MICROBIAL ECOLOGY RESEARCH TRENDS

Microbial Ecology Research Trends

Thijs Van Dijk
Editor

Nova Biomedical Books
New York

Library of Congress Cataloging-in-Publication Data

Microbial ecology research trends / Thijs Van Dijk (editor).
p. ; cm.
Includes bibliographical references and index.
ISBN 978-1-60456-179-1 (hardcover)
1. Microbial ecology. I. Dijk, Thijs Van.
[DNLM: 1. Environmental Microbiology. QW 55 M6268 2008]
QR100.M516 2008
579'.17--dc22

2007049412

Published by Nova Science Publishers, Inc. ✦ New York

Contents

Preface

Microbial ecology is the relationship of microorganisms with one another and with their environment. It concerns the three major domains of life — Eukaryota, Archaea, and Bacteria — as well as viruses. Microorganisms, by their omnipresence, impact the entire biosphere. They are present in virtually all of our planet's environments, including some of the most extreme, from acidic lakes to the deepest ocean, and from frozen environments to hydrothermal vents. Microbes, especially bacteria, often engage in symbiotic relationships (either positive or negative) with other organisms, and these relationships affect the ecosystem. One example of these fundamental symbioses are chloroplasts, which allow eukaryotes to conduct photosynthesis. Chloroplasts are considered to be endosymbiotic cyanobacteria, a group of bacteria that are thought to be the origins of aerobic photosynthesis. Some theories state that this invention coincides with a major shift in the early earth's atmosphere, from a reducing atmosphere to an oxygen-rich atmosphere. This book presents new and important research in the field.

Chapter I - Various techniques have been used to evaluate microbial metabolic activity in natural environments. In recent years, tracer additions of radiolabeled substrates coupled with short term, environmentally-relevant incubation experiments have become standard for assessing overall microbial growth rates as well as utilization rates for specific components of the bioavailable carbon pool. Using these techniques, microbial activities have been determined over a wide range of spatial and temporal scales in a variety of estuarine settings, providing valuable insight into carbon cycling in these dynamic systems. In this paper, the authors focus on patterns of microbial carbon consumption in estuarine sediments and relate these to the utilization of polyaromatic hydrocarbons (PAHs) as a recalcitrant, yet microbially available component of the carbon pool. Drawing on extensive field studies and those published in the literature, they relate microbial metabolic activities to the utilization of specific carbon species, discuss the partitioning of carbon between bacterial production and respiration (utilization efficiency), address the substrate-activity relationship for utilization of recalcitrant organic carbon and identify seasonal effects on the utilization of natural and anthropogenic carbon pools. These studies elucidate the advantages, disadvantages and caveats of relating microbial consumption of bulk labile organic carbon to the utilization of specific components and improve our ability to predict the fate of organic contaminants in estuarine systems.

Chapter II - Many animals and plants establish symbiotic associations with benevolent microorganisms. In particular, most of insects that feed exclusively on restricted diets, such as plant sap, vertebrate blood or woody material, usually possess symbiotic microorganisms inside their body. Recent studies have suggested that more than 50% of insect species may be benefited from such endosymbiotic partners. Among these insects, members of the Heteroptera, known as true bugs or stinkbugs, exhibit an extraordinary diversity in their microbial symbiotic system. Such diversity is exemplified by: (i) physiological impact of the symbiotic associations on the host fitness ranging from parasitic to mutualistic; (ii) interdependence of the symbiotic associations ranging from facultative to obligate; (iii) location of the symbiont ranging from extracellular in gut cavity to intracellular in specialized cells; (iv) among the extracellular associations, specialized symbiotic structures such as midgut sacs and tubular outgrowths, whose cavity harbors specific symbiotic bacteria, varying in their number and arrangement; and (v) mechanisms for symbiont transmission ranging from vertical to environmental. Here, the current biological knowledge of the diverse stinkbug-bacteria symbioses is reviewed, which will provide novel insights into and useful model systems for understanding of the host-symbiont interactions and evolution.

Chapter III - Using various feeding strategies ciliates consume all sizes of autotrophic cells from bacteria to the larger dinoflagellates or diatoms. In addition they consume other heterotrophic protists. In this way every component of prokaryotic and eukaryotic nanoproducer biomass can be transferred to higher trophic levels. Predator ciliates provide an intermediate level in the microbial loop and to obtain adequate information on their biology is important from an ecological point of view. On the other hand, effort spent in attempting to deepen our understanding of predation among protozoa are completely justified by the basic importance of this process. Protozoa, indeed, were not only the first primary consumers in the primeval Oceans, but the first predators as well. Such a new trophic niche is quite important, due to the two consequences it leads to: 1) it creates new empty spaces for new organisms to settle in; 2) it triggers a sort of evolutionary competition between predators and prey as to their morpho-functional acquisitions. So predation in protozoa offers a wide range of examples of more or less species specific and more or less direct cell interactions leading the organism to specific morphological or behavioral responses. Indeed if predators developed a variety of hunting strategies, the potential prey evolved defensive responses. Nevertheless there are not models exhaustively examined to obtain an overall picture of this phenomenon considered from a more specific sinecological point of view. This chapter deals with predation in ciliates. After an introduction to the various feeding strategies of ciliates and a survey of the ciliate predator-prey systems already studied by other authors, the model *Amphileptus* preying on *Euplotes*, is presented in more detail as a prototype of ciliate raptorial feeding behavior. Our attention is successively focused on: a) the ultrastructure of predator toxicysts (the hunting devices); b) the electrophysiological and ultrastructural effects of toxicysts discharge by *Amphileptus* on *Euplotes*; c) the behavioral patterns of the predator (prey recognition from a distance; nearby recognition; engulfment process); d) the peculiar digestion process; e) the defensive strategies of the prey.

Chapter IV - Laboratory cultures have been extensively used in the past as a tool to study the behavior of model organisms subject to a variety of environmental conditions and hence to extrapolate the type of behavior that could be expected from similar microorganisms living in nature. The information obtained using this approach has been crucial for our current understanding of the functional role of certain microbial groups in the environment, but falls short when facing the complexity of interactions operating in real microbial communities.

The main drawback associated with the use of laboratory cultures lays on the fact that conditions in such cultures are usually homogeneous and strongly selective. Batch cultures contain initial concentrations of growth substrates much higher than the growth-limiting threshold thus selecting for organisms with very high specific growth rates. Continuous cultures overcome this limitation by forcing the organisms to grow at low substrate concentrations, well within the growth-limiting range. As a result, organisms with a high affinity for the limiting substrate are selected. None of these cases allows the establishment of a complex microbial community.

The development of complexity in laboratory reactors requires the existence of a certain degree of heterogeneity, either in time or in space, which allows organisms with different requirements to find their niche and avoid competitive exclusion.

In an attempt to provide such heterogeneity the authors have developed a type of bioreactor where solid surfaces within the reactor allow the structuring of the microbial community in space. Analysis of microbial diversity within the reactor using culture-independent molecular techniques reveals the existence of complex assemblages of bacteria diverse both taxonomically and functionally. While the taxonomic composition of these assemblages seems to vary depending on the origin of the inoculum, the functional composition remains quite constant. The results indicate that the use of relatively simple laboratory reactors incorporating a small degree of heterogeneity allows the development of a significant level of biological complexity and can constitute a valuable tool to study the factors determining microbial diversity in natural environments.

Chapter V - The detection, enumeration and fate of *Escherichia coli* in the aquatic systems are on the priority objectives for researches as well as public and health administrations. The role of this bacterium as universal indicator of faecal pollution and of pathogen presence in waters justifies the interest in its study. The method directed to detection and enumeration of this indicator microorganism are gathered and standardized in the prevailing regulations concerning water quality (International Standard Office, American Protection Agency, etc) and are based on their ability to grow in culture. Out of the intestine, in a hostile environment such the aquatic systems, *E. coli* adopts the viable but nonculturable (VBNC) state. In this state, *E. coli* loses its ability to grow on culture media but remains metabolically active. Apparently, this fact nullifies the regulations for its detection justifying the renewed interest in the biological significance of VBNC phenomena and its relation with the survival of microorganisms in the aquatic systems. It is proposed that VBNC state is a strategy for the survival which implies the recovery of growth ability after reversion of the adverse conditions. Another theory hints this state to be a previous step for cellular death. In this review it is presented a global vision of works studying the proposed theses and it is analyzed the ability to resuscitate or to resuming culturability after the entry into the VBNC state, the participation of molecular signals inducing the entry into and the exit from this state, as well as, the genes implied in the overall process through proteomic studies. Moreover, studies analyzing relationship with protozooplankton are too reviewed, due to the VBNC phenomena can alter the ecological relationship between *E. coli* and predator microorganisms from aquatic systems. This approach provides an ecological vision, very close to reality for repercussions of entry into the VBNC state during *E. coli* survival in aquatic systems.

Chapter VI - In this chapter the authors summarize the current information regarding alteration of aquaporin-encoding genes by the arbuscular mycorrhizal (AM) symbiosis, trying to correlate the modulation of aquaporin genes with the enhanced tolerance to water deficit conferred by the AM symbiosis. The results obtained so far show that the effects of the symbiosis on PIP gene expression depends on the own intrinsic properties of the osmotic stress. Under drought stress conditions, the AM symbiosis usually decreases or anticipates the decrease of PIP gene expression. Under salt stress, the trend is just the opposite since the AM symbiosis enhanced the expression of most of the PIP genes analyzed. The regulation of PIP gene expression under cold stress is less evident since one of the genes analyzed was down-regulated by the AM symbiosis, another was up-regulated and two genes were not affected by the symbiosis under such conditions. It seems also that the effects of the AM symbiosis on PIP gene expression depends on the endogenous levels of ABA in the host plant. In any case, the induction or inhibition of particular aquaporins by AM symbiosis should result in a better regulation of plant water status and contribute to the global plant resistance to the stressful conditions as evidenced by their better growth and water status under conditions of water deficit.

Chapter VII - Recently, it has been proposed that plants use pathogenic aerobic and anaerobic bacteria, as well as pathogenic fungi that harbor spines, thorns and prickles, for defense from herbivores, especially vertebrates. These sharp defensive organs introduce various pathogenic agents into the body of the herbivores through the outer defensive layer of the skin in a type of biological warfare. This paper provides a review of our data regarding the various bacterial taxa found on spines, as well as the medical literature regarding infections by bacteria and fungi related to spine injuries. In addition, the authors propose that many plant species which have internal microscopic defensive sharp structures, such as raphides (calcium oxalate needles) and sharp silica bodies, may also introduce pathogenic microorganisms into the tissues of herbivores through the microscopic wounds that these defensive structures induce, especially in the mouth and digestive system. The internal microscopic spines can wound large herbivores and also insects and other very small herbivores that usually attack plants in between spines and thorns. The large spines and the microscopic sharp structures may introduce not only the microorganisms that harbor them, but also those that pre-exist on the skin or inside the digestive system of the herbivores and on the surface of non-spiny consumed plant parts. A majority of the spiny plants visually advertise their spiny nature, a characteristic known as aposematism (warning coloration). Since the pathogenic microorganisms may sometimes be much more dangerous than the physical wounds inflicted by the spines, the authors propose that the cooperation with the pathogenic microorganisms contributes to the evolution of aposematism in spiny plants (including those with raphides and silica needles) and possibly also in spiny animals.

Chapter VIII - Heavy metals are a significant source of pollution in soils that have been demonstrated to exert significant toxic effects on soil microbial communities. Here, the authors investigate the impact of the heavy metals lead and zinc on the bacterial community of an upland metallophytic grassland the UK, contaminated by historical lead mining activity.

The authors focus on the total microbial diversity and that of the Actinobacteria, a significant component of the aerobic soil community that demonstrate a broad metabolic activity and represent useful indicators of soil health.

Two experimental approaches were adopted, the first employed microcosms containing soil from the site of an abandoned lead mine spoil heap, bacterial diversity was investigated using denaturing gradient gel electrophoresis (DGGE) of PCR amplified 16S rRNA DNA with universal primers to amplify the total community and more specific Actinobacteria primers. The Actinobacteria specific PCR primers showed that this taxon represented a diverse and metabolically active element of the microbial community of lead contaminated upland soil. Subsequently, the authors investigated how the stability of the bacterial communities responded to an increase in stress by adding additional Pb and Zn to the soil. There is significant interest in determining whether communities under stress are more or less robust in the face of additional environmental stress. Our data was equivocal, the total community diversity responded negatively to additional heavy metal input into the soil, whereas, the Actinobacteria community did not show any substantial change in its diversity.

A second experiment characterised the bacterial soil community along a transect of a metallophytic grassland site that had significant differences in the Pb, Zn, organic C and N content across it. In this study the metabolically active bacterial population was studied using reverse transcription of mRNA from16S rRNA gene. Universal primers showed a low diversity of metabolically active bacteria, and both organic C and heavy metal concentrations in the soil seemed to impact on the community diversity. In contrast Actinobacteria showed higher diversities and seemed to be able to maintain this diversity even in soils with low organic C and high heavy metal concentrations suggesting that they are competitive in this stressed environment with other soil flora. The data presented here may prove useful in determining the effect of heavy metal concentrations on soil microbial diversity and community stability, particularly in managing rare metallophytic grassland sites and more widely in remediating contaminated soils.

Chapter IX - Many planktonic microbes exhibit nearly global distributions. Apparent adaptability to wide ranges of environmental conditions in cosmopolitan species could involve physiological diversity among individuals. However, variability between individuals among microorganisms is rarely considered. The authors examined individual variability in the common marine ciliate *Euplotes vannus*. Digestion was followed in individual cells subjected to different levels of ambient food concentration. Average digestion times were not significantly different because, in the absence of food or in the presence of super-saturating food concentrations, there was a large variability between individuals. Some ciliates completed food vacuole processing under 1 h while others took as long as 17 h. The range of digestion times, which increased under extreme conditions, may correspond to range of growth rates or growth potentials among individuals.

Chapter X - On the basis of screening 19 strains of nitrogen-fixing and 9 strains of phosphorous-mobilizing bacteria, isolated from rhizosphere of cotton, alfalfa and sugar-beet, two active strains - producers of biosurfactants (BS) - have been isolated and identified as *Pseudomonas aeruginosa CH5* and *Bacillus subtilis 26.* The value of surface tension of cultural broth (CB) of the mentioned bacteria varied within the limits 57.2 – 58.0 mN/m. The optimal cultivation conditions for active production of BS by *P. aeruginosa CH5* and *B.*

subtilis 26 have been determined, at which the surface tension of CB had value 36.6 – 40.5 mN/m. From CB of *P. aeruginosa CH5* biomass, grown in BEB (beef-extract broth) with addition of sunflower or cotton oil, the BS complex (0.2-2.1%) has been selected. In composition of BS no less than three rhamnolipids and lipophilic pigments of phenazine nature have been found, the amount of which essentially depended on type of vegetative oil in nutrient medium. CB, BS and water-soluble pigments of *P. aeruginosa CH5* displayed growth-stimulant activity towards seeds and seedlings of above-mentioned agricultural crops and antagonistic activity to *Verticillium dahliae, Rhizoctonia solani* and *Fusarium oxysporum* phytopathogenic fungi. This activity remained after BS autoclaving (121°C, 1 atm.), enduring at pH 5.5-8.5 and NaCl concentration up to 20 g/L. CB, BS and water-soluble pigments were stable during storage in usual conditions more than 9 months.

Chapter XI - The ecosystem research in a temperate rain forest was not fully understood. Thus, the aims of this chapter are illustrated the changes of low molecular weight organic acids (LMWOAs), stable isotope ratios of soil organic matter, forest fire influenced the changes of vegetation communities, nitrogen transformation and microbial populations study in rhizosphere and bulk soils, selecting Ta-Ta-Chia long-term ecological research (LTER) site, Taiwan in a temperate rain forest as an example. Low molecular weight organic acids and the nature of organic matter were determined by gas chromatograph (GC) and CPMAS ^{13}C nuclear magnetic resonance analysis. Changes in the meadow-forest boundary at this LTER site detected by stable isotope ratios of soil organic matter of $\delta^{13}C$ analysis of plants and soils. The buried bag incubation method was used to study temporal pattern of net N mineralization and nitrification in soils. Several types of microorganisms were counted for their spatial and temporal variations. The recovery percentage of LMWOAs determination by GC shows a high correlation of $r^2 = 0.99$. All recovery of LMWOAs was above 83.6%. The proportion of organic functional groups in the rhizosphere and bulk soils quantified by ^{13}C NMR analysis followed the general order: alkyl-C > O-alkyl-C > N-alkyl-C > acetal-C > aromatic-C > carboxylic-C > phenolic-C. $\delta^{13}C_{PDB}$ values in pedons 1 and 2 tended to decrease with increasing with soil depth. The values of EB, E and Bw horizons were close to $\delta^{13}C_{PDB}$ values of C_3 plants. The data indicated that SOM was a mixture of *transmorrisonensis* (C_4), *nittakeyamensis* (C_3) and *Tsuga* (C_3) plant residues.$\delta^{13}C_{PDB}$ values obtained from deeper meadow pedons 1 and 2 of soil layers indicated that the vegetation community has changed over the years. *M. transmorrisonensis* (C_4) plants replaced C_3 plants (e.g., *Tsuga*). Extractable NO_3^- concentrations ranged from 0.2 to 2.1 mg N kg^{-1} soil in the meadow zone and from 0.2 to 13.0 mg N kg^{-1} soil in the forest zone. No seasonal changes in soil nitrate concentration were present. Extractable NH_4^+ concentrations ranged from 1.0 to 12.4 mg N kg^{-1} soil in the meadow zone and from 2.8 to 25.0 mg N kg^{-1} soil. The NH_4^+ concentrations showed highly temporal variations. The net of mineralization and nitrification rates showed spatial and temporal variations, depending on microbial activities and populations. Bacteria were most abundant, followed by cellulolytic microbes, phosphate-solubilizing microbes, nitrogen-fixing microbes, fungi and actinomycetes. Microbial populations showed high correlation with microbial biomass carbon and organic acid contents.

In: Microbial Ecology Research Trends
Editor: Thijs Van Dijk, pp. 1-38

ISBN: 978-1-60456-179-1

Chapter I

Evaluating PAH Biodegradation Relative to Total Bacterial Carbon Demand in Coastal Ecosystems: Are PAHs Truly Recalcitrant?

Thomas J. Boyd, David C. Smith, Jude K. Apple, Leila J. Hamdan, Christopher L. Osburn and Michael T. Montgomery

Abstract

Various techniques have been used to evaluate microbial metabolic activity in natural environments. In recent years, tracer additions of radiolabeled substrates coupled with short term, environmentally-relevant incubation experiments have become standard for assessing overall microbial growth rates as well as utilization rates for specific components of the bioavailable carbon pool. Using these techniques, microbial activities have been determined over a wide range of spatial and temporal scales in a variety of estuarine settings, providing valuable insight into carbon cycling in these dynamic systems. In this paper, we focus on patterns of microbial carbon consumption in estuarine sediments and relate these to the utilization of polyaromatic hydrocarbons (PAHs) as a recalcitrant, yet microbially available component of the carbon pool. Drawing on extensive field studies and those published in the literature, we relate microbial metabolic activities to the utilization of specific carbon species, discuss the partitioning of carbon between bacterial production and respiration (utilization efficiency), address the substrate-activity relationship for utilization of recalcitrant organic carbon and identify seasonal effects on the utilization of natural and anthropogenic carbon pools. These studies elucidate the advantages, disadvantages and caveats of relating microbial consumption of bulk labile organic carbon to the utilization of specific components and improve our ability to predict the fate of organic contaminants in estuarine systems.

I. Introduction

Estuaries are dynamics environments with considerable physical, chemical and biological forcing factors, as freshwater mixes into the sea. Their complex nature makes understanding biogeochemical cycles at both spatial and temporal scales particularly challenging. The heterogeneity of the sources and types of organic matter entering estuarine systems makes predicting microbial utilization of individual structural components difficult at best. Given the natural complexity, it is difficult to predict the residence time and biodegradability of organic contaminants introduced into estuaries from human activities. As human impact on coastal systems continues to increase (*e.g.* Small and Nicholls 2003), it will be ever more imperative to better our understanding of contaminant turnover in sensitive estuarine systems.

The ability of microbes to degrade organic contaminants such as PAHs has been known for decades [*e.g.* Sisler and ZoBell 1947] and an excellent general review of environmental factors controlling marine hydrocarbon biodegradation is provided by Atlas [1981]. In theory, organic contaminants should represent "just another" substrate available for microbial growth and respiration. Most persistent organic compounds bear some chemical resemblance to natural compounds, thus allowing at least limited biodegradation. And, as been demonstrated empirically, most organic compounds are biodegradable provided favorable environmental conditions. Hence, considerable effort has been made in relating chemical structure to degradability for a number of organic contaminant classes. For cultured microorganisms, structure-activity relationships have been developed. Initially, converting laboratory or theoretical structure-activity relationships to field settings proved difficult. *Wolfe et al.* [1980] found a reasonable correlation between second order biodegradation rate constants in natural water samples and second order alkaline hydrolysis rates for phthalates and several pesticides in distilled water. Although not the first such study to link structure to biodegradability, Wolfe *et al.* [1980] used natural water samples, thus attempting to include the effects of physicochemical forcing factors on biodegradation in the field. Research in structure-activity relationships has refined models and generated computer algorithms to predict not only biodegradation potential, but also environmental risk and potential degradation pathways [Boethling *et al.* 1989, Kompare 1998, Dearden 2002, Jaworska *et al.* 2003, Hou *et al.* 2003]. Despite these gains, Wammer and Peters [2005] showed a very narrow range of biodegradation rates based solely on structure (i.e. without environmental forcing factors such as bioavailability), indicating that environmental forcing factors are a critical component in understanding biodegradation rates.

At the molecular level, chemical structure dictates whether a given enzyme system can catabolize an organic molecule. The family of enzymes expressed in response to PAHs often fall into one of two broad categories of multi-enzymes (upper and lower degradation pathways) [*c.f.* Dagley 1987]. A strain that is expressing genes responsible for naphthalene degradation is also typically expressing genes useful for degrading other lower molecular weight aromatic hydrocarbons. Thus subtle differences in chemical structure would be of little importance in determining hydrocarbon metabolism rates unless the compound was so novel as not to be recognized by any of the enzymes of either the upper or lower pathway. Chemical structure can also be important in metabolic efficiency as PAHs that are more thermodynamically stable may require more cellular energy to metabolize and give the degrading strain a disadvantage when competing for nutrients and oxygen with natural organic matter (NOM) degraders.

A factor dictating biodegradation is contaminant concentration. It is widely accepted that a threshold concentration exists under which biodegradation is not metabolically favorable [Boethling and Alexander 1979a]. The threshold concentration is believed to either represent a level where no net energy can be obtained from the substrate or a level insufficient to stimulate the synthesis and deployment of catabolic enzymes [Wiggins *et al.* 1987, Aelion *et al.* 1987, Wiggins and Alexander 1988]. Exceeding the threshold concentration, one might expect a greater rate of biodegradation [Boethling and Alexander 1979b, Alexander 1985]. Pre-exposure of populations to low and high levels of contaminants may therefore impact subsequent biodegradation [Aelion *et al.* 1989]. Hwang et al. [1989] suggested that different enzyme systems may be responsible for contaminant biodegradation at different concentrations and that enzymes with different kinetics are expressed depending on contaminant concentration and pre-exposure. Adaptation appears to be only one aspect of the story as contaminant-adapted cells eventually loose the ability to degrade compounds at lower concentrations [Roch and Alexander 1997]. Because various enzyme systems working at an array of saturation values are generally responsible for contaminant biodegradation in soils, water and sediments, kinetics necessarily depend on pre-exposure and contaminant concentrations [Scow *et al.* 1986, Wiggins *et al.* 1987, Wiggins and Alexander 1988, Scow *et al.* 1989].

Another parameter impacting biodegradation of organic contaminants in soils and sediments is contaminant bioavailability. Contaminants aged in soil and sediments become less extractable and amenable to biodegradation the longer the aging process continues. Hatzinger and Alexander [1995] demonstrated decreased nitrophenol and phenanthrene biodegradation after extensive sediment or soil aging. Mineralization of contaminants generally decreased as a function of the aging time, indicating a decrease of bioavailability with increased aging [Hatzinger and Alexander 1995]. While the exact means of bioavailability loss for every contaminant in every soil and sediment type is unknown, many explanations exist. A recent review has proposed several general "cases" to conceptualize sorption in soils and sediments [Luthy *et al.* 1997]. In this conceptual model, different geosorbents were evaluated in terms of sorption kinetics, activation energy, heat of sorption, competitive binding sites, steric effects, and solvent extractability. While conceptual models have helped define general processes impacting bioavailability, Chung and Alexander [1998] were unable to definitively model either solvent extractability or bioavailability despite using 16 different soil types. There is evidence that metal-organic complexes may be partially responsible for sediment-PAH sorption [Eschenbach *et al.* 1998]. This phenomenon would be especially applicable to estuarine sediments as estuarine mixing greatly impacts the ionic composition of sedimentary material. Reichenberg and Mayer [2006] suggest two different aspects control bioavailability; the accessible quantity – that being available (or which can become available), and the chemical activity – the ability of the "local" environment to yield the contaminant to resident microorganisms. In soils and sediments, this may of less importance when OM utilization may be dominated by surface-associated strains. A common conceptual model is presented by Johnsen *et al.* [2005 – Figure 3]. In this model, microbes must come in contact with the diffusion boundary dictated by particle sequestration and the particle-to-water diffusion coefficient of the sorbed contaminant. If microbes actively attach to particles or in fact to the PAH molecules as shown by Wick et al [2002], the diffusion boundary layer may be minimal, giving attached bacteria a selective advantage for biodegradation. If not, biodegradation through intermediary dissolution into the aqueous phase would be assumed for free-living bacteria [Reid *et al.* 2000, Johnsen *et al.* 2005, Johnsen and Karlson 2007].

While geosorbents have received much of the experimental and modeling attention, the role of black carbon (soot) and colloids on contaminant bioavailability have also been assessed. Black carbon constitutes approximately 4% of the organic carbon in soils, and about 9% in sediments [Cornelissen and Gustafsson 2005]. It is believed that hydrophobic contaminants are more strongly bound to black carbon than mineral-based geosrobents [Jonker *et al.* 2005]. In marine systems, soot also appears to be a dominant sorbent for PAH [Gustafsson and Gschwend 1997, Lohmann *et al.* 2005]. In estuarine porewaters, colloids effectively sorbed 25-52% of added PAHs within roughly one day [Chin and Gschwend 1992]. Surprisingly few direct studies have targeted the impact of colloidal sorption with PAH biodegradation. Meredith et al [1998] observed no change in PAH biodegradation based after addition of colloidal humic acid as a sorbent. In contrast, Vacca et al [2005] reported enhanced biodegradation of PAH sorbed to humic acid colloids, postulating that bacteria were able to utilize PAH without previous partition from the colloidal to aqueous phase. One can imagine competing forces in natural systems, with some facilitating and others hindering the biodegradation of hydrophobic organic contaminants. Empirical data on the persistence of "degradable" concentrations of PAH clearly demonstrate that bioavailability has a large impact on the fate of contaminants in soils and sediments [*c.f.* Johnsen and Karlson, 2007, Johnsen et al., 2005, Reid et al., 2000].

Given that hydrocarbons represent an excellent source of energy, routinely made use of by humans, it should come as no surprise that hydrocarbon contaminants also represent a potential source of energy and carbon for microorganisms. Although bacteria in nature seldom grow as they do in laboratory cultures, it is obvious that given ample organic material and nutrients, copiotrophic bacteria will grow and reproduce rather than utilize the organic material for energy alone. In nature, the ability of bacteria to find both ample substrate and nutrients is rare and most likely episodic. Microorganisms generally live in a state between growth and starvation [*c.f.* Kjelleberg et al., 1987]. Large-scale hydrocarbon contamination represents a "feast" in terms of carbon and energy to impacted environments (if toxicity effects are minimal). Without nutrients, principally nitrogen (for proteins and nucleic acids) and phosphorous (for ATP and nucleic acids), organisms are unable to significantly increase biomass, thus biodegradation is limited. Microorganisms typically try to preferentially increase biomass over utilizing carbon for maintenance energy. Cornelissen and Sijm [1996] reported that while 47-83% of available energy is generally utilized for growth, aromatic hydrocarbon biodegradation (toluene and PAHs) showed much lower growth efficiencies (9-26% used for growth). Boyd et al. [2001] observed relatively high growth efficiencies for benzene (average 66%) and toluene utilization (average 59%) in groundwater under anaerobic conditions. In this study, no correlation was found between benzene or toluene utilization and *in situ* nutrients (NO_3^-, NH_4^+, PO_4^{3+}). Similarly, Guerin and Jones [1989], found no correlation between PAH biodegradation and nutrient concentrations in estuarine sediments.

While it stands to reason that nutrient additions should stimulate biodegradation of hydrocarbons, demonstrating it in the field has been problematic. Paris and Rogers [1986] found virtually no phenol biodegradation enhancement following nutrient enrichment. Jones and Alexander [1988] however observed a shortened acclimation period for nitrophenol utilization when nutrients were abundant. Acclimation, not only to substrates, but perhaps also to elevated nutrient concentrations, appears to have a significant impact on altering the microbial assemblage, stimulating "degraders" to become a more prominent percentage of the total population. In attempting to stimulate organic contaminant degradation by nutrient addition, Steffensen and Alexander [1995] showed that altering an environment to stimulate contaminant biodegradation may serve to stimulate "other" organic matter biodegradation, not biodegradation of the contaminant of interest. This phenomenon was observed in a gasoline-contaminated groundwater site where a groundwater circulation system designed to introduce atmospheric oxygen to anaerobic groundwater greatly stimulated overall bacterial production, but virtually eliminated benzene and toluene utilization which had been occurring anaerobically [Boyd *et al.* 2001]. In terms of stimulating biodegradation, numerous attempts have been made to augment contaminated sites with nutrients in an attempt to enhance microbial catabolic activity. Abundant successful applications in microcosms and mesocosms have been described, but definitive field demonstrations are few [Venosa *et al.* 1996, Macnaughton *et al.* 1999, Venosa *et al.* 2002, Röling *et al.* 2004, Jiménez *et al.* 2006]. Crude petroleum biodegradation appears overall to be stimulated by nutrient additions [Venosa *et al.* 1996, Wrenn *et al.* 1997], PAH and aromatic compound degradation appears less so [Garcia-Blanco *et al.* 2007]. While, biodegradation kinetics may not be altered, addition of fertilizer has been shown to change the microbial community composition [*c.f.* Ogino et al. 2001]. As we discuss in the following section, changes in community composition may be an important precursor to specific organic contaminant degradation.

While nutrient additions have shown limited effectiveness in stimulated contaminant biodegradation, the addition of specific hydrocarbon-degrading bacteria (bioaugmentation) has been even more problematic. For the most part, *in situ* microbial consortia generally out compete added microbes [*c.f.* Chapelle, 1999]. As mentioned above, it appears that "degraders" become a more active component of the assemblage after environments are loaded with organic contaminants, but it has been difficult to correlate an increase in degrader abundance or overall bacterial abundance to enhanced contaminant utilization [*e.g.* Hickman and Novak, 1989]. In general, bioaugmentation using specific microbial strains has shown promise for chlorinated solvent bioremediation [Ellis *et al.* 2000, Morrill *et al.* 2005, Da Silva *et al.* 2006] and some successes have been documented in microcosms [Jouanneau *et al.* 2005, Yu *et al.* 2005] and engineered remediation systems [Trably *et al.* 2003, Domde *et al.* 2007]. While added organisms have been detected in contaminated environments after bioaugmentation [Mishra *et al.* 2001], *in situ* bioaugmentation for PAH bioremediation appears to be a reasonably ineffective treatment [Johnsen *et al.* 2007].

One underlying theme common to soil hydrocarbon plumes and aquatic sediments is the utilization of molecular oxygen to a point where contaminants tend to remain only in localized anaerobic zones. For instance, in groundwater, the majority of the plume may be anaerobic and high in contaminants, while the edges are low in contaminant and are still oxic. In sediments, this generally occurs below some depth where organic carbon utilization exceeds oxygen diffusion. While anaerobic PAH degradation has been documented under sulfate-reducing [Coates *et al.* 1996, Coates *et al.* 1997] and nitrate-reducing environments [McNally *et al.* 1998], a general consensus is that anaerobic biodegradation is extremely slow under natural conditions [*e.g.* Quantin *et al.* 2005]. Dissolved oxygen appears to be a major factor driving PAH biodegradation in estuarine sediments and aquifers [Bauer and Capone 1985, Brubaker and Stroo 1992, Montgomery *et al.* 2002, Boyd *et al.* 2005a].

While there are many trends have been observed regarding general hydrocarbon and specifically PAH biodegradation control, a comprehensive study covering multiple environments and describing the impact of multiple forcing factors does not exist. In general PAH are typically considered an “exception” rather than just another component of the carbon pool available to microorganisms as a source of carbon and energy. This chapter reports on a series of studies completed from 1997 through 2004 in seven different estuarine systems. In these studies, PAH biodegradation was measured during short-term tracer incubations along with concurrently-measured bacterial production, temperature, dissolved oxygen and other environmental forcing factors. We begin the discussion of our findings with sections on the what is known regarding factors controlling natural organic matter (NOM) biodegradation in estuarine systems and the methods used to assess overall microbial activities and specific carbon (*i.e.* contaminant) biodegradation. We then discuss PAH relative to NOM utilization and what is known of the interplay between utilization of labile and refractory organic matter. The results of our studies along with others that highlight underlying principles of PAH biodegradation in estuarine systems are then described. Finally, a conceptual framework for assessing biodegradation is proposed. It is hoped that this cross-system approach to assessing PAH biodegradation in estuarine sediments leads to an improved fundamental understanding of both PAH and perhaps other recalcitrant organic matter biodegradation.

II. Assessing Microbial Activities

The first careful measurement of PAH utilization at environmentally realistic levels in estuarine sediments was performed by Herbes and Schwall [1978]. This study was particularly relevant as they measured not only PAH mineralization, but also its incorporation into cellular biomass and soluble metabolic intermediates. In order to account for variations in ambient concentration of PAH, radiolabeled PAHs were added at various concentrations so that saturation kinetics could be determined. The authors also measured the utilization of radiolabeled PAH over time in order to investigate degradation kinetics. In subsequent studies, ^{14}C-labeled substrates were used in short-term incubations to assess spatial (salinity through and estuary) and temporal (seasonal) variation in aromatic hydrocarbon biodegradation [Pfaender and Bartholomew 1982, Bartholomew and Pfaender 1983] These studies were significant for two reasons. First, as in Herbes and Schwall’s studies, both respiration and incorporation were measured allowing calculation of degradation efficiency (although they were not reported), and second, the authors assumed added radiotracers were in excess of any ambient contaminant. A significant methodological unknown when analyzing radiolabeled substrate mineralization is the efficiency of biodegradation. If one has a minimal number of samples, cellular incorporation can be readily assayed [*e.g.* Herbes and Schwall 1978]. By "digitizing" graphs in this paper, growth efficiencies were calculated as ~38% for naphthalene and ~78% for anthracene utilization. Typically, mineralization (i.e. conversion to CO_2) is the only measured parameter in ^{14}C-substrate addition studies. Clearly, unknown growth efficiencies would introduce errors in substrate utilization estimates.

In addition to efficiency considerations, the studies of Shiaris [1989a, 1989b] highlight another difficulty with assessing contaminant degradation using ^{14}C-labled substrates: the impact of isotopic dilution. Tracer-based biodegradation studies generally rely on one of two assumptions: 1) the radiotracer is added at a concentration far exceeding the ambient concentration, or 2) the radiotracer is added at a concentration that is negligible compared to the ambient concentration. In the first scenario, the microbial assemblage is presented with a "new" substrate to degrade and therefore must synthesize the metabolic capability to do so. This is partially the cause for lags observed when measuring biodegradation rates [*e.g.* Shimp and Pfaender 1987, Aelion *et al.* 1987]. Assumption two is preferable in terms of understanding the *in situ* rate of substrate biodegradation. The idea in this scenario is that organisms will be minimally impacted by a tracer addition of substrate (*i.e.* < 10% of the ambient pool), and will continue their metabolism without up-regulating protein synthesis to degrade the contaminant of interest. Common practice is also to end the incubation in as short a time as possible so that bottle effects are minimized (*i.e.* organisms do not utilize energy and carbon synthesizing proteins strictly to deal with a new environment). A compication for assumption two is that one must measure the ambient concentration of the same compound added as a radiotracer. This allow recalculation of specific activity based on the total pool of substrate in question. The importance of calculating isotopic dilution has been reported for ecological studies [*e.g.* King and Berman 1984], but is largely disregarded in the PAH biodegradation literature. It is a critical component in performing experiments with isotopic tracers as it can dramatically impact the final rate measurements. In addition, because measurements are generally conducted with contaminated groundwater, soils or sediments, ambient concentrations almost certainly impact the analysis. For instance, if one examines the experiment of Boethling and Alexander [1979] in which radiolabeled substrates were added to stream water at concentrations ranging over six orders of magnitude, the impact of an appreciable ambient contaminant load would change the results markedly. The rate of CO_2 production from ^{14}C-labeled 2,4-D ranged 6 orders of magnitude. If exposed stream water were used for the experiments and had an ambient concentration of 2.2 $\mu g\ mL^{-1}$, recalculating the rates using isotopic dilution would only give two orders of magnitude variation in respiration rates. The work of Shiaris [1989a, 1989b], provided detailed methods for calculating and using isotopic dilution in reporting PAH mineralization rates.

Another critical experimental unknown when measuring ^{14}C-labeled substrate mineralization is the stoichiometry of the contaminant biodegradation process. For PAHs this may be lead to considerable uncertainty as some compounds are available as uniformly labeled (UL), for instance naphthalene, while others are labeled only on one of the ring positions, like fluoranthene-3-^{14}C. So, for instance, if one adds 10 molecules of UL-^{14}C-naphthalene to an incubation, then recovers 10 molecules of $^{14}CO_2$, it is unknown if ten naphthalene molecules were partially degraded, one molecule was completely degraded to CO_2, or some intermediate number of naphthalenes were mineralized. In this scenario, over 90% error in degradation rate estimates is possible. Although this seems unmanageable, there are several phenomena that might mitigate experimental error. There are only several major pathways that have been observed for breakdown of aromatic compounds [*c.f.* Dagley 1987]. If one were to allow biodegradation to occur for only a brief period, one might expect buildup of "known" degradation intermediates as certain pathway steps require more activation energy than others leading to transient buildup of certain intermediates. Short incubation times fit perfectly with tracer study theory, wherein a negligible amount of ^{14}C-labeled substrate is added relative to the ambient concentration. If done properly, a short incubation time coupled with a small number of labeled substrate molecules relative to ambient molecules, one might expect to drive the stoichiometry towards each $^{14}CO_2$ coming from one UL substrate molecule. For this reason, our biodegradation studies assume one molecule of ^{14}C-labeled naphthalene is degraded for each $^{14}CO_2$ molecule recovered [*c.f.* Boyd *et al.* 2005].

Short-term tracer incubations are however susceptible to another uncertainty. When tracer substrate is added to soils or sediments, one cannot be certain there is instantaneous homogenous distribution of isotopic tracer with the ambient pool of substrate. Tracer theory relies on equilibrium mixing of added and substrate. Radiolabeled substrates are generally added in solvent diluent. The impact of solvent addition has been studied extensively [*c.f.* Herbes and Schwall 1978, Bauer and Capone 1985] and can be minimized to have no statistical effect. However, considerable effort has been given to understanding the "aging" of ^{14}C-labeled substrates in sediment slurries [*c.f.* Aptiz et al. 1999]. Reid *et al* [1998] found that the most rapid equilibrium for added radiolabeled PAH tracers occurred when soils or sediments were dried before mixing. While this might allow more homogenous distribution of added isotope, it is impractical for performing rapid microbiological assays given the deleterious effect it would have on natural sediment microorganisms. For lower molecular weight PAHs (up to three rings), sorption and desorption to sediment slurries was found to be linear and rapid and was concluded to have little impact on biodegradation rate measurements [Chandra *et al.* 1996]. For studies presented here, ^{14}C-labeld substrates were added to polystyrene tubes after minimal dilution in solvent. The solvent was then allowed to evaporate prior to inoculation with sediment slurry. PAH sorption to the incubation vessel was assayed and averaged >90% remaining after aqueous rinses.

In previous work by the current authors and others a strong relationship between availability of dissolved oxygen (DO) and PAH biodegradation has been noted [see Introduction section]. Because oxygen comprises ~21% of the atmosphere, it can become a contaminating presence when performing biodegradation studies to determine *in situ* rates. One difficulty in performing incubations is an accurate knowledge of sediment DO content of collected sediments prior to incubation. Without this information, it is impossible to acclimate the incubation headspace to the appropriate oxygen partial pressure. Incubations reported here were performed in 100 mL culture tubes using surficial sediment material. One milliliter of filtered natural water collected immediately above the sediment surface was added to make sediment slurries. We calculated that oxygen introduced to the tube during sample processing would likely penetrate only ~2 mm into the overlay water used to make the slurries [Boyd *et al.* 2005a] and thus not significantly influence the results. Bastviken and Tranvik [2001] conducted a study to determine if atmospheric oxygen could bias estimates of bacterial production in aerobic and anaerobic waters. They concluded it was possible to accurately measure leucine incorporation in anoxic samples without using an anaerobic hood to transfer and incubate samples.

Given the uncertainties inherent in measuring PAH biodegradation, it seems of utmost importance to perform experiments with consistent methodology. For the reasons discussed, it is appropriate to design experiments such that true tracer levels of PAH are added to determine *in situ* rates, rather than determine "potential" by adding PAH to pristine sediment or water. This seems intuitive as the driver to measure PAH utilization rates is almost universally related to detectable levels in the environment. Although the uncertainties in methodologies may be considerable, one might predict that consistent application of a particular experimental protocol should provide largely systematic error such that cross-season or cross-system comparisons prove valid even if actual rates are not completely accurate (Table 1).

III. Estuarine Microbial Activities

Bacteria dominate the biomass of estuarine microbial communities and mediate almost every ecologically relevant biogeochemical process [Day *et al.* 1989]. Understanding environmental factors that regulate the microbial metabolism are fundamental to understanding natural and contaminant carbon cycling in estuarine ecosystems. Manipulative experiments and field studies alike have identified significant effects of temperature [Apple *et al.* 2006], dissolved organic carbon concentrations (DOC) [Shiah and Ducklow 1994a], inorganic nutrients [Smith and Kemp 2003], nutrient stoichiometry [Goldman *et al.* 1987], salinity [del Giorgio and Bouvier 2002], and organic substrate source and quality [Revilla *et al.* 2000, Amon *et al.* 2001] on microbial carbon metabolism. These various factors differentially influence pathways of carbon metabolism.

Temperature and Salinity

Temperature and salinity are two of the most important factors influencing biotic and abiotic processes in estuarine ecosystems, with temperature exhibiting a universal effect on the range and magnitude of biological processes [Malone *et al.* 1988, Smith and Kemp 1995, Lomas *et al.* 2002, Apple *et al.* 2006] and salinity being a fundamental force shaping biogeochemical processes, species distribution and water-column chemistry [Day *et al.* 1989]. Recent analysis of data from over 50 estuarine systems corroborates the primacy of temperature and salinity in shaping the variability in microbial processes and water column chemistry [Apple *et al.* 2007].

Table 1. Locations, substrates, mineralization rates and turnover times for estuarine sediments

		Mineralization Rate (μg C kg^{-1} d^{-1})0					Turnover Time (d)				
Estuary	Substrate	Stations	Average	Median	Range		Stations	Average	Median	Range	
					low	high				low	high
Charleston, SC	Naphthalene	157	31.7	2.29	0	567	41	218	73	3.7	3680
	Phenanthrene	185	30.9	0.41	0	935	118	2586	526	1.1	28716
	Fluoranthene	185	191	0.34	0	16730	107	1500	517	0.6	16347
Chesapeake Bay, VA	Naphthalene	66	6.5	2.07	0	45	32	191	8.1	0.07	2498
	Phenanthrene	65	3.9	2.96	0	92	54	1470	431	0.32	22388
	Fluoranthene	62	18.4	1.08	0	854	49	723	195	5.6	7264
Narragansett Bay, RI	Naphthalene	15	1.51	1.1	0.13	0.48	0	NA	NA	NA	NA
	Phenanthrene	12	0.31	0.22	0	0.92	12	225	122	5.1	756
	Fluoranthene	13	1.36	1.13	0	2.87	13	102	41	5	412
Pearl Harbor, HI	Naphthalene	37	93.3	1.26	0	603	35	1229	702	34	6950
	Phenanthrene	37	74.1	1.56	0	895	35	14590	918	17	125495
	Fluoranthene	37	168	1.54	0	2911	30	2164	565	8	34125
San Francisco Bay, CA	Naphthalene	24	1.23	0.5	0	8.8	3	8.3	3.9	1	20
	Phenanthrene	17	11.8	4.1	0	91.4	11	8	1.47	0.17	62.4
	Fluoranthene	20	0.65	0.2	0	5.4	16	342	184	3.2	2088
San Diego Bay, CA	Naphthalene	22	2.5	0.91	0	10.9	0	NA	NA	NA	NA
	Phenanthrene	22	1.3	0.4	0	8.6	20	1006	185	8	8342
	Fluoranthene	22	4.8	2.97	0	42	20	1922	796	20	11592
Delaware Bay, PA	Naphthalene	198	2.2	0.36	0	48.5	164	1470	131	0.01	21000
	Phenanthrene	196	4.3	1.50	0	60	185	440	125	2.2	4220
	Fluoranthene	185	6.1	0.41	0	223	183	973	367	0.19	15600

Table 2. Concentration of labile and refractory DOC in Chesapeake Bay and Delaware Bay in µg C L^{-1}. L1 refers to DOC utilized by bacterioplankton during aerobic respiration in less than 5 days. L2 refers to DOC utilized by bacterioplankton during aerobic respiration in less than 20 days. ROM refers to refractory DOM requiring greater than 20 days to be oxidized

Date	Location	Depth (m)	Lat	Long	DOM (total)	L1-DOM	L2-DOM	ROM (DOC -(L1+L2))	% Labile (L1)	% Labile (L2)	% Labile (Sum L1+L2)
Sep-03	Tidal Fresh Chesapeake Bay	1.0	39.4531	-76.0222	2950	346	539	2064	12	18	30
Sep-03	Tidal Fresh Chesapeake Bay	3.0	39.4531	-76.0222		318	397				
Sep-03	Tidal Fresh Chesapeake Bay	6.0	39.4531	-76.0222		363	433				
Sep-03	Mesohaline Chesapeake Bay	1.0	37.8345	-76.1815	2750	253	423	2074	9	15	25
Sep-03	Mesohaline Chesapeake Bay	12.0	37.8345	-76.1815		148	309				
Sep-03	Mesohaline Chesapeake Bay	17.0	37.8345	-76.1815		131	376				
Sep-03	Polyhaline Chesapeake Bay	1.5	36.9602	-75.9970	1880	206	341	1333	11	18	29
Sep-03	Mesohaline Delaware Bay	1.0	39.2387	-75.3157		250	19				
Sep-03	Mesohaline Delaware Bay	5.0	39.2387	-75.3157		374	502				
Sep-03	Mesohaline Delaware Bay	10.0	39.2387	-75.3157		421	361				
Mar-04	Oligohaline Delaware Bay	2.0	39.5014	-75.5530	2500	794	1096	610	32	44	76
Mar-04	Mesohaline Delaware Bay	2.0	39.3824	-75.4779	2000	646	892	463	32	45	77
Mar-04	Mesohaline Delaware Bay	5.9	39.3824	-75.4779		933	1288				
Mar-04	Mesohaline Delaware Bay	10.4	39.3824	-75.4779		495	684				
Mar-04	Tidal Fresh Chesapeake Bay	1.9	39.4403	-75.9961	1790	566	782	442	32	44	75
Mar-04	Tidal Fresh Chesapeake Bay	5.0	39.4403	-75.9961		698	964				
Mar-04	Mesohaline Chesapeake Bay	1.9	37.8712	-76.1579	2140	617	851	672	29	40	69
Mar-04	Mesohaline Chesapeake Bay	16.5	37.8712	-76.1579		570	787				
Mar-04	Mesohaline Chesapeake Bay	26.5	37.8712	-76.1579		505	697				
Mar-04	Mesohaline Potomac River	1.9	38.1715	-76.5875	1810	635	878	297	35	48	84
Mar-04	Mesohaline Potomac River	8.0	38.1715	-76.5875		572	790				
Mar-04	Mesohaline Potomac River	14.2	38.1715	-76.5875		539	744				
Mar-04	Polyhaline Chesapeake Bay	2.0	37.2324	-76.1295	1888	446	616	826	24	33	56
Mar-04	Polyhaline Chesapeake Bay	4.0	37.2324	-76.1295		271	378				
Mar-04	Polyhaline Chesapeake Bay	9.0	37.2324	-76.1295		403	556				
									24	34	58

*L1 and L2 calculated from 5 and 20 day BOD data respectively assuming a ratio of xxx

Temperature plays a fundamental role in regulating the activity and growth of all microorganisms [Rose 1967, Madigan *et al.* 2003]. The effect of temperature on cellular processes in non-psychrophylic cultured bacteria has been well documented, with a general consensus that metabolic rates approximately doubles for each 10°C increase in temperature [Morita 1974]. The temperature effects on natural bacterioplankton communities in temperate and sub-tropical estuaries are not as well understood but have been the subject of numerous studies [Hoch and Kirchman 1993, Sampou and Kemp 1994, Shiah and Ducklow 1994a, Pomeroy *et al.* 1995, Felip *et al.* 1996, Raymond and Bauer 2000, Staroscik and Smith 2004], the overwhelming majority of which have focused on the temperature dependence of bacterial growth rates and production (BP) alone.

In general, these studies share two fundamental conclusions. First, that the temperature dependency of bacterial growth and production is stronger at lower temperatures, and second, that temperature effects are often modulated by other environmental conditions, namely the availability of inorganic nutrients and the quality and quantity of organic matter substrates available for growth. Three studies of seasonal PAH biodegradation in estuarine sediments showed found no direct correlation between the former and *in situ* temperature [Shiaris 1989, Montgomery *et al.* 2002, Boyd *et al.* 2005a]. The rate of PAH delivery to the sediments may vary during the season, providing a constant flux [*c.f.* Arzayus *et al.* 2001, Pohlman *et al.* 2002]. Accumulation may be impacted not only by PAH degraders, but also by lower activities of benthic invertebrates that rework sediments allowing DO and nutrient influx and removal of wastes [Schaffner *et al.* 1997].

Estuarine circulation creates an environment where dramatic changes in salinity may occur on relatively small spatial and temporal scales. In this regard, changes in salinity have the potential to influence microbial metabolic processes, especially for microorganisms faced with the need to maintain internal osmotic state. del Giorgio and Bouvier [2002] documented increased mortality, decreased cell growth and single-cell activity in the estuarine turbidity maxima, attributing this to physiological stress associated with changing ionic strength. Salinity may also have an direct effect on phylogenetic composition of resident bacterial communities [Barcina *et al.* 1997], which in turn influences microbial metabolic processes. For example, shifts in both phylogenetic composition and the abundance of highly-active cells have been documented along salinity gradients in a wide range of estuaries [del Giorgio and Scarborough 1995, Crump *et al.* 1999, Pinhassi *et al.* 1999, Bouvier and del Giorgio 2002, Cottrell *et al.* 2006] and have been linked to the variability and magnitude of community-level metabolic processes [Yokokawa *et al.* 2004]. Concurrent changes in metabolism and salinity may also be related to changes in resource availability, as the supply and quality of organic matter varies dramatically along the estuarine gradient. In their study of York River estuary, Schultz and Ducklow [2000] observed distinct differences in the type of DOM used by bacterioplankton at different salinity ranges, while Schultz et al. [2003] report a strong correlation between bacterial metabolism and salinity, with highest values of specific growth at the freshwater endmember. Unlike the effect of temperature, the role of salinity in the regulation of microbial metabolism represents the effect of a complex combination of underlying mechanisms that are not altogether understood. In the only experiments we know of to assess salinity impacts, Shiaris [1989] found no relationship between *in situ* salinity and PAH utilization in Boston Harbor.

Dissolved Oxygen Concentrations

Microbial degradation of organic matter occurs most effectively in the presence of oxygen, thus the availability of dissolved oxygen (DO) plays an extremely important role in the longevity and cycling of organic matter in aquatic systems. Organic matter loading to estuarine ecosystems typically exceeds the input of oxygen that would be necessary to convert it all into CO_2, resulting in the accumulation of unconsumed organic matter, particularly in the sediments. Researchers have found that the oxygen demand of 1ml of sediment in the Georgia coastal zone during summer was equivalent to the total amount of dissolved oxygen in almost 1 L of overlying water [Day *et al.* 1989]. Oxygen consumption in the water column alone is lower, yet still substantial and at rates typical of estuarine systems can drive the water column hypoxic (i.e. < 2mg L^{-1}) in 5-7 days [Jonas 1997]. The availability of dissolved oxygen is also determined to a large extent by ambient water temperatures, as temperature regulates the solubility of oxygen in water and mediates the magnitude of the very metabolic processes (discussed above) that consume oxygen. Sediment oxygen consumption (SOC) rates generally increase with increasing temperature until bottom water concentrations of dissolved oxygen fall below 2.5 mg L^{-1}, at which point SOC rates become oxygen limited [Cowan and Boynton 1996]. Likewise, dissolved oxygen concentrations tend to be higher in colder months when solubility is higher and pelagic and benthic respiration is lower [Hopkinson, Jr. and Smith 2005]. In this regard, temperature and oxygen interact to regulate the relationship between oxygen availability and microbial metabolism. The primary mechanism is the requirement of oxygen as a terminal electron acceptor in aerobic respiration, with temperature modulating the availability of oxygen directly through the effect on solubility and indirectly via stimulation of oxygen-consuming microbial metabolic processes. As briefly discussed in the introduction, DO appears to have a profound effect on PAH biodegradation in estuarine sediments – seemingly to the exclusion of most other environmental forcing factors [Bauer and Capone 1985, Hurst *et al.* 1996, Boyd *et al.* 2005a].

Types and Lability of NOM in Estuarine Environments

The term "labile organic matter" is often applied to OM utilized for bacterial re-growth in 1-2 weeks. Thus LOM represents the net lability of the highly-labile pool (L1) and that of the intermediate lability (L2) pool [Søndergaard and Middelboe 1995, del Giorgio and Davis 2003]. Although it is tempting to routinely combine the L1 and L2 pools, first order organic matter decay rates (*k*) for these substrate pools are often significantly different. Hendrickson, et al. [2007] demonstrated that *k* values range from 0.094 to 0.001 day^{-1} depending on the source of the organic mater. del Giorgio and Davis [2003] identified similar differences in the shape of slopes describing the consumption of organic matter, with evidence of rapid consumption of the highly-labile fraction relative to that of intermediate lability. However, following the consumption of DOM over longer periods, (i.e. 7-10d) across a wide range of estuaries reveals higher proportion of semi-labile DOM and underscores the importance of this DOM in carbon cycling in estuarine systems. Thus, the most labile substrates in aquatic systems appear to originate from autochthonous sources, specifically phytoplankton [Münster and Chróst 1990, Canuel and Martens 1993, Søndergaard and Middelboe 1995, Connolly and Coffin 1995], and exhibits the highest *k* value of all classes of organic matter routinely observed in estuarine systems [Hendrickson *et al.* 2007]. However, microbial carbon consumption is supplemented by a pool of semi-labile DOM, especially in estuarine systems.

The relative lability of the OM pool exhibits seasonal patterns which are linked to kinetics of supply of allochthonous and autochonous sources [Hamdan and Jonas 2006, Hendrickson *et al.* 2007]. In the Chesapeake Bay and Delaware Bay, during Fall of 2003, on average, the water column L1 pool accounted for 9 to 12% of the total DOM pool (Table 2). Conversely, during spring 2004, the water column L1 pool accounted for 24-34% of total DOM indicating that the supply of autochonous DOM resulting from phytoplankton primary production greatly influenced the lability of the bulk pool. A similar pattern was observed in the mesohaline Potomac River, where the concentration of LDOM was often highest during early spring and late summer in surface waters in conjunction with a bimodal chlorophyll a maximum [Hamdan and Jonas 2006].

Seasonal cycles in sediment LOM concentration and composition have also been observed in Cape Lookout Bight, NC [Canuel and Martens 1993], and San Francisco Bay [Lesen 2006] which reflect autochonous production and supply of LOM to the water column. Deposition of particulate matter to sediments resulting from phytoplankton primary production is the dominant source of LOM in many locations [Zimmerman and Canuel 2001, Sobczak *et al.* 2002, Stepanauskas *et al.* 2005]. In Cape Lookout Bight LOM accounts for approximately 21% of the total sediment OM pool, higher than the average observed in aquatic systems, and results primarily from phytoplankton-derived particulate organic matter. Organic matter derived from terrestrial sources is a minor component of sediment LOM in this location [Canuel and Martens 1993, Sobczak *et al.* 2002, Stepanauskas *et al.* 2005]. A study of San Francisco Bay by Sobczak et al. [2002] revealed that LDOM contributes significantly to the LOM pool and at times equals or exceeds LPOM. Unlike most locations, data from San Francisco indicate that LDOM in sediments results from deposition and hydrolysis of LPOM but also from direct import of LDOM from the water column. This indicates that LDOM import is anomalous in this location compared to other estuaries, however, in general, DOM import to estuarine sediments is rarely studied.

Pulses of LOM or phytoplankton biomass in the water column result in a time lagged (depending on hydraulic residence) depositional event of LOM to sediments. The deposition of LOM may have a significant and predictable impact on mineralization rates of OM as well as contaminants in estuarine sediments. Differential bacterial utilization of OM types impact the resulting composition of the bulk OM in sediments. The rapid uptake of the L1 and L2 fraction will result in an increasingly refractory OM pool [Canuel and Martens 1993, Hendrickson *et al.* 2007] and thus results in a complex mixture of increasingly humic material. Microbial utilization of the remaining (bulk) material is poorly understood due to limited knowledge of chemical structure and substrate utilization rates of the R fraction of OM [Münster 1993, Hendrickson *et al.* 2007]. The presence of high concentrations of ROM in marine and estuarine sediments will enhance sediment affinity for hydrophobic anthropogenic substrates such as PAHs [Xing 1997, Lueking *et al.* 2000, Granberg *et al.*

2005, Parrish *et al.* 2005]. However, an increase in the bulk organic matter content (L1, L2 and R) may decrease the magnitude of sorption of PAHs to sediment particles and thus render PAHs more bioavailable.

Influence of LOM on Mineralization of PAHs

In sediments with elevated concentration of labile substrates, relative to OM poor locations, the utilization of refractory compounds, such as PAH may be reduced due to selective utilization of more labile compounds[Hendrickson *et al.* 2007]. However, some studies indicate that the uptake of recalcitrant OM may actually be enhanced by the presence of organic LOM or other organic compounds [Turnewitsch *et al.* 2007]. Some studies have investigated the impact of OM additions on the bioavailability (to bacteria) of specific PAHs (e.g., naphthalene, fluoranthene, and pyrene). In one study, non-PAH contaminated marine sediments were amended with a 2% increase in total organic carbon resulting from terrestrial lignin (R source) and lysed microalgae (*Tetraselmis* sp.) (LOM source), the latter of which was high in fatty acids and amino acids (L1 and L2) [Granberg *et al.* 2005]. To the same samples, 30 $\mu g\ g^{-1}$ pyrene was added to determine the effect of OM additions on pyrene bioavailability. In surface (oxic) sediments amended with LOM, microbial pyrene mineralization (measured as $^{14}CO_2$ evolution) was 31% higher than observed in unamended sediments and 55% higher than observed in R amended sediments. The addition of LOM may have enhanced mineralization rates due to the addition of both carbon and nitrogen, the latter which may have been the a limiting nutrient. However, the addition of ROM resulted in lower mineralization rates compared to both the LOM-amended and control treatments. Bonding of pyrene to lignin leading to lower bioavailability was used to explain this finding. Xing [1997] likewise observed that sorption of naphthalene to soil organic matter increased as the aromaticity of the soil increased, and relativity lability of the OM pool decreased.

Selck et al. [2005] conducted a experiment to determine the relative bioavailability of fluoranthene after sediments were amended with ROM (terrestrial lignin) and LOM (*Tetraselmis* sp. Lysate). After 45 days of exposure, fluoranthene contaminated sediments amended with LOM had an 80% decline in total organic carbon indicating increased mineralization rates of both fluoranthene as well as bulk OM compared to unamended controls. Conversely, organic carbon content moderately increased in ROM treated sediments. After 45 days of exposure, the lowest concentrations of fluoranthene and highest concentrations of daughter products of fluoranthene were detected in sediments enriched with LOM. Interestingly, in the presence of LOM, the toxicity of fluoranthene to *Amphiura filiformis* (brittle star) increased, likely due to the production of daughter products during microbial degradation.

The studies described above indicate that bacterial mineralization of PAHs in sediments containing LOM may occur more rapidly than in sediments enriched with ROM due to both physicochemical and biological interactions between substrate, contaminant and sediment [Xing 1997, Lueking *et al.* 2000, Granberg *et al.* 2005, Selck *et al.* 2005]. A survey of estuarine sediments in Delaware Bay near Philadelphia, PA showed no correlation between bulk sediment DOM and PAH mineralization (naphthalene, phenanthrene and fluoranthene)

[Boyd *et al.* 2005a]. However, sediment OM was not fractionated to determine the proportion of labile and recalcitrant material. Such information would greatly enhance understanding of PAH mineralization in natural estuarine sediments, specifically after seasonal depositional events of autochonous OM.

The period that LOM concentrations in the water column and sediments in temperate estuaries are highest (late spring to early summer) [Canuel and Martens 1993, Canuel and Zimmerman 1999, Zimmerman and Canuel 2001, Stepanauskas *et al.* 2005] coincides with annual maximum increases in ambient sediment temperature and the onset of seasonal water column hypoxia in some estuaries [Shiah and Ducklow 1994b, Hamdan and Jonas 2006]. Some reports indicate that temperature increases have a minor impact on PAH mineralization rates, and instead the availability of molecular oxygen has the strongest impact on regulating PAH mineralization [Boyd *et al.* 2005b]. The data above indicates that seasonal patterns in PAH mineralization may also be attributable to increased bioavailability of PAHs due to physical-chemical interactions with fresh LOM or stimulation of microbial communities of PAH degraders due to the presence of fresh LOM.

IV. PAH Biodegradation in Estuaries

Ambient PAH concentration is the net result of PAH flux into the media minus chemical and biological removal. In bioremediation studies, PAH concentrations have been measured in sediments over time and if the concentration of a given compound changes little, then it is assumed that the PAH is poorly biodegraded or recalcitrant. However, this determination cannot truly be made in the absence of knowledge of the amount of flux or biodegradation rate. Although flux is difficult to measure in natural systems, there are two reports comparing PAH flux to biodegradation rates in coastal sediments [Pohlman *et al.* 2002, Chadwick *et al.* 2006]. Chadwick *et al.* [2006] concluded that PAH flux to the sediments of San Diego Bay (sediment trap study) was largely negated by PAH biodegradation (radiotracer additions) in bioturbated surface sediments. Other transport processes (e.g. groundwater seepage, diffusion) were several orders of less significant in affecting ambient PAH concentration. Pohlman *et al.* [2002] came to a similar conclusion when examining seasonal PAH transport on particles into the Naval Reserve Basin in Philadelphia though they found that PAH metabolism was a larger component of total heterotrophic organic matter degradation during the winter months (higher oxygen levels).

In six of the seven ecosystems studied (Table 1), average naphthalene mineralization rates were less rapid than either those for phenanthrene or fluoranthene or both. A similar phenomenon has also been reported by Rockne and Strand [2001]. Though we did not find a linear relationship between PAH concentration and mineralization rate, we did find the same general trends of higher mineralization rates being associated with sediments with higher ambient PAH concentrations (Figures 1-7). This trend is more evident is ecosystems with a greater range of ambient PAH concentrations (1 to 10,000 $\mu g\ g^{-1}$), such as Pearl Harbor, HI (Figure 1). PAH flux may be more related to ambient PAH concentration in sediment at the extremes, that is, relatively pristine environments with little anthropogenic impacts would likely have both low ambient PAH concentrations (below 1 $\mu g\ g^{-1}$) and low flux (and likewise low mineralization). Heavily impacted ecosystems (PAH concentration > 100 μg g^{-1}) might be expected to have such high flux that the microbial system is overwhelmed and a high ambient PAH concentration may be maintained despite relatively high mineralization rates. In sediments in most anthropogenically influenced ecosystems (PAH concentration = 1-100 μg g^{-1}), the PAH inputs could be episodic and so PAH flux and ambient concentration would be more uncoupled.

To further investigate the relationship between ambient PAH concentration and PAH biodegradation, we took data from published reports of flask studies of bioremediation treatments. Because PAH mixtures used in the studies typically have different concentrations of the various PAH compounds, the initial concentration of the individual PAH compound was regressed with the amount of that compound that had biodegraded over the course of the treatment.

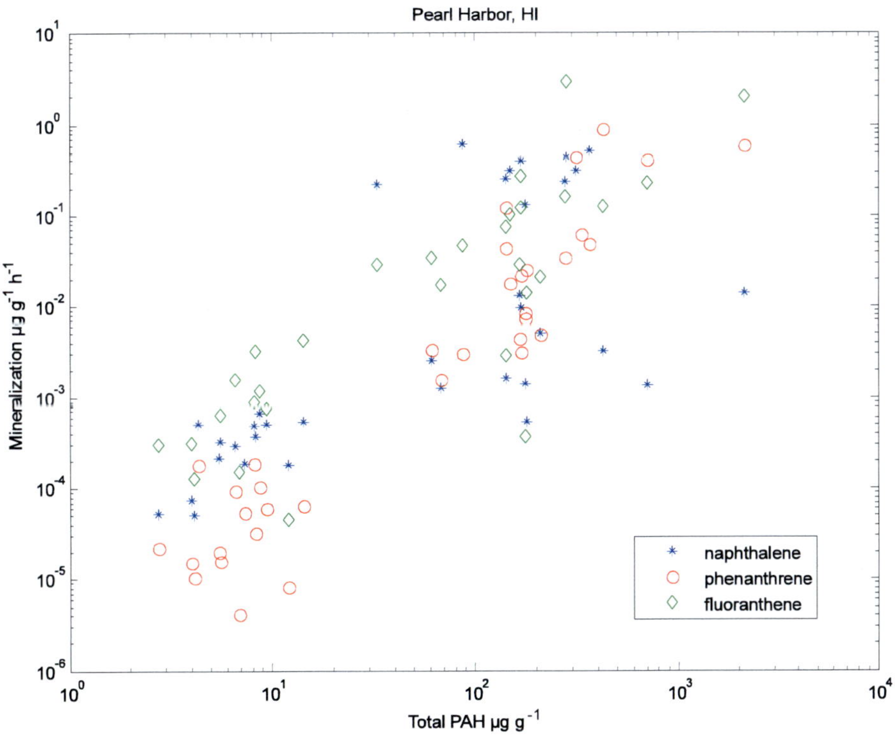

Figure 1. log Total PAH related to log mineralization of naphthalene, phenanthrene and fluoranthene, Pearl Harbor, HI.

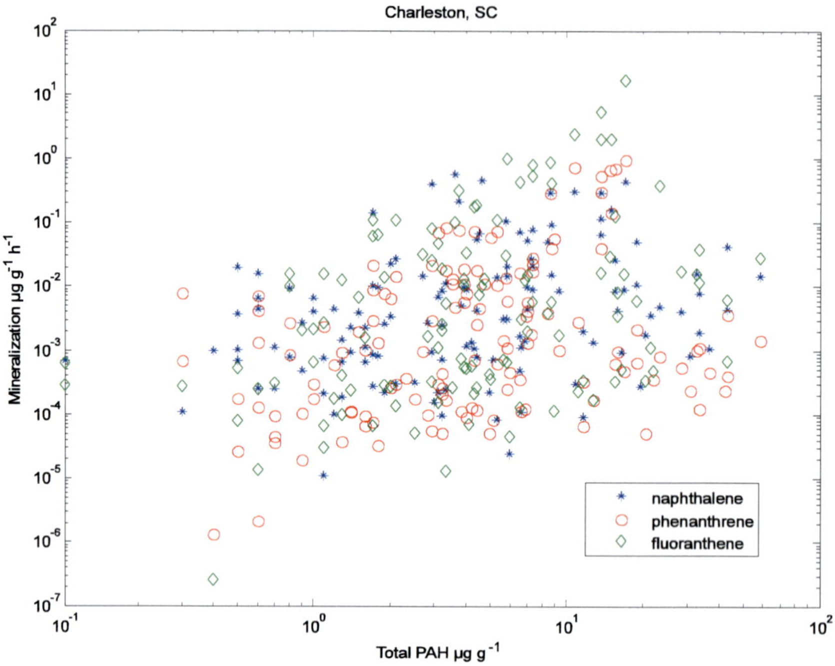

Figure 2. log Total PAH related to log mineralization of naphthalene, phenanthrene and fluoranthene, Charleston Harbor, SC.

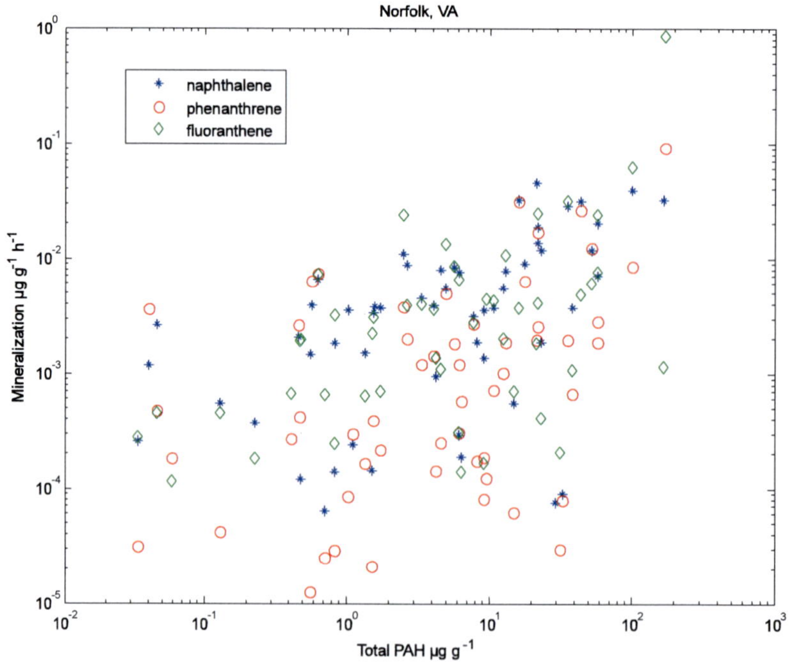

Figure 3. log Total PAH related to log mineralization of naphthalene, phenanthrene and fluoranthene, Norfolk Harbor, VA.

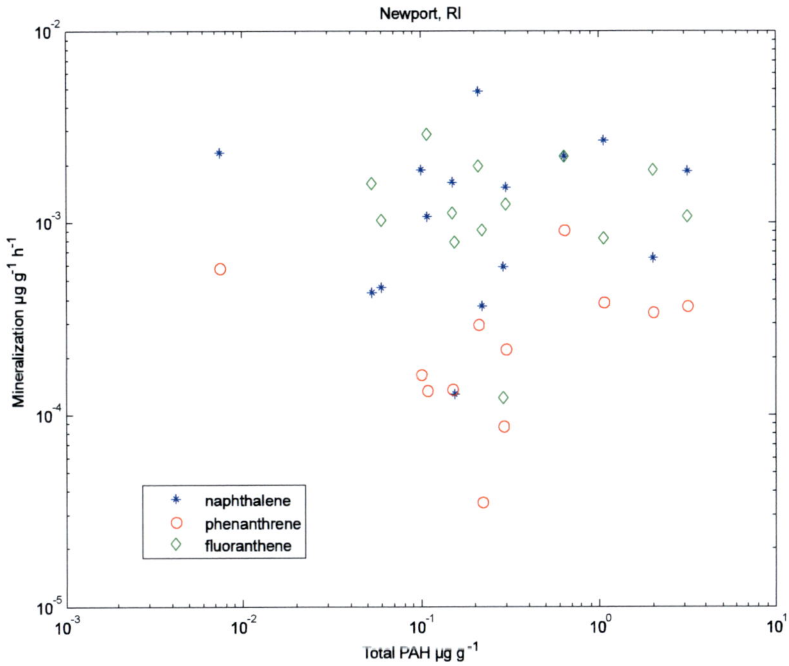

Figure 4. log Total PAH related to log mineralization of naphthalene, phenanthrene and fluoranthene, Newport Harbor, RI.

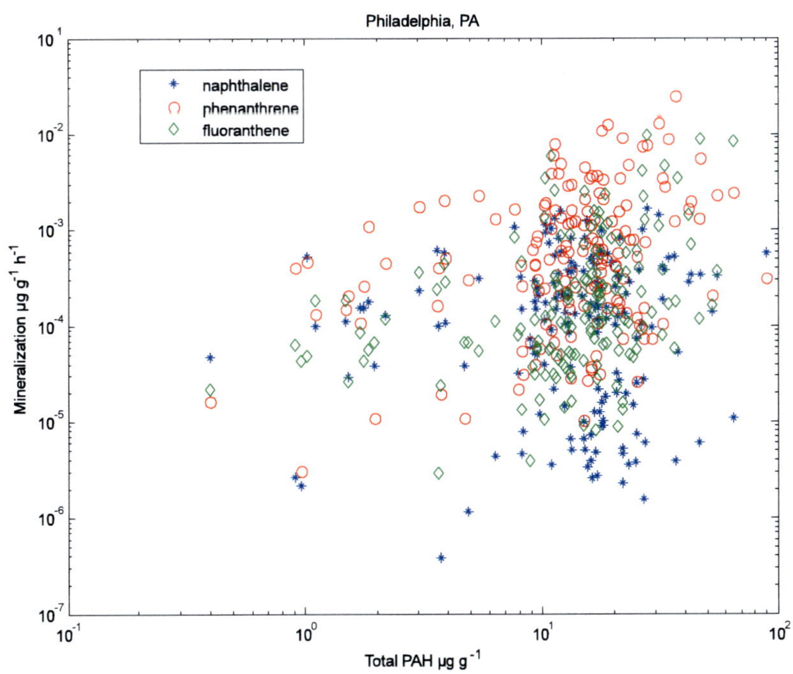

Figure 5. log Total PAH related to log mineralization of naphthalene, phenanthrene and fluoranthene, Delaware Bay and Schuylkill River, Philadelphia, PA.

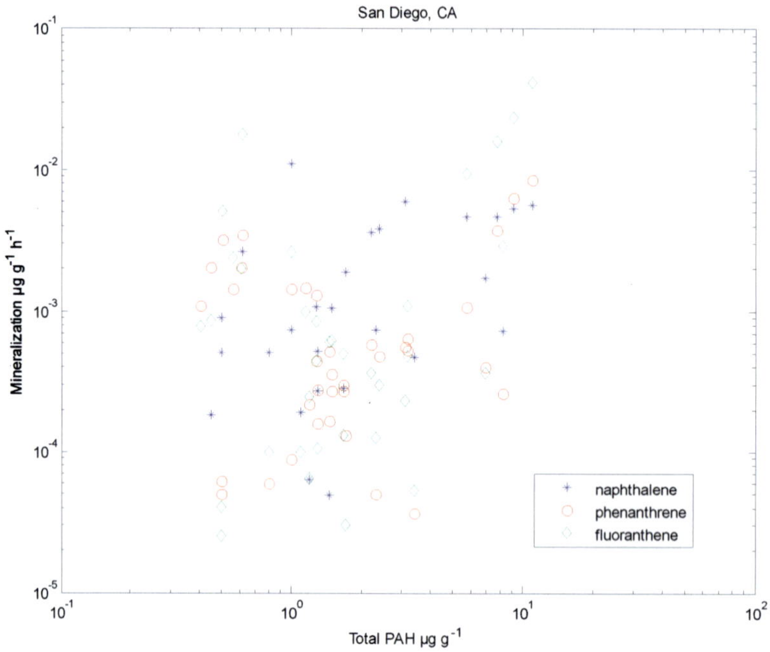

Figure 6. log Total PAH related to log mineralization of naphthalene, phenanthrene and fluoranthene, San Diego Bay, CA.

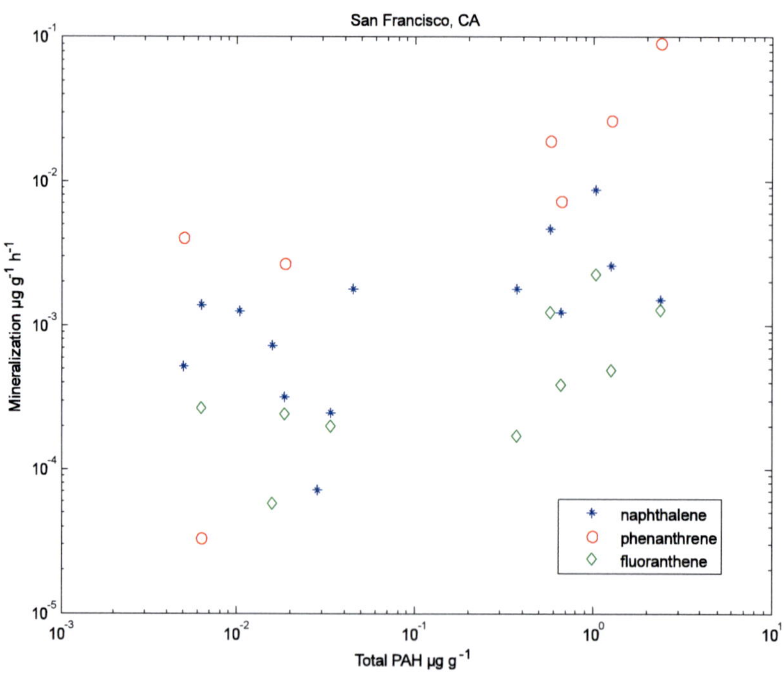

Figure 7. log Total PAH related to log mineralization of naphthalene, phenanthrene and fluoranthene, San Francisco Bay, CA.

If chemical structure of the individual compound was more important to biodegradation than initial concentration, then there would be a non-linear relationship between initial concentration and amount biodegraded. Data calculated from Talley *et al.* [2002], Lantz *et al.* [1997], and Ghosh *et al.* [2003] showed linear relationships (r^2 = 0.96, 0.98, 0.93, respectively) between initial PAH concentration and the amount of the individual PAH that was biodegraded (Figure 8). This suggests that in these bioremediation treatments, chemical structure was not a determinant of whether a compound was recalcitrant to biodegradation.

The relationship was even stronger (r^2 = 1.0) in Ko *et al.* [1995] data comparing the concentration of individual PAHs in sediment trap samples with those that eventually make it through the biodegradation process in the water column and nepheloid layer into the sediment (Figurc 8). This relationship even held when comparing unbranched phenanthrene (C0) to branched isomers (C1-C5) (r^2 = 0.99, Figure 8) [Dutta and Harayama 2000]. Ratio changes among parent and branched isomers of PAHs are often given as evidence of biodegradation or weathering of the hydrocarbons. Interestingly, the C6 and C7 isomers were biodegraded much less than was predicted based on their initial concentration suggesting that only the C0:C6 or C0:C7 may be useful as a relative measure of biodegradation.

Impact of Temperature and Salinity

While salinity may play a part in controlling NOM utilization, there was no evidence for salinity control on sediment production or PAH mineralization in this study ($r^2 < 0.20$, $P < 0.05$).

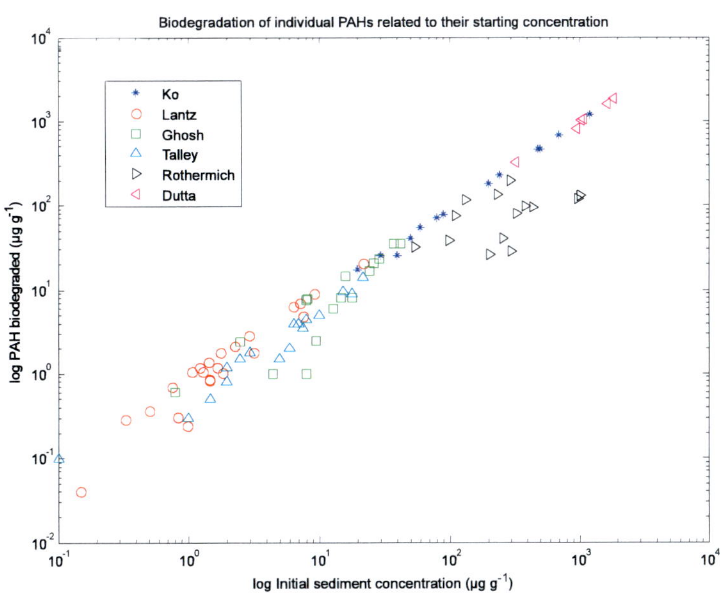

Figure 8. Initial concentration related to degraded amount with 2-5 ring PAHs spiked into sediment. Linear regressions: $y = 0.60x - 0.43$; $r^2 = 0.97$, [Talley *et al.* 2001]; $y = 0.94x - 0.25$; $r^2 = 0.98$, [Lantz *et al.* 1997]; $y = 0.88x - 2..7$; $r^2 = 0.93$, [Ghosh *et al.* 2003]; $y = 0.98x - 9.1$; $r^2 = 1.0$, [Ko *et al.* 1995]; $y = 1.0x - 37$; $r^2 = 0.99$, [Dutta and Harayama 2000].

Data for each site were subjected to linear regression both in terms of raw data and as seasonal averages where available [see Boyd *et al.* 2005 for methods]. Our results are similar to those of Shiaris [1989a]. Salinity, being a conservative tracer, should indicate what proportion of oceanic to estuarine end-member resides in a given parcel of water. Though OM lability varies between freshwater, estuarine and seawater end-members, sediment PAH mineralization and microbial activities (bacterial production, [Smith and Azam 1992])could not be related to mixing processes. In estuarine systems exposed to tidal salinity fluctuations, microorganisms should be less likely to undergo metabolic shifts due to salnity changes. Early reports suggested changes in salnity might be responsible for flocculation of particles and eventual settlement [Sholkovitz 1976], thus perhaps providing additional substrate to the sediments when mixing is intense. It is now widely accepted that estuarine mixing is less universal in dictating particle deposition in most of an estuary [Fisher *et al.* 1998]. Organisms would thus be less likely to ramp up their metabolism when sensing higher ionic strength. Finally, chronic deposition of particle-associated PAHs might likely obscure any fine-scale salinity or temperature relation to microbial PAH degradation in

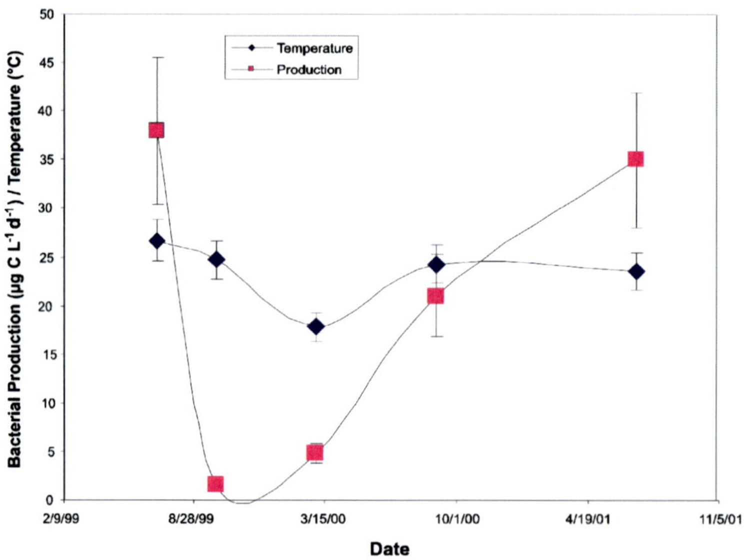

Figure 9. Average temperature and bacterial production in an fuel-impacted tidally-influenced groundwater well (Norfolk, VA).

As with salinity, we were unable to correlate any microbial activity index with temperature ($r^2 < 0.38$, $P < 0.05$). This was somewhat surprising for bacterial production, which is known to vary seasonally with temperature. In order to gain a higher resolution, we sampled a groundwater well which was contaminated with fuel hydrocarbons (significant levels of PAHs) and known to be tidally influenced. The well was located at the Norfolk Naval Station in VA (USA). Sampling was performed every two hours for two days. The well was so sampled five times over a two year period. Over the entire sampling period bacterial production was found to loosely track seasonal temperature shifts (Figure 9). Similar temperature effects were not seen with naphthalene, phenanthrene or fluoranthene mineralization. PAH mineralization was highest in March, May and October, representing cooler months. As with sediments in Delaware Bay near Philadelphia, we hypothesize higher temperatures stimulate overall heterotrophic production which is fueled by more labile carbon substrates negating a direct relationship between production and PAH mineralization [Boyd *et al.* 2005a].

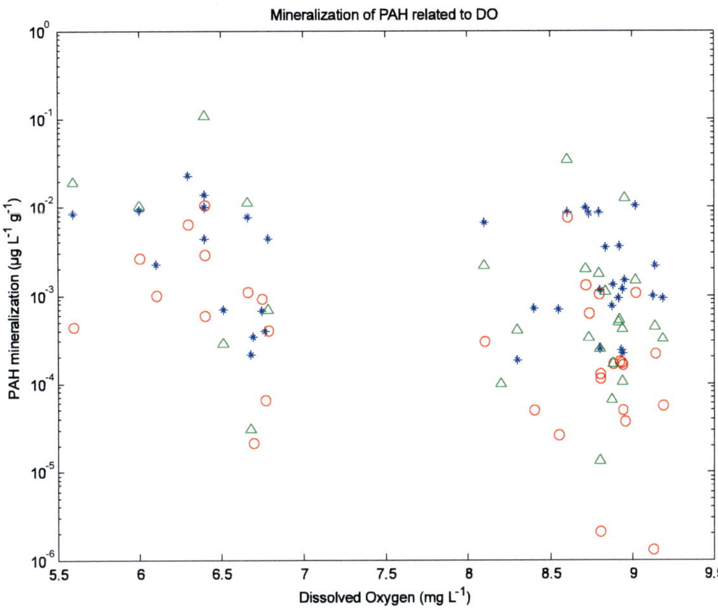

Figure 10. Relationship between sediment PAH mineralization rate and bottom water DO in Charleston Harbor estuary.

Impact of Dissolved Oxygen

PAHs are known to be degraded under anoxic conditions, however much of the burden to the sediments comes from chronic deposition of particle-associated PAHs [Ko and Baker 1995, Arzayus *et al.* 2001, Pohlman *et al.* 2002, Countway *et al.* 2003, Ko *et al.* 2003]. We are aware of only one study which examined the influence of multiple forcing factors on PAH biodegradation in estuarine sediments which encompassed all temperate seasons [Pohlman *et al.* 2002, Boyd *et al.* 2005a]. In this study of the Delaware Bay and Schuylkill River, a positive relationship was observed for naphthalene, phenanthrene and fluoranthene mineralization and ambient concentration, although linear regressions were not significant (Figure 5). The influence of temperature and DO were evaluated and it was found that DO was the primal forcing factor dictating mineralization rate in surface sediments. The overall conclusion was that during cooler months, when overall heterotrophic activity is suppressed, relatively cool, highly oxygenated water allows for maximum PAH biodegradation rates. At these times of year, deposition is low, so PAH mineralization can approximate PAH settling. During warmer months, when high overall production utilizes DO toward hypoxia in surface sediments, PAH mineralization is low and relatively high deposition rates increase net PAH accumulation in surface sediments.

Most of the other studies described in this chapter spanned several seasons, but only in Charleston, SC and Chesapeake Bay (Norfolk) sites were seasonal data available. In the case of Charleston, DO was not available from shipboard sensors, however for several of the field campaigns, representative numbers were available from South Carolina Department of Health and Environmental Control's (www.scdhec.net) monitoring network. PAH mineralization was regressed against DO, but we found no linear relationship as in Delaware Bay (Figure 10). While this was somewhat surprising, it is clear from the data that DO never approached levels of hypoxia during any of our samplings. In the Delaware Bay, two summer seasons showed considerable hypoxia and we observed a significant "inflection" in the relationship between PAH mineralization and DO at about 70% saturation. Because the Delaware Bay work spanned DO down to ~30 % saturation where little PAH mineralization occurred, it is clear that at this site DO plays a major role in facilitating PAH biodegradation.

Unlike the Delaware Bay work, DO did not correlate with PAH biodegradation or bacterial production in Chesapeake Bay sediments (Norfolk). Much like relationships between PAH concentration and degradation, correlations were not significant, but a general trend showing higher rates coincident with higher DO was evident (Figure 11 as an example). Seasonal measurements of temperature, salinity, DO, and PAH mineralization were averaged for the seven seasonal samplings in Norfolk. Unlike the Delaware Bay samples, correlations were insignificant for all potential forcing factors ($r^2 < 0.25$, $P < 0.05$). Data from the higher resolution sampling at the groundwater well showed a clearer trend toward higher DO relating to higher PAH mineralization (Figure 12). Correlation coefficients for all three substrates were higher, but still not significant ($r^2 < 0.45$, $P < 0.05$). While the relationship was strong in the Delaware Bay samples, it cannot be stated that a statistically significant relationship exists between DO and PAH mineralization a t the other sites included in this study.

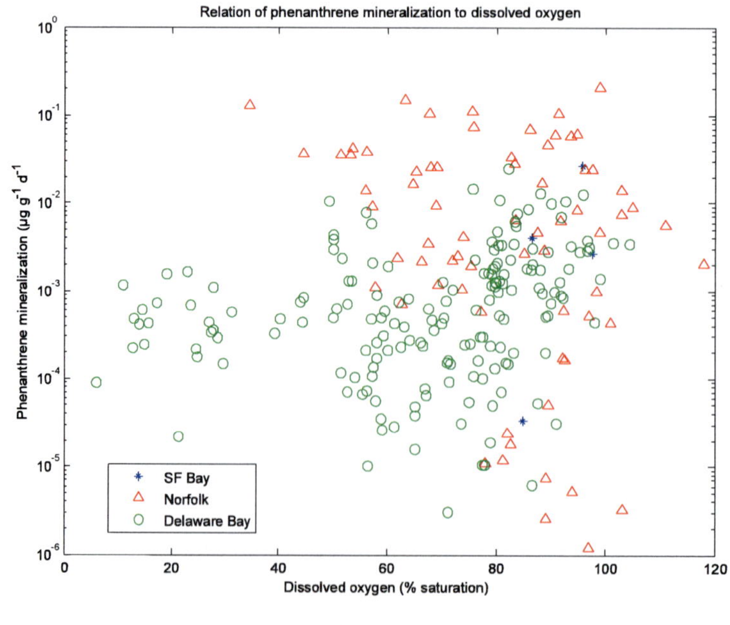

Figure 11. Relationship between sediment phenanthrene mineralization rate and bottom water DO in labeled estuary.

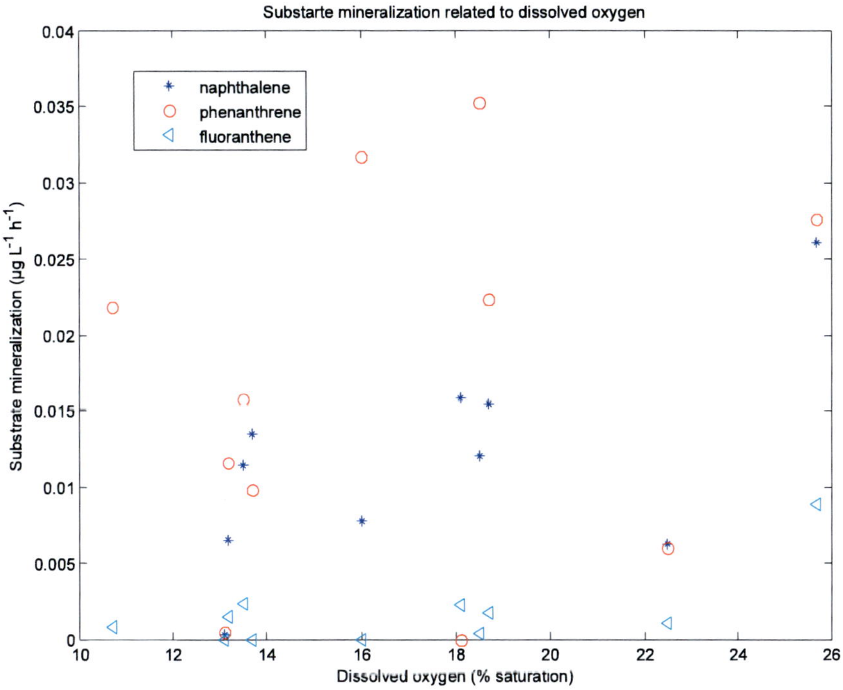

Figure 12. PAH mineralization related to DO over tidal cycles in fuel-contaminated groundwater.

Impact of Recalcitrant DOM

In order to determine the impact of recalcitrant OM on sediment bacterial production and PAH mineralization, lignin phenols were measured for selected sites [Goñi and Montgomery 2000]. If lignin behaves as a proxy for ROM in estuaries (as a component of humic material), it's sedimentary content should have either a positive (by stimulating degraders to become a larger proportion of the overall microbial assemblage) or negative (by competing with PAHs for microbial attack) impact on PAH biodegradation. Previous studies have linked the addition of labile OM with increases in PAH biodegradation, while addition of ROM hinders PAH biodegradation. We have found no correlation with overall heterotrophic production (which should be a proxy for availability of LOM) and PAH mineralization in any of our data. In natural systems, LOM should out-compete PAHs for available limiting nutrients and DO when the microbial assemblage is active (i.e. during warmer periods in temperate estuaries). As with others, we have observed coherence between ambient temperature and bacterial production [*c.f.* Boyd *et al.* 2005, Figure 14], although statistical regressions have generally been insignificant.

In San Francisco Bay and Charleston Harbor sediments, there was a very loose negative relationship between PAH mineralization and total lignin content for naphthalene and fluoranthene mineralization (Figure 13) while no relationship was observed for phenanthrene mineralization. In Norfolk Harbor sediments, we observed no or a very loose positive relationship between total lignin content and sediment PAH mineralization (Figures 14). In all cases, it was not possible to correlate sediment PAH mineralization with sediment lignin content (r^2 < 0.25, P < 0.05). The presence of lignin in sediments likely provides an environmental stimulus to harbor if not deploy the enzymes for aromatic hydrocarbon degradation. We hypothesized that a concentration-dependent biodegradation response similar to what is observed with total PAH content would also be observed with lignin. At this point, it is not possible to assess the impact of *in situ* ROM on PAH biodegradation in the environment, though ROM additions appear to inhibit PAH utilization in laboratory experiments.

V. Conclusion

In this study we have relied on large datasets to visualize trends in PAH mineralization and some of the environmental forcing factors which might impact its rate. PAHs represent a refractory energy and carbon source that is degradable by estuarine sediment microbes (and likely particle-associated or attached water column bacteria). PAH were degraded in every site presented here, making biodegradation potential a ubiquitous phenomenon. If broad specificity enzyme systems are expressed in response to sediment PAHs, it is likely that PAH structure has less control over PAH biodegradation rate as *in situ* PAH concentration. Ample concentration is likely necessary to stimulate biodegradation. Two to five ring PAHs generally showed a linear response to biodegradation based on their starting concentration(s). Sediment PAH concentration appears to be the major factor found to control PAH biodegradation. Higher concentrations stimulate microorganisms to switch from utilizing labile substrates to PAHs.

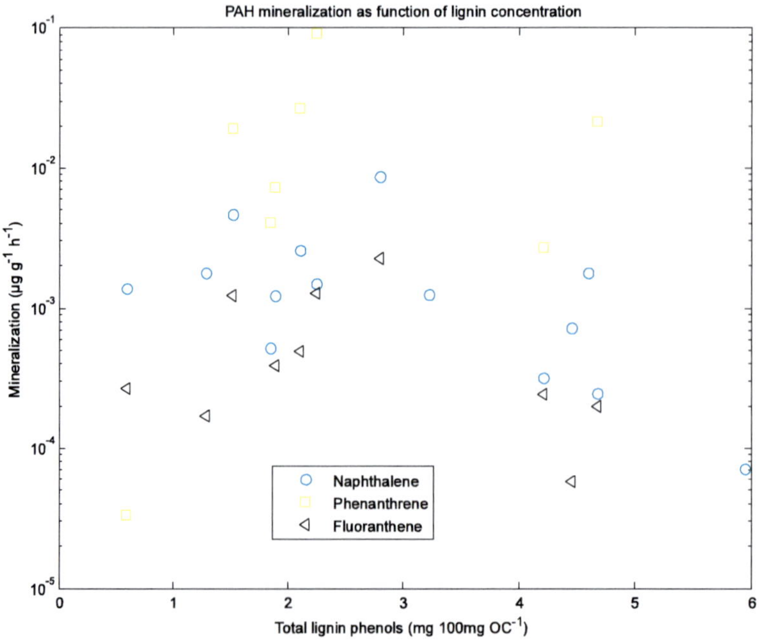

Figure 13. Relationship between total sediment lignin phenols and PAH mineralization, San Francisco Bay.

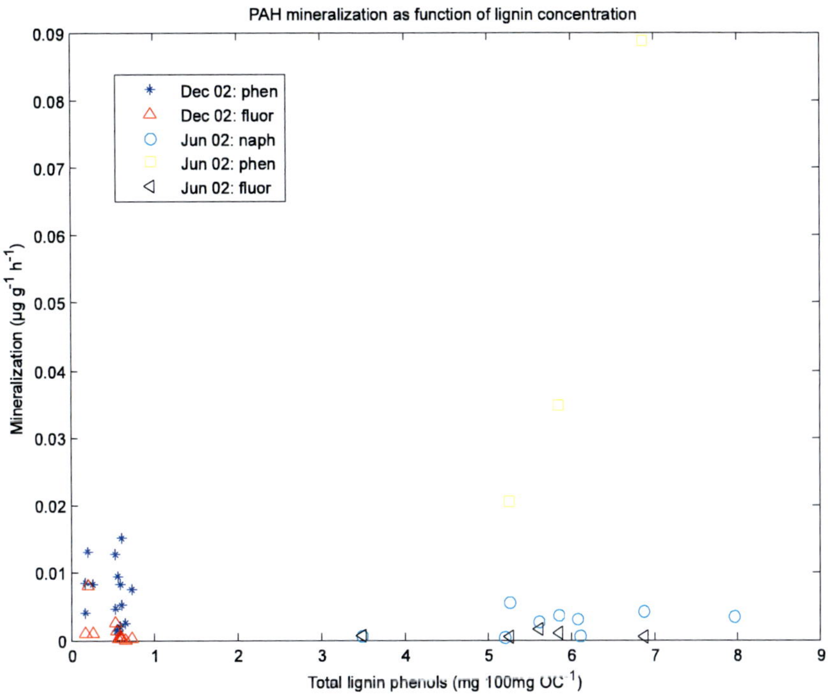

Figure 14. Relationship between total sediment lignin phenols and PAH mineralization, Norfolk Harbor, Chesapeake Bay.

PAH biodegradation in surface sediments was also found to be influenced by DO availability. In cooler months, when DO is relatively high, but overall heterotrophic production is temperature depressed, PAH mineralization may be highest. This observation was demonstrated markedly in Delaware Bay [Boyd *et al.* 2005a] and to a lesser degree in a tidally-influenced fuel-impacted groundwater plume. However, DO was not determined to be a major controlling factor for PAH biodegradation in Chesapeake Bay sediments near Norfolk, VA. Temperature and salinity were not determined to be controlling factors in PAH biodegradation, although temperature may control overall heterotrophic activity in temperate estuaries. Recalcitrant organic matter represented by lignin might impact PAH biodegradation by both up-regulating the synthesis of catabolic enzymes and competing with PAH for those enzymes. We observed both positive and negative relationships between lignin content and PAH mineralization. A needed component for exploring sediment PAH degradation is a more thorough understanding of ROM and PAH fluxes. All of our studies have focused on industrial waterways where field sampling represents only a snapshot in time. Mass balance estimates suggest relatively high particle-bound PAH fluxes to the sediments [Arzayus *et al.* 2001, Pohlman *et al.* 2002, Countway *et al.* 2003]. Regardless of which factors control rates, biodegradation must account for much of the difference between total PAH deposition and ultimate PAH burial in estuarine sediments.

Acknowledgements

The opinions and assertions contained herein are not to be construed as official or reflecting the views of the US Navy or the Federal Government at large. Mention of tradenames, or commercial products does not constitute endorsement or recommendation for use. The authors would like to thank the Captains and Crews of the *R/V Cape Henlopen*, *R/V Pt. Sur*, *R/V Barnes*, *R/V Cape Hatteras* and the *R/V Ecos* for logistical support for this work.

References

Aelion, C. M., D. C. Dobbins, and F. K. Pfaender. 1989. Adaptation of aquifer microbial communities to the biodegradation of xenobiotic compounds: Influence of substrate concentration and preexposure. *Environmental Toxicology and Chemistry* 8:75-86.

Aelion, C. M., C. M. Swindoll, and F. K. Pfaender. 1987. Adaptation to and Biodegradation of Xenobiotic Compounds by Microbial Communities from a Pristine Aquifer. *Applied and Environmental Microbiology* 53:2212-2217.

Alexander, M. 1985. Biodegradation of organic chemicals. *Environmental Science and Technology* 19:106-111.

Amon, R., H. P. Fitznar, and R. Benner. 2001. Linkages among the bioreactivity, chemical composition, and diagenetic state of marine dissolved organic matter. *Limnology and Oceanography* 46:287-297.

Apitz, S. E., E. Arias, S. A. Clawson, e. W. Lin, R. J. Melcher, and B. B. Hemmingsen. 1999. The development of a sterile, PAH-spiked, aged marine sediment for biodegradation experiments: chemical results. *Organic Geochemistry* 30:891-900.

Apple J.K., Boyd T.J. and Smith E.M. Temperature, salinity and the coherence of BP and chlorophyll-a in estuarine systems. *Journal of Coastal Research*. (submitted).

Apple, J. K., P. A. del Giorgio, and W. M. Kemp. 2006. Temperature regulation of bacterial production, respiration, and growth efficiency in a temperate salt-marsh estuary. *Aquatic Microbial Ecology* 43:243-254.

Arzayus, K. M., R. M. Dickhut, and E. A. Canuel. 2001. Fate of atmospherically deposited polycyclic aromatic hydrocarbons (PAHs) in Chesapeake Bay. *Environmental Science and Technology* 35:2178-2183.

Atlas, R. M. 1981. Microbial degradation of petroleum hydrocarbons: An environmental perspective. *Microbiological Reviews* 45:180-209.

Barcina, I., P. LeBaron, and J. Vives-Rego. 1997. Survival of allochthonous bacteria in aquatic systems: a biological approach. *FEMS Microbiology Ecology* 23:1-9.

Bartholomew, G. W., and F. K. Pfaender. 1983. Influence of spatial and temporal variations on organic pollutant biodegradation rates in an estuarine environment. *Applied and Environmental Microbiology* 45:103-109.

Bastviken, D., and L. Tranvik. 2001. The leucine incorporation method estimates bacterial growth equally well in both oxic and anoxic lake waters. *Applied and Environmental Microbiology* 67:2916-2921.

Bauer, J. E., and D. G. Capone. 1985. Degradation and mineralization of the polycyclic aromatic hydrocarbons anthracene and naphthalene in intertidal marine sediments. *Applied and Environmental Microbiology* 50:81-90.

Boethling, R. S., and M. Alexander. 1979a. Effect of concentration of organic chemicals on their biodegradation by natural microbial communities. *Applied and Environmental Microbiology* 37:1211-1216.

Boethling, R. S., and M. Alexander. 1979b. Microbial degradation of organic compounds at trace levels. *Environmental Science and Technology* 13:989-991.

Boethling, R. S., B. Gregg, R. Frederick, N. W. Gabel, S. E. Campbell, and A. Sabljic. 1989. Expert systems survey on biodegradation of xenobiotic chemicals. *Ecotoxicology and Environmental Safety* 18:252-267.

Bouvier, T. C., and P. A. del Giorgio. 2002. Compositional changes in free-living bacterial communities along a salinity gradient in two temperate estuaries. *Limnology and Oceanography* 47:453-470.

Boyd, T. J., M. T. Montgomery, B. J. Spargo, D. C. Smith, R. B. Coffin, C. A. Kelley, and J. G. Mueller. 2001. Effects of oxygenation on hydrocarbon biodegradation in a hypoxic environment. *Bioremediation Journal* 5:145-157.

Boyd, T. J., M. T. Montgomery, J. K. Steele, J. W. Pohlman, S. R. Reatherford, B. J. Spargo, and D. C. Smith. 2005a. Dissolved oxygen saturation controls PAH biodegradation in freshwater estuary sediments. *Microbial Ecology* 49:226-235.

Boyd, T. J., D. M. Wolgast, I. Rivera-Duarte, O. Holm-Hansen, C. D. Hewes, A. Zirino, and D. B. Chadwick. 2005b. Effects of dissolved and complexed copper on heterotrophic bacterial production in San Diego Bay. *Microbial Ecology* 49:353-366.

Brubaker, G. R., and H. F. Stroo. 1992. *In situ* bioremediation of aquifers containing polyaromatic hydrocarbons. *J.Hazard.Mater.* 32:163-177.

Canuel, E. A., and C. S. Martens. 1993. Seasonal-Variations in the Sources and Alteration of Organic-Matter Associated with Recently-Deposited Sediments. *Organic Geochemistry* 20:563-577.

Canuel, E. A., and A. R. Zimmerman. 1999. Composition of particulate organic matter in the southern Chesapeake Bay: *Sources and reactivity. Estuaries* 22:980-994.

Chadwick D.B., Kirtay V., Blake A.C., Arias E., Davidson B., Guerrero J., Apitz S.E., Germano J., Maa J., Smith C., Paulsen R., Montgomery M.T., Ziebis W., Gieskes J. and Groves J. (2006) Pathway Ranking for In-place Sediment Management (CU1209): Final Site II Report - Pearl Harbor. CU1209, 1-857.

Chandra, S. D., C. H. Ward, and J. B. Hughes. 1996. Biodegradation of sorbed fluorene in sediment slurries. Hazard.Waste Hazard.Mater. 13:375-385.

Chapelle, F. H. 1999. Bioremediation of petroleum hydrocarbon-contaminated ground water: The perspectives of history and hydrology. *Ground water* 37:122-132.

Chin, Y.-P., and P. M. Gschwend. 1992. Partitioning of polycyclic aromatic hydrocarbons to marine porewater organic colloids. *Environmental Science and Technology* 26:1621-1626.

Chung, N., and M. Alexander. 1998. Differences in sequestration and bioavailability of organic compounds aged in dissimilar soils. *Environmental Science and Technology* 32:855-860.

Coates, J. D., R. T. Anderson, and D. R. Lovley. 1996. Oxidation of Polycyclic Aromatic Hydrocarbons under Sulfate-Reducing Conditions. *Applied and Environmental Microbiology* 62:1099-1101.

Coates, J. D., J. Woodward, J. Allen, P. Philp, and D. R. Lovley. 1997. Anaerobic Degradation of Polycyclic Aromatic Hydrocarbons and Alkanes in Petroleum-Contaminated Marine Harbor Sediments. *Environmental Science and Technology* 63:3589-3593.

Connolly, J. P., and R. B. Coffin. 1995. Model of carbon cycling in planktonic food webs. *Journal of Environmental Engineering* 121:682-690.

Cornelissen, G., and Ö. Gustafsson. 2005. Importance of unburned coal carbon, black carbon, and amorphous organic carbon to phenanthrene sorption in sediments. *Environmental Science and Technology* 39:764-769.

Cornelissen, G., and D. T. H. M. Sijm. 1996. An energy budget model for the biodegradation and cometabolism of organic substances. *Chemosphere* 33:817-830.

Cottrell, M. T., R. R. Malmstrom, V. Hill, A. E. Parker, and D. L. Kirchman. 2006. The metabolic balance between autotrophy and heterotrophy in the western Arctic Ocean. Deep-Sea Research Part I: *Oceanographic Research Papers* 53:1831-1844.

Countway, R. E., R. M. Dickhut, and E. A. Canuel. 2003. Polycyclic aromatic hydrocarbon (PAH) distributions and associations with organic matter in surface waters of the York River, VA Estuary. *Organic Geochemistry* 34:209-224.

Cowan, J. L. W., and W. R. Boynton. 1996. Sediment-water oxygen and nutrient exchanges along the longitudinal axis of Chesapeake Bay: Seasonal patterns, controlling factors and ecological significance. *Estuaries* 19:562-580.

Crump, B. C., E. V. Armbrust, and J. A. Baross. 1999. Phylogenetic Analysis of Particle-Attached and Free-Living Bacterial Communities in the Columbia River, Its Estuary, and the Adjacent Coastal Ocean. *Applied and Environmental Microbiology* 65:3192-3204.

Da Silva, M. L. B., R. C. Daprato, D. E. Gomez, J. B. Hughes, C. H. Ward, and P. J. J. Alvarez. 2006. Comparison of bioaugmentation and biostimulation for the enhancement of dense nonaqueous phase liquid source zone bioremediation. *Water Environment Research* 78:2456-2465.

Dagley, S. 1987. Lessons from biodegradation. Annual Reviews of Microbiology 41:1-23.

Day J. W. J., C. A. S. Hall, W. M. Kemp, and A. Yanez-Arancibia. 1989. Estuarine Ecology. John Wiley and Sons, New York.

Dearden, J. C. 2002. Prediction of environmental toxicity and fate using quantitative structure-activity relationships (QSARs). *Journal of the Brazilian Chemical Society* 13:754-762.

del Giorgio, P. A., and T. C. Bouvier. 2002. Linking the physiologic and phylogenetic successions in free- living bacterial communities along an estuarine salinity gradient. *Limnology and Oceanography* 47:471-486.

del Giorgio P. A., and J. Davis. 2003. Patterns in dissolved organic matter lability and consumption across aquatic ecosystems. Elsevier Science, San Diego.

del Giorgio, P. A., and G. Scarborough. 1995. Increase in the proportion of metabolically active bacteria along gradients of enrichment in freshwater and marine plankton:

implications for estimates of bacterial growth and production rates. *Journal of Plankton Research* 17:1905-1924.

Domde, P., A. Kapley, and H. J. Purohit. 2007. Impact of bioaugmentation with a consortium of bacteria on the remediation of wastewater-containing hydrocarbons. *Environmental Science and Pollution Research* 14:7-11.

Dutta, T. K., and S. Harayama. 2000. Fate of crude oil by the combination of photooxidation and biodegradation. *Environmental Science and Technology* 34:1500-1505.

Ellis, D. E., E. J. Lutz, J. M. Odam, R. J. Buchanan, C. L. Bartlett, M. D. Lee, M. R. Harkness, and K. A. DeWeerd. 2000. Bioaugmentation for accelerated *in situ* anaerobic bioremediation. *Environmental Science and Technology* 34:2254-2260.

Eschenbach, A., R. Wienverg, and B. Mahro. 1998. Fate and stability of nonextractable residues of [^{14}C]PAH in contaminated soils under environmental stress conditions. *Environmental Science and Technology* 32:2585-2590.

Felip, M., M. L. Pace, and J. J. Cole. 1996. Regulation of planktonic bacterial growth rates: The effects of temperature and resources. Microbial Ecology 31:15-28.

Fisher, T. R., J. D. Hagy, and E. Rochelle-Newall. 1998. Dissolved and particulate organic carbon in Chesapeake Bay. *Estuaries* 21:215-229.

Garcia-Blanco, S., A. D. Venosa, M. T. Suidan, K. Lee, S. Cobanli, and J. R. Haines. 2007. Biostimulation for the treatment of an oil-contaminated coastal salt marsh. *Biodegradation* 18:1-15.

Ghosh, U., J. R. Zimmerman, and R. G. Luthy. 2003. PCB and PAH speciation among particle types in contaminated harbor sediments and effects on PAH bioavailability. *Environmental Science and Technology* 37:2209-2217.

Goldman, J. C., D. A. Caron, and M. R. Dennet. 1987. Regulation of gross growth efficiency and ammonium regeneration in bacteria by C:N ratio. *Limnology and Oceanography* 32:1239-1252.

Goñi, M. A., and S. Montgomery. 2000. Alkaline CuO oxidation with a microwave digestion system: Lignin analyses of geochemical samples. *Analytical Chemistry* 72:3116-3121.

Granberg, M. E., R. Hansen, and H. Selck. 2005. Relative importance of macrofaunal burrows for the microbial mineralization of pyrene in marine sediments: impact of macrofaunal species and organic matter quality. *Marine Ecology-Progress Series* 288:59-74.

Guerin, W. F., and G. E. Jones. 1989. Estuarine ecology of phenanthrene-degrading bacteria. *Estuarine Coastal and Shelf Science* 29:115-130.

Gustafsson, Ö., and P. M. Gschwend. 1997. Soot as a strong partition medium for polycyclic aromatic hydrocarbons in aquatic systems. *Molecular Markers in Environmental Geochemistry* 671:365-381.

Hamdan, L. J., and R. B. Jonas. 2006. Seasonal and interannual dynamics of free-living bacterioplankton and microbially labile organic carbon along the salinity gradient of the Potomac River. *Estuaries and Coasts* 29:40-53.

Hatzinger, P. B., and M. Alexander. 1995. Effect of aging of chemicals in soil on their biodegradability and extractability. *Environmental Science and Technology* 29:537-545.

Hendrickson, J., N. Trahan, E. Gordon, and Y. Ouyang. 2007. Estimating relevance of organic carbon, nitrogen, and phosphorus loads to a blackwater river estuary. *Journal of the American Water Resources Association* 43:264-279.

Herbes, S. E., and L. R. Schwall. 1978. Microbial transformation of polycyclic aromatic hydrocarbons in pristine and petroleum-contaminated sediments. *Applied and Environmental Microbiology* 35:306-316.

Hickman, G. T., and J. T. Novak. 1989. Relationship between subsurface biodegradation rates and microbial density. *Environmental Science and Technology* 23:525-532.

Hoch, M. P., and D. L. Kirchman. 1993. Seasonal and interannual variability in bacterial production and biomass in a temperate estuary. Marine Ecology - Progress Series 98:283-295.

Hopkinson C. S., Jr., and E. M. Smith. 2005. Estuarine respiration: an overview of benthic, pelagic and whole system respiration. Oxford University Press, Oxford.

Hou, B. K., L. P. Wackett, and L. B. M. Ellis. 2003. Microbial pathway prediction: A functional group approach. *Journal of Chemical Information and Computer Sciences* 43:1051-1057.

Hurst, C. J., R. C. Sims, J. L. Sims, D. L. Sorensen, J. E. McLean, and S. Huling. 1996. Polycyclic aromatic hydrocarbon biodegradation as a function of oxygen tension in contaminated soil. *J.Hazard.Mater.* 51:193-208.

Jaworska, J. S., R. S. Boethling, and P. H. Howard. 2003. Recent developments in broadly applicable structure-biodegradability relationships. *Environmental Toxicology and Chemistry* 22:1710-1723.

Jiménez, N., M. Viñas, J. Sabaté, S. Díez, J. M. Bayona, A. M. Solanas, and J. Albaiges. 2006. The *Prestige* oil spill. 2. Enhanced biodegradation of a heavy fuel oil under field conditions by the use of an oleophilic fertilizer. *Environmental Science and Technology* 40:2578-2585.

Johnsen,A.R., and U.Karlson 2007. Diffuse PAH contamination of surface soils: environmental occurrence, bioavailability, and microbial degradation. *Applied Microbiology and Biotechnology in press.*

Johnsen, A. R., S. Schmidt, T. K. Hybholt, S. Henriksen, C. S. Jacobsen, and O. Andersen. 2007. Strong impact on the polycyclic aromatic hydrocarbon (PAH)-degrading community of a PAH-polluted soil but marginal effect on PAH degradation when priming with bioremediated soil dominated by mycobacteria. *Applied and Environmental Microbiology* 73:1474-1480.

Johnsen, A. R., l. Y. Wick, and H. Harms. 2005. Principles of microbial PAH-degradation in soil. Environmental Pollution 133:71-84.

Jonas, R. B. 1997. Bacteria, dissolved organics and oxygen consumption in salinity stratified Chesapeake Bay, an anoxia paradigm. *Am.Zool.* 37:612-620.

Jonker, M. T. O., S. B. Hawthorne, and A. A. Koelmans. 2005. Extremely Slowly Desorbing Polycyclic Aromatic Hydrocarbons from Soot and Soot-like Materials: Evidence by Supercritical Fluid Extraction. *Environmental Science and Technology* 39:7889-7895.

Jouanneau, Y., J. C. Willison, C. Meyer, S. Krivobok, N. Chevron, G. Blake, and J. L. Besombes. 2005. Stimulation of pyrene mineralization in freshwater sediments by bacterial and plant bioaugmentation. *Environmental Science and Technology* 39:5729-5735.

King, G. M., and T. Berman. 1984. Potential effects of isotopic dilution on apparent respiration in ^{14}C heterotrophy experiments. Marine Ecology - Progress Series 19:175-180.

Kjelleberg, S., M. Hermansson, P. Marden, and G. W. Jones. 1987. The transient phase between growth and nongrowth of heterotrophic bacteria, with emphasis on the marine environment. *Annual Reviews of Microbiology* 41:25-49.

Ko, F.-C., and J. E. Baker. 1995. Partitioning of hydrophobic organic contaminants to resuspended sediments and plankton in the mesohaline Chesapeake Bay. *Mar.Chem.* 49:171-188.

Ko, F.-C., L. P. Sanford, and J. E. Baker. 2003. Internal recycling of particle reactive organic chemicals in the Chesapeake Bay water column. *Marine Chemistry* 81:163-176.

Kompare, B. 1998. Estimating environmental pollution by xenobiotic chemicals using QSAR (QSBR) models based on artificial intelligence. *Water Science and Technology* 37:9-18.

Lantz, S. E., M. T. Montgomery, W. W. Schultz, P. H. Pritchard, B. J. Spargo, and J. G. Mueller. 1997. Constituents of an organic wood preservative that inhibit the fluoranthene-degrading activity of *Sphingomonas paucimobilis* strain EPA505. Environmental Science and Technology 31:3573-3580.

Lesen, A. E. 2006. Sediment organic matter composition and dynamics in San Francisco Bay, California, USA: Seasonal variation and interactions between water column chlorophyll and the benthos. *Estuarine Coastal and Shelf Science* 66:501-512.

Lohmann, R., J. K. MacFarlane, and P. M. Gschwend. 2005. Importance of black carbon to sorption of native PAHs, PCBs, and PCDDs in Boston and New York, Harbor sediments. *Environmental Science and Technology* 39:141-148.

Lomas, M. W., P. M. Glibert, F.-K. Shiah, and E. M. Smith. 2002. Microbial processes and temperature in Chesapeake Bay: current relationships and potential impacts of regional warming. *Global Change Biology* 8:51-70.

Lueking, A. D., W. L. Huang, S. Soderstrom-Schwarz, M. S. Kim, and W. J. Weber. 2000. Relationship of soil organic matter characteristics to organic contaminant sequestration and bioavailability. *Journal of Environmental Quality* 29:317-323.

Luthy, R. G., G. R. Aiken, M. L. Brusseau, S. D. Cunningham, P. M. Gschwend, J. J. Pignatello, M. Reinhard, S. J. Traina, W. J. Weber Jr, and J. C. Westall. 1997. Sequestration of hydrophobic organic contaminants by geosorbents. *Environmental Science and Technology* 31:3341-3347.

Macnaughton, S. J., J. R. Stephen, A. D. Venosa, G. A. Davis, Y.-J. Chang, and D. C. White. 1999. Microbial Population Changes during Bioremediation of an Experimental Oil Spill. *Applied and Environmental Microbiology* 65:3566-3574.

Madigan M. T., J. M. Martinko, and J. Parker. 2003. Brock Biology of Microorganisms., 10th edition. Prentice Hall, Upper Saddle River, NJ.

Malone, T. C., L. H. Crocker, S. E. Pike, and B. W. Wendler. 1988. Influences of river flow on the dynamics of phytoplankton production in a partially stratified estuary. *Marine Ecology Progress Series* 48:235-249.

McNally, D. L., J. R. Mihelcic, and D. R. Lueking. 1998. Biodegradation of three- and four-ring polycyclic aromatic hydrocarbons under aerobic and denitrifying conditions. *Environmental Science and Technology* 32:2633-2639.

Meredith, C. E., and M. Radosevich. 1998. Bacterial degradation of homo- and heterocyclic aromatic compounds in the presence of soluble/colloidal humic acid. Journal of Environmental Science and Health - Part B Pesticides, Food Contaminants, and Agricultural Wastes 33:17-36.

Mishra, S., J. Jyot, R. C. Kuhad, and B. Lal. 2001. Evaluation of inoculum addition to stimulate in situ bioremediation of oily-sludge-contaminated soil. *Applied and Environmental Microbiology* 67:1675-1681.

Montgomery, M. T., C. L. Osburn, T. J. Boyd, S. Reatherford, and D. C. Smith. 2002. Bacterial degradation of polycyclic aromatic hydrocarbons in surface sediments of coastal ecosystems. *Soil and Sediment Contamination* 11:1021.

Morita R. Y. 1974. Temperature Effects on Marine Microorganisms. University Park Press, Baltimore, MD.

Morrill, P. L., G. Lacrampe-Couloume, G. F. Slater, B. E. Sleep, E. A. Edwards, M. L. McMaster, D. W. Major, and B. Sherwood Lollar. 2005. Quantifying chlorinated ethene degradation during reductive dechlorination at Kelly AFB using stable carbon isotopes. *Journal of Contaminant Hydrology* 76:279-293.

Münster, U. 1993. Concentrations and fluxes of organic-carbon substrates in the aquatic environment. Antonie Van Leeuwenhoek 63:243-274.

Münster U., and R. J. Chróst. 1990. Origin, composition, and microbial utilization of dissolved organic matter. Springer-Verlag, New York.

Ogino, A., H. Koshikawa, T. Nakahara, and H. Uchiyama. 2001. Succession of microbial communities during a biostimulation process as evaluated by DGGE and clone library analyses. J.Appl.Microbiol. 91:625-635.

Paris, D. F., and J. E. Rogers. 1986. Kinetic concepts for measuring microbial rate constants: Effects of nutrients on rate constants. *Applied and Environmental Microbiology* 51:221-225.

Parrish, Z. D., M. K. Banks, and A. P. Schwab. 2005. Assessment of contaminant lability during phytoremediation of polycyclic aromatic hydrocarbon impacted soil. *Environmental Pollution* 137:187-197.

Pfaender, F. K., and G. W. Bartholomew. 1982. Measurement of aquatic biodegradation rates by determing heterotrophic uptake of radiolabeled pollutants. *Applied and Environmental Microbiology* 44:159-164.

Pinhassi, J., F. Azam, U. L. Zweifel, R. A. Long, J. Martinez, U. L. Zweifel, and Å. Hagström. 1999. Coupling between bacterioplankton species composition, population dynamics, and organic matter degradation. *Aquatic Microbial Ecology* 17:13-26.

Pohlman, J. W., R. B. Coffin, C. S. Mitchell, M. T. Montgomery, B. J. Spargo, J. K. Steele, and T. J. Boyd. 2002. Transport, deposition and biodegradation of particle bound polycyclic aromatic hydrocarbons in a tidal basin of an industrial watershed. *Environ.Monit.Assess.* 75:155-167.

Pomeroy, L. R., J. E. Sheldon, W. M. Sheldon, and F. Peters. 1995. Limits to growth and respiration of bacterioplankton in the Gulf of Mexico. *Marine Ecology Progress Series* 117:259-268.

Quantin, C., E. J. Joner, J. M. Portal, and J. Berthelin. 2005. PAH dissipation in a contaminated river sediment under oxic and anoxic conditions. *Environmental Pollution* 134:315-322.

Raymond, P. A., and J. E. Bauer. 2000. Bacterial consumption of DOC during transport through a temperate estuary. *Aquatic Microbial Ecology* 22:1-12.

Reid, B. J., K. C. Jones, and K. T. Semple. 2000. Bioavailability of persistent organic pollutants in soils and sediments - A perspective on mechanisms, consequences and assessment. *Environmental Pollution* 108:103-112.

Reid, B. J., G. L. Northcott, K. C. Jones, and K. T. Semple. 1998. Evaluation of spiking procedures for the introduction of poorly water soluble contaminants into soil. *Environmental Science and Technology* 32:3224-3227.

Revilla, M., A. Iriarte, I. Madariaga, and E. Orive. 2000. Bacterial and phytoplankton dynamics along a trophic gradient in a shallow temperate estuary. *Estuarine Coastal and Shelf Science* 50:297-313.

Roch, F., and M. Alexander. 1997. Inability of bacteria to degrade low concentrations of toluene in water. *Environmental Toxicology and Chemistry* 16:1377-1383.

Rockne, K. J., and S. E. Strand. 2001. Anaerobic biodegradation of naphthalene, phenanthrene, and biphenyl by a denitrifying enrichment culture. *Water Research* 35:291-299.

Röling, W. F. M., M. G. Milner, D. M. Jones, F. Fratepietro, R. P. J. Swannell, F. Daniel, and I. M. Head. 2004. Bacterial Community Dynamics and Hydrocarbon Degradation during a Field-Scale Evaluation of Bioremediation on a Mudflat Beach Contaminated with Buried Oil. *Applied and Environmental Microbiology* 70:2603-2613.

Rose A. H. 1967. Thermobiology. Academic Press Inc., London, UK.

Sampou, P., and W. M. Kemp. 1994. Factors regulating plankton community respiration in Chesapeake Bay. *Marine Ecology Progress Series* 110:249-258.

Schaffner, L. C., R. M. Dickhut, S. Mitra, P. W. Lay, and C. Brouwer-Riel. 1997. Effects of physical chemistry and bioturbation by estuarine macrofauna on the transport of hydrophobic organic contaminants in the benthos. *Environmental Science and Technology* 31:3120-3125.

Schultz Jr, G. E., and H. W. Ducklow. 2000. Changes in bacterioplankton metabolic capabilities along a salinity gradient in the York River estuary, Virginia, USA. *Aquatic Microbial Ecology* 22:163-174.

Schultz Jr, G. E., E. D. White III, and H. W. Ducklow. 2003. Bacterioplankton dynamics in the York River estuary: primary influence of temperature and freshwater inputs. *Aquatic Microbial Ecology* 30:135-148.

Scow, K. M., S. K. Schmidt, and M. Alexander. 1989. Kinetics of biodegradation of mixtures of substrates in soil. *Soil Biol.Biochem.* 21:703-708.

Scow, K. M., S. Simkins, and M. Alexander. 1986. Kinetics of mineralization of organic compounds at low concentrations in soil. *Applied and Environmental Microbiology* 51:1028-1035.

Selck, H., M. E. Granberg, and V. E. Forbes. 2005. Impact of sediment organic matter quality on the fate and effects of fluoranthene in the infaunal brittle star Amphiura filiformis. *Marine Environmental Research* 59:19-45.

Shiah, F.-K., and H. W. Ducklow. 1994a. Temperature and substrate regulation of bacterial abundance, production and specific growth-rate in Chesapeake Bay, USA. *Marine Ecology - Progress Series* 103:297-308.

Shiah, F.-K., and H. W. Ducklow. 1994b. Temperature regulation of heterotrophic bacterioplankton abundance, production, and specific growth rate in Chesapeake Bay. *Limnology and Oceanography* 39:1243-1258.

Shiaris, M. P. 1989. Phenanthrene mineralization along a natural salinity gradient in an urban estuary, boston-harbor, massachusetts. *Microbial Ecology* 18:135-146.

Shiaris, M. P. 1989. Seasonal biotransformation of naphthalene, phenanthrene, and benzo[A]Pyrene in surficial estuarine sediments. *Applied and Environmental Microbiology* 55:1391-1399.

Shimp, R. J., and F. K. Pfaender. 1987. Effect of adaptation to phenol on biodegradation of monosubstituted phenols by aquatic microbial communities. *Applied and Environmental Microbiology* 53:1496-1499.

Sholkovitz, E. R. 1976. Flocculation of dissolved organic and inorganic matter during the mixing of river water and seawater. Geochimica et Cosmochimica Acta 40:831-845.

Sisler, F. D., and C. E. Zobell. 1947. Microbial utilization of carcinogenic hydrocarbons. *Science* 106:521-522.

Small, C., and R. J. Nicholls. 2003. A Global Analysis of Human Settlement in Coastal Zones. *Journal of Coastal Research* 19:584-599.

Smith, D. C., and F. Azam. 1992. A simple, economical method for measuring bacterial protein synthesis rates in seawater using ^{3}H-leucine. *Mar.Microb.Food Webs* 6:107-114.

Smith, E. M., and W. M. Kemp. 1995. Seasonal and regional variations in plankton community production and respiration for Chesapeake Bay. *Marine Ecology - Progress Series* 116:217-231.

Smith, E. M., and W. M. Kemp. 2003. Planktonic and bacterial respiration along an estuarine gradient: responses to carbon and nutrient enrichment. *Aquatic Microbial Ecology* 30:251-261.

Sobczak, W. V., J. E. Cloern, A. D. Jassby, and A. B. Muller-Solger. 2002. Bioavailability of organic matter in a highly disturbed estuary: The role of detrital and algal resources. *Proc.Natl.Acad.Sci.U.S.A.* 99:8101-8105.

Søndergaard, M., and M. Middelboe. 1995. A cross-system analysis of labile dissolved organic-carbon. *Marine Ecology - Progress Series* 118:283-294.

Staroscik, A. M., and D. C. Smith. 2004. Seasonal patterns in bacterioplankton abundance and production in Narragansett Bay, Rhode Island, USA. *Aquatic Microbial Ecology* 35:275-282.

Steffensen, W. S., and M. Alexander. 1995. Role of competition for inorganic nutrients in the biodegradation of mixtures of substrates. *Applied and Environmental Microbiology* 61:2859-2862.

Stepanauskas, R., M. A. Moran, B. A. Bergamaschi, and J. T. Hollibaugh. 2005. Sources, bioavailability, and photoreactivity of dissolved organic carbon in the Sacramento-San Joaquin River Delta. *Biogeochemistry* 74:131-149.

Talley, J. W., U. Ghosh, S. G. Tucker, J. S. Furey, and R. G. Luthy. 2002. Particle-scale understanding of the bioavailability of PAHs in sediment. *Environmental Science and Technology* 36:477-483.

Trably, E., D. Patureau, and J. P. Delgenes. 2003. Enhancement of polycyclic aromatic hydrocarbons removal during anaerobic treatment of urban sludge. *Water Science and Technology* 48:53-60.

Turnewitsch, R., B. Domeyer, and G. Graf. 2007. Experimental evidence for an effect of early-diagenetic interaction between labile and refractory marine sedimentary organic matter on nitrogen dynamics. *Journal of Sea Research* 57:270-280.

Vacca, D. J., W. F. Bleam, and W. J. Hickey. 2005. Isolation of Soil Bacteria Adapted To Degrade Humic Acid-Sorbed Phenanthrene. *Applied and Environmental Microbiology* 71:3797-3805.

Venosa, A. D., M. T. Suidan, K. Lee, S. E. Cobanli, S. Garcia-Blanco, and J. R. Haines. 2002. Bioremediation of oil-contaminated coastal freshwater and saltwater wetlands. Pages 139-148 *in*.

Venosa, A. D., M. T. Suidan, B. A. Wrenn, K. L. Strohmeier, J. R. Haines, B. L. Eberhart, D. King, and E. Holder. 1996. Bioremediation of an Experimental Oil Spill on the Shoreline of Delaware Bay. *Environmental Science and Technology* 30:1764-1775.

Wammer, K. H., and C. A. Peters. 2005. Polycyclic aromatic hydrocarbon biodegradation rates: A structure-based study. *Environmental Science and Technology* 39:2571-2578.

Wick, L., A. De Munain, D. Springael, and H. Harms. 2002. Responses of Mycobacterium sp. LB501T to the low bioavailability of solid anthracene. *Applied Microbiology and Biotechnology* 58:378-385.

Wiggins, B. A., and M. Alexander. 1988. Role of chemical concentration and second carbon sources in acclimation of microbial communities for biodegradation. *Applied and Environmental Microbiology* 54:2803-2807.

Wiggins, B. A., S. H. Jones, and M. Alexander. 1987. Explanations for the acclimation period preceding the mineralization of organic chemicals in aquatic environments. *Applied and Environmental Microbiology* 53:791-796.

Wolfe, N. L., D. F. Paris, W. C. Steen, and G. L. Baughman. 1980. Correlation of microbial degradation rates with chemical structure. Environmental *Science and Technology* 14:1143-1144.

Wrenn, B. A., M. T. Suidan, K. L. Strohmeier, B. L. Eberhart, G. J. Wilson, and A. D. Venosa. 1997. Nutrient transport during bioremediation of contaminated beaches: Evaluation with lithium as a conservative tracer. *Water Research* 31:515-524.

Xing, B. 1997. The effect of the quality of soil organic matter on sorption of naphthalene. *Chemosphere* 35:633-642.

Yokokawa, T., T. Nagata, M. T. Cottrell, and D. L. Kirchman. 2004. Growth rate of the major phylogenetic bacterial groups in the Delaware estuary. *Limnology and Oceanography* 49:1620-1629.

Yu, K. S. H., A. H. Y. Wong, K. W. Y. Yau, Y. S. Wong, and N. F. Y. Tam. 2005. Natural attenuation, biostimulation and bioaugmentation on biodegradation of polycyclic aromatic hydrocarbons (PAHs) in mangrove sediments. *Marine Pollution Bulletin* 51:1071-1077.

Zimmerman, A. R., and E. A. Canuel. 2001. Bulk organic matter and lipid biomarker composition of Chesapeake Bay surficial sediments as indicators of environmental processes. *Estuarine Coastal and Shelf Science* 53:319-341.

In: Microbial Ecology Research Trends
Editor: Thijs Van Dijk, pp. 39-63
ISBN: 978-1-60456-179-1

Chapter II

Diversity of Bacterial Symbiosis in Stinkbugs

Yoshitomo Kikuchi*[1], *Takahiro Hosokawa and Takema Fukatsu
Institute for Biological Resources and Functions,
National Institute of Advanced Industrial Science and Technology (AIST), Japan

Abstract

Many animals and plants establish symbiotic associations with benevolent microorganisms. In particular, most of insects that feed exclusively on restricted diets, such as plant sap, vertebrate blood or woody material, usually possess symbiotic microorganisms inside their body. Recent studies have suggested that more than 50% of insect species may be benefited from such endosymbiotic partners. Among these insects, members of the Heteroptera, known as true bugs or stinkbugs, exhibit an extraordinary diversity in their microbial symbiotic system. Such diversity is exemplified by: (i) physiological impact of the symbiotic associations on the host fitness ranging from parasitic to mutualistic; (ii) interdependence of the symbiotic associations ranging from facultative to obligate; (iii) location of the symbiont ranging from extracellular in gut cavity to intracellular in specialized cells; (iv) among the extracellular associations, specialized symbiotic structures such as midgut sacs and tubular outgrowths, whose cavity harbors specific symbiotic bacteria, varying in their number and arrangement; and (v) mechanisms for symbiont transmission ranging from vertical to environmental. Here, the current biological knowledge of the diverse stinkbug-bacteria symbioses is reviewed, which will provide novel insights into and useful model systems for understanding of the host-symbiont interactions and evolution.

1 Corresponding author. Mailing address: Institute for Biological Resources and Functions, National Institute of Advanced Industrial Science and Technology (AIST), Tsukuba Central 6, 1-1-1 Higashi, Tsukuba, Ibaraki 305-8566, Japan. Phone: +81-29-861-8612; Fax: +81-29-861-6080; E-mail: y-kikuchi@aist.go.jp.

Introduction

Symbiosis, or "living together", generally means close associations between different organisms. The nature of symbiotic associations varies on the basis of interactions between the partners. In mutualism, for example, the outcome of the interactions is beneficial to both of the partners. In parasitism, one partner exploits the other in a more or less selfish manner. Among a diverse array of symbiotic associations, the most cohesive forms are found in those called "endosymbiosis", wherein one partner called symbiont lives inside the body, or sometimes inside the cells, of the other called host. On account of their predominance and diversity, bacteria are regarded as leading players in such endosymbiotic associations. Sometimes the bacterial associates are detrimental or even lethal for the host organisms, and are regarded as "pathogens". Meanwhile, a number of bacterial associates greatly benefit the host organisms with their diverse metabolic abilities, resulting in physiological, ecological and evolutionary advantages of the host organisms. For example, leguminous plants are able to grow under nitrogen-deficient conditions, and the ability is conferred by *Rhizobium* spp. and allied nodule symbionts that fix atmospheric dinitrogen and provide the hosts with nitrate (reviewed in Gualtieri and Bisseling 2000). Recent development of molecular tools such as diagnostic PCR, quantitative PCR, molecular phylogenetic analysis and *in situ* hybridization has revolutionized the whole area of the microbial ecology, which opened a new window to the fascinating world of endosymbiosis.

Symbiotic associations with bacteria are known from a broad range of animals, plants and other organisms (Margulis and Fester 1991; Ruby et al. 2004), among which insects probably comprise the largest group that takes advantage of symbiotic bacteria (Bourtzis and Miller 2003; Buchner 1965; Moran 2007). The insects that feed exclusively on restricted diets, such as plant sap, vertebrate blood or woody material, usually possess well-developed endosymbiotic systems, where the symbionts are involved in provision of essential nutrients and/or digestion of food materials for their host (Douglas 1989). The symbiosis between aphids and their intracellular bacteria *Buchnera* is rated among the ultimate forms of mutualistic associations. Almost all aphids possess huge cells, called bacteriocytes or mycetocytes, that are specialized for harboring the symbiotic bacteria. *Buchnera* provides the host aphid with essential amino acids that are deficient in phloem sap. The symbiont is passed from mother to offspring by an elaborate mechanism called transovarial transmission, in which the symbiont directly infects the embryos in the maternal body (Braendle et al. 2003; Mira and Moran 2002; Miura et al. 2003). The symbiont phylogeny mirrors the host phylogeny, indicating strict vertical transmission over evolutionary time (Clark et al. 2000; Moran et al. 1993). During the long-term intimate association, the genome of *Buchnera* has been drastically reduced, and thus the symbiont cannot survive outside the host cells (Shigenobu et al. 2000). In turn, the host aphid suffers retarded growth, mortality and sterility when deprived of the symbiont, indicating an obligate reliance on each other. As it were, aphid and *Buchnera* are no longer two independent organisms but regarded as an almost inseparable biological entity.

Such highly interdependent and co-speciating host-symbiont associations have been also identified in other insects, such as *Wigglesworthia* in tsetse flies (Aksoy 2000; Chen et al. 1999), *Baumannia* in sharpshooters (Moran et al. 2005; Takiya et al. 2006; Wu et al. 2006),

Carsonella in psyllids (Thao et al. 2000), *Tremblaya* in mealybugs (Thao et al. 2002), *Blochmannia* in carpenter ants (Sauer et al. 2000), *Nardonella* in weevils (Lefevre et al. 2004), and many others. For the past decade, these endosymbiotic systems have significantly contributed to our understanding of host-symbiont associations in insects. However, it has been elusive how these sophisticated symbiotic systems have evolved from more primitive stages of insect-microbe associations, mainly because intermediate stages of host-symbiont associations are lacking in these insect groups.

Stinkbugs belong to the Heteroptera, a taxonomically and phylogeneticaly well-defined insect group involving a diverse array of bacterial endosymbiosis. In this article, the current biological knowledge of the diverse stinkbug-bacteria symbioses is reviewed, which will provide novel insights into and useful model systems for understanding of the host-symbiont interactions and evolution.

Heteroptera: Taxonomy, Habitats, Feeding Types, and Bacterial Symbioses

Insects belonging to the Heteroptera are commonly called "stinkbugs", because they have a habit to exude foul-smelling fluid when disturbed. Many stinkbugs have relevance to various aspects of human life and activity, including a large number of crop pests like chinch bugs and notorious medical and sanitary pests like bedbugs (Schaefer and Panizzi 2000; Schuh and Slater. 1995). The Heteroptera, a term for Greek "different wings", is distinguished from its sister group, the Homoptera, in that their forewings have both membranous and hardened portions, while both insect groups possess a unique shape of sucking mouthparts called proboscis. The Heteroptara consists of more than 38,000 described species (Schuh and Slater. 1995), and is among the largest groups in insects with incomplete metamorphosis. Their habitats and feeding types are impressively diverse. Some species live as parasite of birds and mammals, while others pray on other arthropods, suck mycelia of fungi, or feed on almost all parts of plants, including phloem sap, seeds, and pollen grains. They live in grasslands, under stones, in the soil, and even in spider webs and termite nests. They are also found in the water and on its surface, and furthermore, a few species inhabit the open ocean. Taxonomically, the heteropteran bugs are grouped into seven infraorders: the Enicocephalomorpha, the Dipsocoromorpha, the Gerromorpha, the Nepomorpha, the Leptopodomorpha, the Cimicomorpha, and the Pentatomomorpha (Schuh and Slater. 1995). Phylogenetic analyses on the basis of molecular and morphological data have suggested that the Heteroptera is a well-defined monophyletic group in the Hemiptera (Ouvrard et al. 2000; Schuh 1979; von Dohlen and Moran 1995; Wheeler et al. 1993) and that each of the seven infraorders forms a monophyletic clade, respectively (Schuh 1979; Wheeler et al. 1993).

In the Heteroptera, symbiotic bacteria have been reported from two infraorders, the Cimicomorpha and the Pentatomomorpha (Buchner 1965; Dasch et al. 1984; Glasgow 1914). Buchner (1965) pointed out the general pattern that insects living on nutritionally unbalanced diets such as woody material, plant sap or vertebrate blood tend to possess well-developed endosymbiotic system with microorganisms whereas carnivorous and herbivorous insects feeding on animal or plant tissues do not. The same also applies to heteropteran insects:

stinkbugs feeding exclusively on unbalanced diets such as plant sap or vertebrate blood possess symbiotic bacteria, while predatory and mycophagous species carry no symbionts. In phytophagous stinkbugs, however, not only sap feeders but also a number of seed feeders carry endosymbiotic bacteria. Since plant seeds seem to be more nutritious and balanced than plant sap like phloem sap, the symbiotic associations in seed feeders possibly suggest an exceptional instance of the Buchner's theory in heteropteran insects.

In the Cimicomorpha, blood-sucking species representing the families Reduviidae (assassin bugs) and Cimicidae (bedbugs) harbor endosymbiotic bacteria in gut cavity, inside gut epitherial cells, or in a specialized symbiotic organ called bacteriome or mycetome (Buchner 1965; Dasch et al. 1984; Glasgow 1914). Meanwhile, in the family Miridae, the largest group in the Cimicomorpha and embracing notorious pest species of various crops, no obvious symbionts have been recorded probably because almost all mirids feed on both plants and prey (Eubanks et al. 2003; Schuh and Slater. 1995). In the Pentatomomorpha, almost all members, except for the predatory Asopinae and the mycophagous Aladoidae, are phytophagous, most of which possess symbiotic bacteria in their alimentary tract. In these phytophagous species, conspicuous sacs or outgrowths, called caeca or crypts, are usually developed at a midgut posterior region (midgut 4th section), whose lumen is filled with a number of specific bacterial symbiont cells (Buchner 1965; Dasch et al. 1984; Glasgow 1914). In some species of the Cimicomorpha and the Pentatomomorpha, it has been experimentally demonstrated that elimination of the symbiont causes retarded growth and nymphal mortality of the host insects (Abe et al. 1995; Buchner 1965; Chang 1974; Fukatsu and Hosokawa 2002; Hill et al. 1976; Hosokawa et al. 2006; Huber-Schneider 1957; Müller 1956; Schorr 1957), suggesting important biological roles of the symbiont for the host. In addition to these beneficial symbionts, facultative or parasitic bacterial endosymbionts of the genus *Wolbachia*, which are known to cause a variety of reproductive aberrations in many arthropods (O'Neill et al. 1997), are commonly detected from these heteropterans groups (Kikuchi and Fukatsu 2003).

Diversity of Bacterial Symbiosis in the Heteroptera

A diverse array of symbiotic associations with bacteria have been known from the Heteroptera. Biological features of the associations are also quite diverse, in terms of physiological impact on the host fitness, interdependence of the symbiotic associations, location of the symbiont, formation of symbiotic organs, transmission mechanisms, and others (Table 1).

Table 1. Bacterial symbiosis in the Heteroptera: host feeding habit, symbiont taxonomical affiliation, interaction and association types, symbiont culturability, symbiont localization, and transmission mechanism[a]

Host stinkbug	Infraorder: Family (Subfamily)	Feeding habit	Symbiont	Interaction type	Association type	Symbiont culturability	Symbiont localization	Transmission mechanism
Rhodnius prolixus	Cimicomorpha: Reduviidae (Triatominae)	Haematophagous	*Rhodococcus rhodnii* (Actinobacteria)	Mutualistic	Facultative	○	Extracellular	Coprophagy
Cimex lecturalius	Cimicomorpha: Cimicidae	Haematophagous	Unknown	Mutualistic	Obligate	-	Intracellular	Transovarial transmission
Megacopta punctatissima	Pentatomomorpha: Plataspidae	Phytophagous	*Ishikawaella capsulata* (γ -Proteobacteria)	Mutualistic	Obligate	-	Extracellular	Capsule transmision
Riptortus clavatus	Pentatomomorpha: Alydidae	Phytophagous	*Burkholderia* sp. (β -Proteobacteria)	Mutualistic	Facultative	○	Extracellular	Environmental acquisition
Plautia stali	Pentatomomorpha: Pentatomidae	Phytophagous	Unknown	Mutualistic	Obligate	-	Extracellular	Egg smearing
Nezara viridula	Pentatomomorpha: Pentatomidae	Phytophagous	γ -Proteobacteria	Commensa-listic	Facultative	-	Extracellular	Egg smearing
Many heteropterans		Diverse	*Wolbachia* (α -Proteobacteria)	Parasitic	Facultative	-	Intracellular	Transovarial transmission

[a] Based on the following studies: *Rhodnius prolixus* (Beard et al. 2002; Harrington 1960; Hill et al. 1976); *Cimex lecturalius* (Chang 1974; Chang 1975; Chang & Musgrave 1973; Hypsa & Aksoy 1997); *Megacopta punctatissima* (Fukatsu& Hosokawa 2002; Hosokawa et al. 2005; Hosokawa et al. 2006; Muller 1956; Schneider 1940); *Riptortus clavatus* (Kikuchi et al. 2005; Kikuchi et al. 2007); *Plautia stali* (Abe et al. 1995); *Nezara viridula* (Prado et al. 2006); and *Wolbachia* symbiosis in diverse stinkbug species (Kikuchi & Fukatsu 2003).

In the Heteroptera, (i) physiological impact of the symbiotic associations on the host fitness ranges from parasitic through commensalistic to mutualistic. (ii) Interdependence of the symbiotic associations also varies, ranging from facultative to obligate. In the facultative associations, the symbiont is often culturable outside the host, and the host often suffers no or relatively low fitness disadvantage in the absence of the symbiont. In the obligate associations, by contrast, neither the symbiont nor the host survives without the partner. (iii) Location of the symbiont ranges from extracellular in gut cavity to intracellular in specialized cells. (iv) Among the extracellular associations, specialized symbiotic structures such as sacs and tubular outgrowths in the midgut, called midgut crypts, which harbor specific symbiotic bacteria in the cavity and vary in their number and arrangement across different taxa, are formed (Figure 1); (Buchner 1965; Glasgow 1914; Miyamoto 1961).

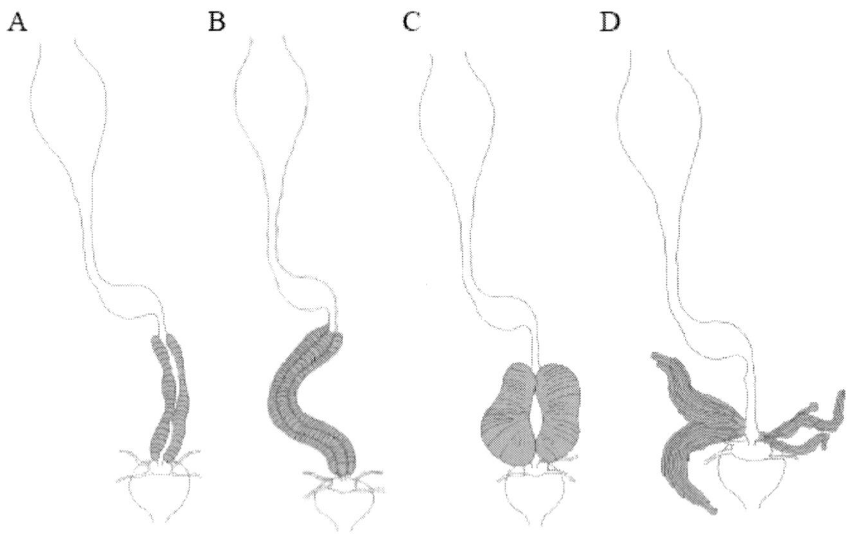

Figure 1. Structural variety of the midgut crypts in the Pentatomomorpha. (A) Two-rows of small crypts in the families Cydnidae, Coreidae, Alydidae, and their relatives. (B) Four-rows of small crypts in the family Pentatomidae and Scutelleridae. (C) A pair of large crypts in the family Acanthosomatidae. (D) Tubular crypts in the family Largidae and its relatives. Crypts are indicated by gray color.

In addition, (v) the mechanisms for symbiont transmission are remarkably diverse in the Heteroptera. From a variety of insects, three types of mechanisms for symbiont transmission have been identified, all of which are of vertical mode (Buchner 1965). Intracellular symbionts like *Buchnera* and *Wolbachia* are typically passed to the next generation via transovarial transmission (Braendle et al. 2003; Frydman et al. 2006; Mira and Moran 2002; Miura et al. 2003), while insects harboring extracellular symbionts have evolved post-hatch symbiont transmission mechanisms, such as superficial bacterial contamination of eggs (egg smearing) and probing of parental bacteria-containing excrement (coprophagy). In the Heteroptera, all of the three vertical transmission manners are known (Buchner 1965). In addition, a unique and elaborate transmission mechanism, called “capsule transmission”, has been reported from the family Plataspidae, in which symbiont-filled particles, called

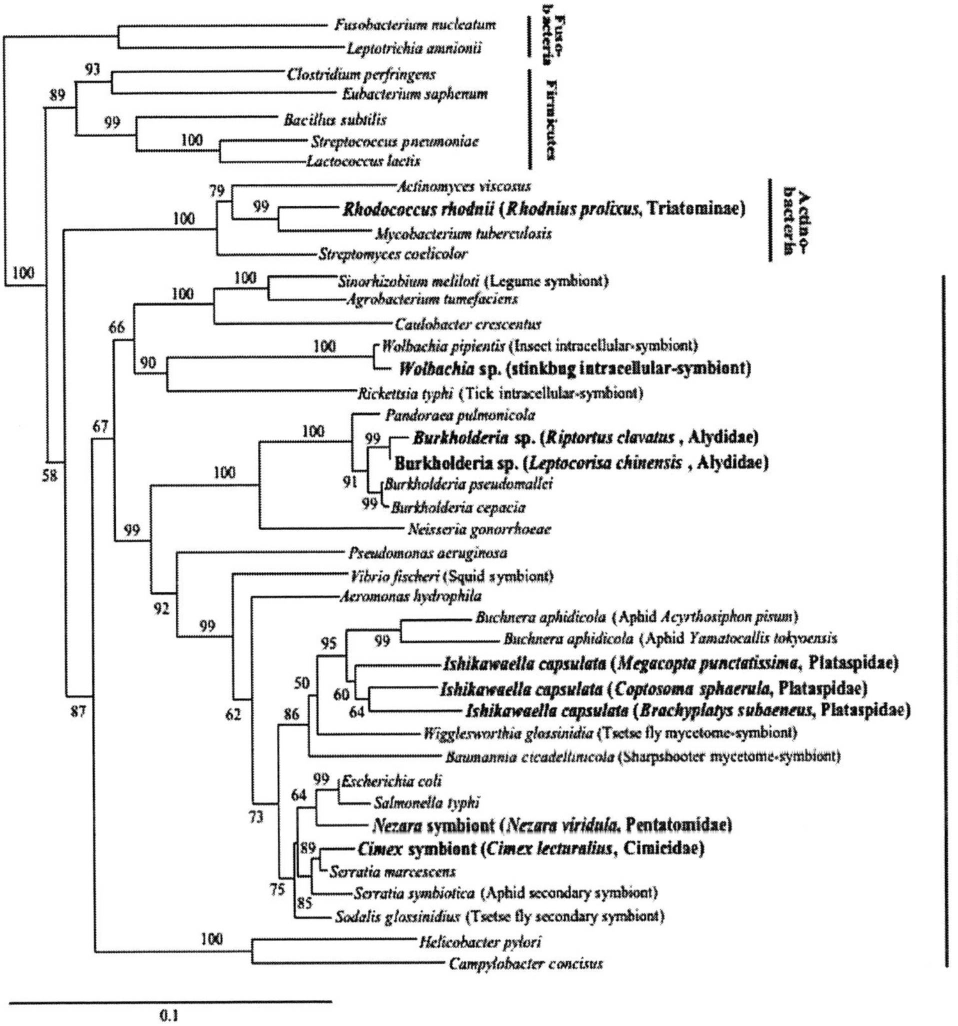

Figure 2. Phylogenetic tree based on the sequence of the 16S rRNA gene (Neighbor-joining analysis with a Kimura's correction; aligned 985 bp). Bacterial phyla are shown on the right. Stinkbug symbionts are represented in boldface and host species and family names are shown in parentheses. Bootstrap values of higher than 50% are depicted at the nodes. The scale bar represents 0.1 changes per base.

"symbiont capsules", are deposited in association with eggs, and hatchlings probe the content of the capsules to acquire the symbiont (Fukatsu and Hosokawa 2002; Hosokawa et al. 2005; Hosokawa et al. 2006; Müller 1956; Schneider 1940). Furthermore, recent studies revealed a novel symbiont transmission mechanism in the broad-headed bug *Riptortus clavatus* (family Alydidae). In this species, the symbiont is not vertically transmitted but acquired from surrounding environment by nymphal insects every generation (Kikuchi et al. 2007). Such a transmission manner, called "environmental transmission", has been reported from symbioses in marine invertebrates like squid-*Vibrio* symbiosis (Nyholm and McFall-Ngai 2004) and in terrestrial plants like legume-*Rhizobium* symbiosis (Denarie et al. 1992), but had been unknown among insects before the discovery in the alydid stinkbug. Recent molecular phylogenetic studies on bacterial symbionts of diverse stinkbugs have revealed that they belong to various lineages of bacterial groups (Figure 2, see below), suggesting these symbiotic associations have evolved independently. In the following sections, we will review each of the model systems of heteropteran endosymbiosis in detail.

Bacterial Symbiosis in Blood Sucking Bugs

Because of medical importance, biological aspects of blood-sucking insects, such as tsetse flies and lice, have been investigated for a long time. Blood feeders have had a strong impact on human history because of their transmission ability of serious diseases, such as malaria, African trypanozomiasis, and rickettsiosis. Despite the recent significant advances in public health and medical science, these insect-born diseases remain afflicting worldwide. It is notable that many of these insects are in obligate association with symbiotic bacteria. Actually, almost all insects that feed exclusively on vertebrate blood throughout their life possess symbiotic bacteria inside their body (Buchner 1965). Several experimental and genomic studies have revealed that the symbiotic bacteria provide their host with B vitamins that are lacking in vertebrate blood (Akman et al. 2002; Nogge 1976; Puchta 1955). In the Heteroptera, two groups of the infraorder Cimicomorpha, triatomine assassin bugs and bedbugs, are blood-feeding and have been reported to harbor symbiotic bacteria.

Triatomine Assassin Bugs

The subfamily Triatominae is a group of the family Reduviidae, known as assassin bugs, and mostly distributed across Central and South America. To date, 111 species and 14 genera have been described in the subfamily (Schuh and Slater. 1995), most of which possess symbiotic bacteria in their alimentary tract, mainly in the hindgut (Buchner 1965; Dasch et al. 1984). These bacteria usually inhabit the lumen of the digestive tract and are occasionally found inside the gut epithelial cells. *Rhodnius prolixus* provides one of the well-characterized symbiotic systems, wherein a gram-positive actinomycete *Rhodococcus rhodonii* has been identified as the gut symbiont (Figure 2) (Dasch et al. 1984; Harrington 1960; Hill et al. 1976). This microorganism is acquired by hatchlings through coprophagy, where nymphs actively ingest feces of other *R. prolixus* individuals. Experimental inhibition of the symbiont acquisition results in high mortality in early nymphal stages (Buchner 1965; Hill et al. 1976), and the fitness damage was recovered by supplementation with B vitamins (Baines 1956), suggesting that the symbiotic bacterium provides the host insect with B-complex vitamins, like *Wigglesworthia* in tsetse flies (Akman et al. 2002; Nogge 1976). Notably, the symbiont *R. rhodonii* is culturable outside the host bug and detectable in environmental soil (Beard et al. 1992; Dasch et al. 1984; Hill et al. 1976). The culturability of the symbiont is exceptional, in contrast to most of insect microbial mutualists that are generally not viable outside their hosts, and is potentially useful for genetic manipulation of the symbiont.

In addition to the beneficial associate *R. rhodonii*, the members of the Triatominae carry a protozoan parasite *Tripanosoma cruzi*, which causes the Chagas disease in human. Almost 10% to 30% of patients of the disease progress to a chronic, debilitating, and life-threatening illness of the cardiac and gastrointestinal systems, and at least 50,000 deaths per year have been attributed to the pathogen in Latin America (Magill and Reed 2000). Prevention of the disease is still difficult, because there is no effective vaccine to the parasite and insecticide resistance easily develops in the vector bugs. By using genetic manipulation of the symbiotic bacterium *R. rhodonii*, Durvasula et al. (1997) tried a "paratransgenetic approach" to control the *Trypanosoma* transmission. "Paratransgenesis" is a recently-proposed research concept, where a host organism is modified physiologically, ecologically or behaviorally not by genetic manipulation of the organism itself but via transgenesis of its symbiotic microorganism (Beard et al. 2002). In the case of *R. prolixus*, the symbiont *R. rhodonii* was genetically modified for expressing cecropin A, a peptide lethal to *Trypanosoma,* and was reintroduced into the triatomine host. Hatchlings of the insect were exposed to an artificial bug's feces, which was named "CRUZIGARD" and containing the genetically modified symbiont cells, and survival rates of *Trypanosoma* were monitored in these insects. These experiments resulted in complete elimination or reduction in number of the parasite in these insects (Durvasula et al. 1997), suggesting that the paratransgenic approach could be certainly effective for controlling the Chagas disease. Although the ecological impact and the safety aspects of this strategy should be further examined, such paratransgenetic approaches have been also considered for other insect-born diseases, such as malaria vectored by mosquitoes (Favia et al. 2007) and African trypanosomiasis mediated by tsetse flies (Aksoy and Rio 2005; Rio et al. 2004).

Bedbugs

Bedbugs, members of the family Cimicidae, are nocturnal insects famous for sneaking into a bedroom for sucking human blood. However, except the common bedbug *Cimex lecturalius* and the tropical bedbug *C. hemipterus*, most members of the Cimicidae are parasites of wild bats (Reinhardt and Siva-Jothy 2007; Schuh and Slater. 1995; Usinger 1966). Although the human bedbugs are not known to vector any pathogens in the wild, laboratory experiments have demonstrated their potential capabilities of carrying a wide variety of pathogens, including trypanosomes, rickettsial bacteria, hepatitis B virus, and HIV (Burton 1963; Reinhardt and Siva-Jothy 2007; Silverman et al. 2001; Usinger 1966). Bedbug infestations are still a big problem in developing countries, while the insects are no longer serious household pests in developed countries since the use of DDT after the beginning of World War II (Usinger 1966). Recently, however, bedbugs have been increasing in number and widespread even in the developed world, often causing troubles in hotels and other short-stay lodges (Reinhardt and Siva-Jothy 2007; Ryan et al. 2004).

Female bedbugs usually possess a pair of mycetomes in their body cavity, while male bedbugs have a pair of mycetomes connected to the duct of testis (Buchner 1965; Usinger 1966). Besides the specialized symbiotic organs, some of the insects were reported to harbor additional symbiotic bacteria inside gut epithelial cells (Usinger 1966). In *C. lecturalius*, the symbiotic bacteria are vertically passed from mother to offspring by transovarial transmission (Buchner 1965). Heat shock treatment of female individuals resulted in morphological abnormalities of the symbiont in the mycetomes and a drastic reduction in the host fecundity (Chang 1974), suggesting that the symbiotic bacteria play a vital role in the insect.

In the mycetomes of *C. lecturalius*, Chang and colleagues found three morphotypes of intracellular symbionts by using transmission electron microscopy: primary, secondary, and guest symbionts (Chang 1975; Chang and Musgrave 1973). They investigated two insect strains, where the primary symbiont was found in both of the strains, the secondary one was observed only in a strain, and the guest one was scarcely observed in the mycetomes but predominantly in the ovary (Chang 1975). Later, a molecular phylogenetic study identified two bacterial symbionts from female ovaries of *C. lecturalius*, a γ-proteobacterial symbiont and *Wolbachia* (Figure 2) (Hypsa and Aksoy 1997), although it remains unclear which of these bacteria correspond to the primary, secondary, or guest symbionts.

Capsule-Transmitted Symbiont in the Family Plataspidae

Diverse insects are associated with symbiotic bacteria, and most of them have some mechanisms for vertical transmission of the symbionts from mother to offspring. However, no insects have evolved a transmission mechanism as unique as that found in stinkbugs of the family Plataspidae. Members of the insect group transmit their symbiotic bacteria through a peculiar mechanism, called “capsule transmission”.

Plataspid Stinkbugs

The pentatomomorphan family Plataspidae consists of 56 genera and 530 species (Schuh and Slater. 1995), which are distributed mainly in the tropics and subtropics of the Eastern Hemisphere but a small number of temperate species are recorded in the Palearctic (Schuh and Slater. 1995). The family is characterized by their large scutellum that completely covers the abdomen, and their body shape is globular or convex (Figure 3 A). Most of plataspid species feed on phloem sap of leguminous plants, and occasionally become pests of crop legumes (Schaefer and Panizzi 2000). In the midgut of the insects, numerous crypts are present, whose lumen is filled with a dense population of specific bacterium.

Capsule Transmission

The mechanism of symbiont transmission is the most fascinating feature in the plataspid symbiosis. When females of the plataspid stinkbugs lay eggs, they also deposit small brownish particles under the egg mass.

Figure 3. Capsule symbiosis in the Plataspidae. (A) Symbiotic (left) and aposymbiotic (right) female adults of the plataspid species, *Megacopta punctatissima*. Aposymbiotic adult exhibits small body size, pale coloration, and soft cuticle. (B) Hatchlings of *M. punctatissima* probing "symbiont capsules". Arrowheads indicate symbiont capsules that contain *Ishikawaella* symbiont. (C) Dissected midguts of the plataspid species, *Brachyplatys subaeneus*. Top, adult female; Bottom, adult male. Asterisks indicate the midgut posterior region, symbiotic organ with numerous bacteria-containing crypts. Arrowheads indicate filamentous tissues, where anterior and posterior parts of the midgut are disconnected.

The particles, called "symbiont capsules", contain a copious amount of the symbiotic bacterium inside (Hosokawa et al. 2005; Schneider 1940). Hatchlings immediately probe the capsules with their proboscis and orally ingest the symbiont from the capsules (Figure 3 B). Such a unique transmission manner allows us to easily generate aposymbiotic nymphs by removing capsules from egg masses. Irrespective of the host species, aposymbiotic nymphs exhibit reduced developmental performance, high mortality, prolonged nymphal period, smaller body size, pale coloration, soft cuticle, and sterility (Figure 3 A) (Fukatsu and Hosokawa 2002; Hosokawa et al. 2006; Müller 1956), suggesting the obligatory nature of the host-symbiont association.

Although there have been no studies demonstrating physiological or biochemical mechanisms by which the symbiont enables the host to develop and reproduce normally, considering the phloem-feeding habit of the stinkbug, the symbiont probably synthesizes essential amino acids and other nutrients for the host insects, as reported in the aphid–*Buchnera* association (Douglas 1989; Douglas 1998; Shigenobu et al. 2000). Future studies should focus on physiological aspects of the symbiosis by using a nutritionally defined artificial diet.

Midgut Morphology

Anatomy of the adult midgut of the plataspid stinkbugs is very unique. The posterior part of the midgut, which bears numerous crypts and contains symbiotic bacteria, is completely disconnected from the anterior part, which has no crypts and is free of the symbiont (Figure 3 C) (Fukatsu and Hosokawa 2002; Hosokawa et al. 2006; Schneider 1940). In the posterior part, the lumen is filled with a great amount of the symbiont cells, and there is no food flow in the section (Hosokawa et al. 2005; Schneider 1940). Namely, the posterior midgut of the stinkbug is transformed into a specialized organ for harboring the pure-cultured symbiont.

Symbiont "*Candidatus* Ishikawaella capsulata"

Electron microscopy showed that the symbiont of the Japanese common plataspid stinkbug, *Megacopta punctatissima*, is a globular-shaped bacterium of 1-3 μm in diameter (Hosokawa et al. 2005). Phylogenetic analysis based on 16S rRNA genes of the symbiotic bacteria derived from seven Japanese plataspid species showed that the symbionts form a monophyletic group in the γ-Proteobacteria (Hosokawa et al. 2006) (Figure 2). For the novel group of the extracellular gut symbionts, the designation "*Candidatus* Ishikawaella capsulata" was proposed (Hosokawa et al. 2006). Notably, the extracellular *Ishikawaella* symbionts are related to the intracellular insect symbionts such as *Wigglesworthia* and *Blochmannia*, and its closest relative is the aphid intracellular symbiont *Buchnera*. Moreover, the plataspid symbionts exhibited several peculiar genetic traits such as (i) AT biased nucleotide composition, (ii) accelerated molecular evolution, and (iii) reduced genome size in comparison with related free-living bacteria (Hosokawa et al. 2006). These genetic traits have been widely observed with obligate intracellular symbiotic bacteria of various insects like *Buchnera* in aphids, *Wigglesworthia* in tsetse flies, *Blochmannia* in ants, *Carsonella* in psyllids, *Baumannia* in sharpshooters and others, and had been thought that the intracellular condition where the symbionts inhabit might be relevant. However, now it became evident that the intracellular habitat is not responsible for the peculiar genetic traits. Alternatively, attenuated purifying selection due to small population size and strong bottleneck is a candidate mechanism that better accounts for the evolutionary patterns found in the symbiont genomes (Moran 1996; Wernegreen 2002).

Host-symbiont phylogenetic congruence has been reported from various insect-microbe endosymbiotic systems wherein obligate intracellular bacteria play an essential role for their host, indicating their stable association over evolutionary time (Baumann 2005). Considering the extracellular habitat of *Ishikawaella* and the transmission mechanism depending on nymphal behavior, the plataspid–*Ishikawaella* association might be expected to be evolutionarily more casual than the intracellular associations. However, the *Ishikawaella* phylogeny showed a perfect congruence with the host stinkbug phylogeny (Figure 4), which strongly suggests that a single bacterial infection in the common ancestor of the plataspid stinkbugs has been stably maintained in the course of the host speciation (Hosokawa et al. 2006).

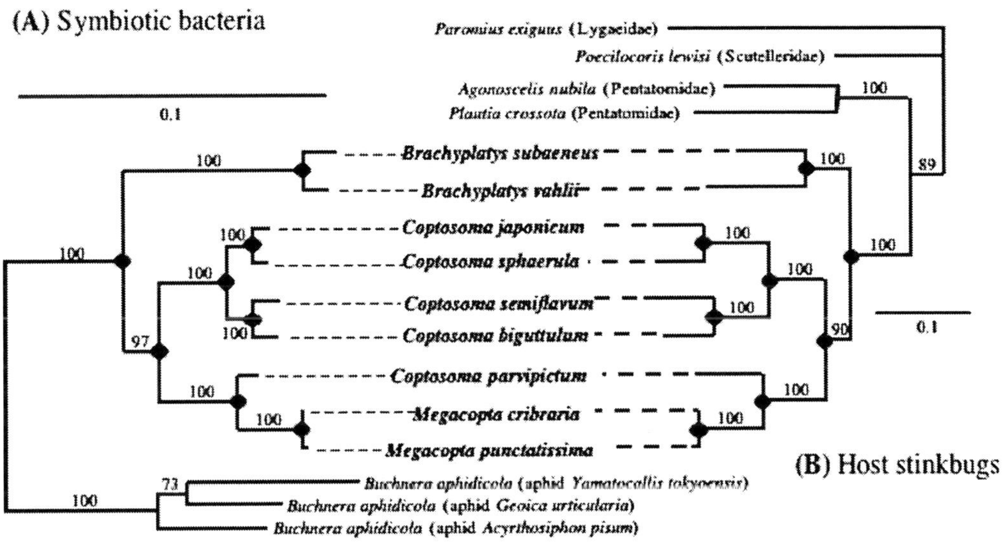

Figure 4. Phylogenetic congruence between the plataspid stinkbugs and their *Ishikawaella* symbionts. (A) A Bayesian phylogeny of *Ishikawaella* symbionts on the basis of eubacterial 16S rRNA gene sequences (1,229 aligned nucleotide sites). In addition *Ishikawaelle* symbionts, three *Buchnera* symbionts were examined as outgroup taxa. (B) A Bayesian phylogeny of the host stinkbugs on the basis of mitochondrial 16S rRNA gene sequences (847 aligned nucleotide sites). In addition to the plataspid stinkbugs, four pentatomomorphan species were examined as outgroup taxa. In parentheses are taxonomic affiliations. Dots indicate co-divergence events inferred by the jungles algorithm (Charleston 1998). Posterior probabilities in the Bayesian analysis depicted at the nodes.

Capsule Symbiosis as a Model System for Studying Obligate Mutualism

A great experimental advantage of the plataspid-*Ishikawaella* symbiosis is that we can easily separate the host and the symbiont without damaging them, despite their obligate mutualistic association. By using the system, we are able to perform unprecedented experimental approaches that are impossible with other obligate intracellular insect-bacterium associations. For example, egg/capsule ratios in an egg mass can be freely altered, by which the levels of maternal investment in symbiont transmission can be experimentally manipulated (Hosokawa et al., 2007a). By exchanging eggs and capsules between populations, species and genera of the insects, the extent of host-symbiont coevolution and coadaptation would be experimentally evaluated (Hosokawa et al., 2007b). Furthermore, pure symbiont DNA for genome sequencing is easily prepared from isolated symbiont capsules. We expect that the plataspid–*Ishikawaella* association will provide an excellent model system for studying insect-microbe symbiosis.

Mutualistic Symbiosis without Vertical Transmission in Alydid Stinkbugs

Diverse mechanisms for symbiont transmission have been identified in insects, most of which are vertical mode, including transovarial transmission, egg smearing, coprophagy, and the unique "capsule transmission". In a number of insect groups, the vertical transmission over evolutionary time leads to host-symbiont coevolution and cospeciation, as mentioned above. However, a bacterial symbiosis recently found in the family Alydidae (Pentatomomorpha) is totally different from such a typical insect-symbiont association in that symbiotic bacteria are not vertically transmitted but acquired from surrounding environment every generation.

Burkholderia Symbiont in Alydid Stinkbugs

Members of the family Alydidae, or broad-headed bugs, are distributed worldwide and include a number of pest species for leguminous and gramineous crops. In Japan, the alydid stinkbugs *Riptortus clavatus* and *Leptocorisa chinensis* are the major pests for soybean and rice, respectively (Schaefer and Panizzi 2000; Tomokuni et al. 1993). Both species possess well-developed midgut crypts in two rows (Figure 5A, B) (Kikuchi et al. 2005). Molecular phylogenetic studies revealed that their crypt-associated symbiotic bacteria belong to the genus *Burkholderia* in the β-Proteobacteria (Figure 2) (Kikuchi et al. 2005).

Surveys of *Burkholderia* infection in host populations showed that the symbiont is maintained in each of the local populations with almost 100% prevalence (Kikuchi et al. 2005), suggesting an intimate host-symbiont association. However, the symbiotic bacteria exhibited a considerable level of genetic variations within and between the species, in terms of 97-100% sequence identities of their 16S rRNA gene. Furthermore, the symbiont phylogeny was not concordant with the host systematics: the symbionts from *R. clavatus* and the symbionts from *L. chinensis* did not form separate clades but constituted an intermingled clade in the *Burkholderia* phylogeny (Kikuchi et al. 2005). The tangled phylogeny despite the specific and prevalent host-symbiont association was an enigma.

No Vertical Transmission but Environmental Acquisition of the *Burkholderia* Symbiont

In *R. clavatus*, inspection of eggs by diagnostic PCR and a series of rearing experiments revealed that the symbiont is neither associated with the eggs nor vertically transmitted (Kikuchi et al. 2007). On the other hand, the *Burkholderia* symbiont was frequently detected in the rhizosphere of soybean, a host plant of the stinkbug, and, surprisingly, almost all nymphs reared on potted soybean plants acquired the symbiont in the absence of their parents (Kikuchi et al. 2007). These results strongly suggested environmental transmission of the *Burkholderia* symbiont in *R. clavatus*.

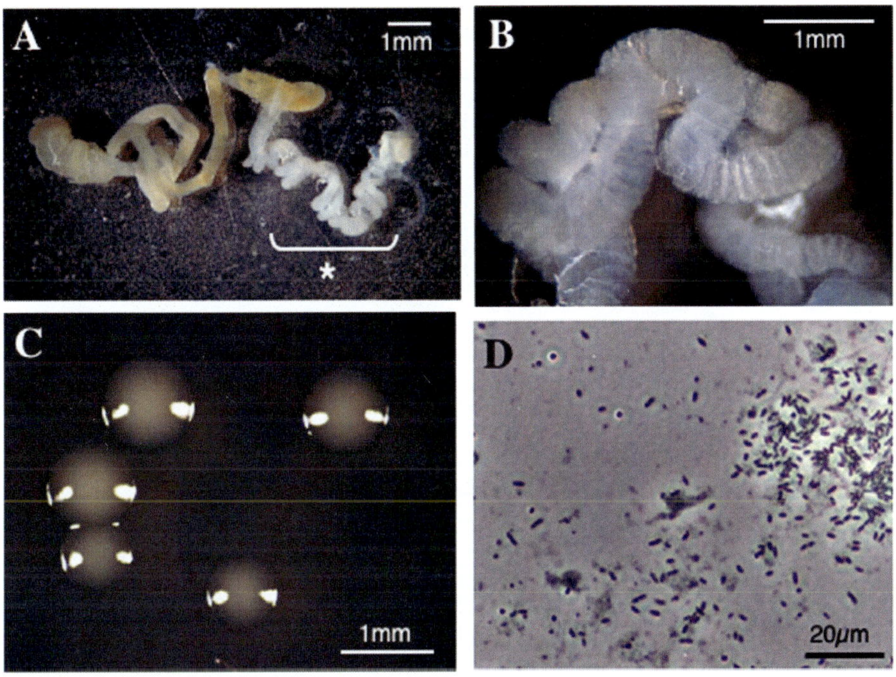

Figure 5. Midgut organization of alydid species, *Riptortus clavatus*, and the *Burkholderia* symbiont isolated from the stinkbug. (A) A dissected midgut of *R. clavatus*. Asterisk indicates the midgut region with crypts. (B) Enlarged image of midgut crypts. (C) Colonies of the *Burkholderia* symbiont on LB (Luria-Bertani) medium. (D) Microscopic image of cultured *Burkholderia* symbiont.

It should be noted that, as might be expected from occurrence of the symbiont in soil environment, a culture strain of the symbiont was successfully isolated from the midgut crypts of the insect (Figure 5 C, D) (Kikuchi et al. 2007). Rearing experiments by using sterilized soybean bottles demonstrated that the cultured symbiont is certainly able to establish a normal infection in the host stinkbug, verifying its environmental transmission.

Beneficial Nature of the *Burkholderia* Symbiont

Comparison between symbiont-infected and uninfected individuals demonstrated that the *Burkholderia* symbiont significantly improves the growth of the stinkbug: in both males and females, body size parameters (body length, thorax width, and abdomen width) of the infected bugs were significantly larger than those of the uninfected ones (Figure 6). For example, the dry weights of the infected individuals were twice as large as those of the uninfected bugs (Kikuchi et al. 2007). These results clearly demonstrated that the *Burkholderia* symbiont plays a beneficial role for the host *R. clavatus*. Meanwhile, effects of the symbiont on host survival was not obvious in that the aposymbiotic individuals produced viable offspring, suggesting the facultative nature of the host-symbiont association (Kikuchi et al. 2007).

Figure 6. Symbiotic (left) and aposymbiotic (right) females of *Riptortus clavatus*. Aposymbiotic adult shows smaller body size compared with symbiotic one.

It remains unclear what the symbiont actually does for the host insect. Members of the genus *Burkholderia* are major soil bacteria that are most commonly found on plant roots and adjacent areas, and in other moist environments (Woods and Sokol 2000). Interestingly, a wide variety of biological activities have been reported from the bacterial group: a number of strains possess dinitrogen-fixing ability (Estrada-De et al. 2001); some strains nodulate the root of leguminous plants (Chen et al. 2003; Moulin et al. 2001); some species are associated with plant leaf galls (Van Oevelen et al. 2002; Van Oevelen et al. 2004); some members promote plant growth and suppress plant diseases (Bevivino et al. 1998) and are used as biofertilizing agents (Vessey 2003). Such biological activities of the bacterial group might be, although speculative, relevant to biological functions of the *Burkholderia* symbiont in the alydid stinkbug.

Alydid-*Burkholderia* Symbiosis as an Insect Analogue of Legume-*Rhizobium* Symbiosis

All these results suggest that the stinkbug-*Burkholderia* symbiosis is an insect analogue of the well-known symbiotic systems between plants and soil bacteria such as legume-*Rhizobium* and alder-*Frankia* relationships: the mutualistic symbiont is present in the soil of the host habitat and is environmentally acquired every generation. Since these plant-associated bacteria are culturable and genetically manipulatable, these model systems have greatly contributed to our understanding of the molecular mechanisms of host-symbiont interactions, such as nutritional exchanges, host-symbiont specificity, development of symbiotic organs and others (reviewd in Gualtieri and Bisseling 2000; Perret et al. 2000). In the alydid-*Burkholderia* system, the symbiont is easily culturable in standard microbiological media. Potentially the symbiont is manipulatable genetically, considering that several transposon-mediated transformation systems have already been established in other *Burkholderia* species (Choi et al. 2006; DeShazer et al. 1997; Hunt et al. 2004; Lefebre and Valvano 2002). Reintroduction of genetically-manipulated symbiont into the host stinkbug could provide unique opportunities to investigate bacterial genes responsible for symbiosis-related phenotypes, such as specificity and stability of infection, bacterial localization in the digestive tract, fitness effects on the stinkbug host, etc. The alydid-*Burkholderia* symbiosis would provide a novel system that enables molecular genetic approaches to the mechanisms underlying the insect-microbc mutualism.

Other Bacterial Symbiosis in the Heteroptera

Bacterial Symbiosis in the Family Pentatomidae

In addition to the bacterial symbioses in the heteropteran insects described above, those in two stinkbugs of the family Pentatomidae have been investigated in detail. One species is the brown-winged green bug *Plautia stali* and the other one is the southern green bug *Nezara viridula*, both of which are economically important pests of various crops (Schaefer and Panizzi 2000; Tomokuni et al. 1993). These two species possess the midgut crypts arranged in four rows, where dense populations of symbiotic bacteria are found (Abe et al. 1995; Prado et al. 2006). Both of the species transmit the symbiont to their offspring via egg smearing, and aposymbiotic insects can be easily generated by superficial treatment of the eggs with ethanol and/or antibiotics (Abe et al. 1995; Prado et al. 2006). Despite such similarities, physiological impact of the symbiosis is different between the species. Aposymbiotic nymphs of *P. stahli* suffered high motility and almost all individuals died until adult molt (Abe et al. 1995), while symbiont-free *N. viridula* insects were able to persist in the laboratory at least for two generations (Prado et al., 2006). These results suggest that the nature of the symbioses might be mutualistic in *P. stali* and rather commensalistic in *N. viridula*.

To date, the symbiotic bacterium has been molecular-phylogenetically characterized in *N. viridula*. Phylogenetic analysis based on 16S rRNA gene sequences revealed that the symbiont of *N. viridula* belongs to the Enterobactericeae in the γ-Proteobacteria (Figure 2) (Prado et al. 2006). AT content of the symbiont's 16S rRNA gene (45.6%) was not similar to those of other insect microbial mutualists (*Buchnera aphidicola*, 49.8%; *Ishikawaella capsulata,* 50.7%), but rather to those of free-living γ-proteobacteria (*Escherichia coli*, 45.6%; *Serratia marcescens*, 45.0%) (Prado et al. 2006), suggesting that the symbiosis in *N. viridula* is of relatively recent origin. Plausibly that may be relevant to the commensalistic nature of the association between the southern green bug and the symbiotic bacterium.

Wolbachia Endosymbionts

In addition to the beneficial symbionts, parasitic/commensaristic endosymbionts of the genus *Wolbachia* have been detected from a wide variety of heteropteran species. Members of the genus *Wolbachia* are intracellular symbionts belonging to the α-Proteobacteria (Figure 2) and inherited through transovarial transmission (Frydman et al. 2006). This group of bacteria has been paid special attention by researchers, because they cause intriguing reproductive aberrations in their arthropod hosts, such as cytoplasmic incompatibility, male killing, feminization, and parthenogenesis (reviewed in O'Neill et al. 1997). On account of the maternal inheritance of *Wolbachia*, these reproductive symptoms are regarded as selfish strategies of the symbionts whereby the frequency of infected females increases in host populations.

Broad surveys of *Wolbachia* infection in the Heteroptera revealed 37% (47/134) of stinkbug species carried the intracellular parasite (Kikuchi and Fukatsu 2003). This infection rate was relatively high compared with other insect groups, such as mosquitoes (28%), butterflies (29%), and oak gallwasps (17%) (Jiggins et al. 2001; Kittayapong et al. 2000; Rokas et al. 2002). Phylogenetic analysis based on *Wolbachia* surface protein (*wsp*) sequences strongly suggested frequent horizontal transmission of the symbionts between host species (Kikuchi and Fukatsu 2003), as shown in other insect groups (McGraw and O'Neill 1999; O'Neill et al. 1997). It remains unclear how the intracellular, transovarially transmitted symbionts can travel across different insect species. In addition, it is also under investigation whether the *Wolbachia* endosymbionts cause reproductive disorders in their stinkbug hosts or not.

Since many heteropterans possess beneficial symbionts, a considerable number of stinkbugs may be multiply infected with two different microbial associates, a primary symbiont and a *Wolbachia* endosymbiont. Here, a kind of interaction and/or conflict may occur between the coexisting symbiotic bacteria. Recent studies on various insect symbioses have identified such competitive interactions between coexisting microbial associates (Goto et al. 2006; Juchault et al. 1992; Oliver et al. 2006; Rio et al. 2006). It will be of great interest to investigate the impact of *Wolbachia* infection not only on the stinkbug hosts but also on the beneficial symbiotic bacteria.

Concluding Remarks

Endosymbiosis is rated among important mechanisms driving the organismal evolution, in which the symbiont provides the host with a novel physiological ability and enables the host to exploit a brand-new ecological niche. For a long time, diverse phenomenal aspects concerning endosymbiotic associations have fascinated many researchers, probably because it appears to counter the classical evolutional theory where only species differentiation is the evolutionary process and also because it can explain the origin of eukaryotic cells (Margulis and Fester 1991). However, it still remains an enigma how such endosymbioses had occurred and established. When we attempt to infer an evolutionary process in the past, there are mainly two types of approaches: one is thorough understanding of a certain model system in depth, and the other is through comparison of diverse systems with reference to their phylogenetic and ecological information. Ultimately, of course, these approaches should be integrated and synthesized. While several insect models like aphids and tsetse flies have enriched our knowledge of the general biology and evolution of endosymbiotic systems, few comparative studies on different symbiotic systems in a certain insect clade have been performed because variations in symbiotic systems are generally quite limited or lacking in a specific insect taxon. As reviewed here, however, the Heteroptera embraces diverse endosymbiotic systems ranging from a strictly co-evolved, obligate association as in plataspid stinkbugs (Fukatsu and Hosokawa 2002; Hosokawa et al. 2005; Hosokawa et al. 2006), to an environmentally transmitted, facultative association as in alydid stinkbugs (Kikuchi et al. 2007; Kikuchi et al. 2005). The immense diversity of bacterial symbiosis in the Heteroptera provides a unique opportunity to thoroughly investigate each of the model systems and also to synthesize those findings in a well-characterized phylogenetic framework of the host insects.

Many animals, including humans, harbor a dense bacterial community in their body, mainly in the digestive tract. Recent studies have revealed that such a microbiota play important roles in the host biology, such as provision of essential nutrients (Backhed et al. 2005), involvement in development and morphpgenesis (Stappenbeck et al. 2002), and priming of host immunity (Mazmanian et al. 2005; Rakoff-Nahoum et al. 2004). Nowadays, these endosymbionts are often considered as "second genome" for the host organism (Relman and Falkow 2001). Despite the seemingly considerable impact of the symbiotic bacteria on the evolution of the host organism, these microscopic partners have been almost ignored in the field of biodiversity. Although a number of biodiversity projects have been promoted for mammals, insects, plants, protozoans, and environmental bacteria (Edwards et al. 2000; Leslie 2007; Wilson 2000), we are still at early beginnings of understanding the almost untouched diversity of endosymbiotic microorganisms. The heteropteran species reviewed here constitute just a tiny fraction of the total of 38,000 stinkbug species ever described. Further surveys of the Heteroptera will no doubt unveil a more diverse and fascinating world of insect symbioses, which would contribute to our understanding of the evolution of endosymbiotic associations.

References

Abe, Y., Mishiro, K. and Takanashi, M. 1995 Symbiont of brown-winged green bug, *Plautia stali* Scott. *Jpn. J. Appl. Entomol. Zool.* 39, 109-115.

Akman, L., Yamashita, A., Watanabe, H., Oshima, K., Shiba, T., Hattori, M. and Aksoy, S. 2002 Genome sequence of the endocellular obligate symbiont of tsetse flies, *Wigglesworthia glossinidia. Nat Genet* 32, 402-407.

Aksoy, S. 2000 Tsetse--A haven for microorganisms. *Parasitol. Today* 16, 114-118.

Aksoy, S. and Rio, R. V. M. 2005 Interactions among multiple genomes: tsetse, its symbionts and trypanosomes. *Insect Biochem. Mol. Biol.* 35, 691-698.

Backhed, F., Ley, R. E., Sonnenburg, J. L., Peterson, D. A. and Gordon, J. I. 2005 Host-bacterial mutualism in the human intestine. *Science* 307, 1915-1920.

Baines, S. 1956 The role of the symbiotic bacteria in the nutrition of *Rhodnius prolixus* (Hemiptera) *J. Exp. Biol.* 33, 533-541.

Baumann, P. 2005 Biology bacteriocyte-associated endosymbionts of plant sap-sucking insects. *Annu. Rev. Microbiol.* 59, 155-189.

Beard, C. B., Cordon-Rosales, C. and Durvasula, R. V. 2002 Bacterial symbionts of the triatominae and their potential use in control of Chagas disease transmission. *Annu. Rev. Entomol.* 47, 123-141.

Beard, C. B., Mason, P. W., Aksoy, S., Tesh, R. B. and Richards, F. F. 1992 Transformation of an insect symbiont and expression of a foreign gene in the Chagas' disease vector *Rhodnius prolixus*. *Am. J. Trop. Med. Hyg.* 46, 195-200.

Bevivino, A., Sarrocco, S., Dalmastri, C., Tabacchioni, S., Cantale, C. and Chiarini, L. 1998 Characterization of a free-living maize-rhizosphere population of *Burkholderia cepacia*: effect of seed treatment on disease suppression and growth promotion of maize. *FEMS Microbiol. Ecol.* 27, 225-237.

Bourtzis, K. and Miller, T. A. 2003 *Insect symbiosis*. Boca Raton, F. L.: CRC press.

Braendle, C., Miura, T., Bickel, R., Shingleton, A. W., Kambhampati, S. and Stern, D. L. 2003 Developmental origin and evolution of bacteriocytes in the aphid-*Buchnera* symbiosis. *PLoS Biol.* 1, e21.

Buchner, P. 1965 *Endosymbiosis of animals with plant microorganisms.* New York, N.Y.: Interscience.

Burton, G. J. 1963 Bedbugs in relation to transmission of human disease. *Public Health Rep.* 78, 513-524.

Chang, K. P. 1974 Effects of elevated temperature on the mycetome and symbiotes of the bed bug *Cimex lectularius* (Heteroptera). *J. Invertebr. Pathol.* 23, 333-340.

Chang, K. P. 1975 Hematophagous insect and haemoflagellate as host for prokaryotic endosymbionts. *Symp. Soc. Exp. Biol.* 29, 407-427.

Chang, K. P. and Musgrave, A. J. 1973 Morphology, histochemistry, and ultrastructure of mycetome and its rickettsial symbiotes in *Cimex lectularius* L. *Can. J. Microbiol.* 19, 1075-1081.

Charleston, M. A. 1998 Jungles: a new solution to the host/parasite phylogeny reconciliation problem. *Math. Biosci.* 149, 191-223.

Chen, W. M., Moulin, L., Bontemps, C., Vandamme, P., Bena, G. and Boivin-Masson, C. 2003 Legume symbiotic nitrogen fixation by ß-proteobacteria is widespread in nature. *J. Bacteriol.* 185, 7266-7272.

Chen, X., Li, S. and Aksoy, S. 1999 Concordant evolution of a symbiont with its host insect species: molecular phylogeny of genus *Glossina* and its bacteriome-associated endosymbiont, *Wigglesworthia glossinidia*. *J. Mol. Evol.* 48, 49-58.

Choi, K. H., DeShazer, D. and Schweizer, H. P. 2006 mini-Tn7 insertion in bacteria with multiple *glmS*-linked *att*Tn7 sites: example *Burkholderia mallei* ATCC 23344. *Nat. Protocol.* 1, 162-169.

Clark, M. A., Moran, N. A., Baumann, P. and Wernegreen, J. J. 2000 Cospeciation between bacterial endosymbionts (*Buchnera*) and a recent radiation of aphids (Uroleucon) and pitfalls of testing for phylogenetic congruence. *Evolution* 54, 517-25.

Dasch, G. A., Weiss, E. and Chang, K. P. 1984 Endosymbionts of insects. In *Bergey's manual of systematic bacteriology*, vol. 1 (ed. N. R. Krieg and J. G. Holt). Baltimore, Md: Williams and Wilkins.

Denarie, J., Debelle, F. and Rosenberg, C. 1992 Signaling and host range variation in nodulation. *Annu. Rev. Microbiol.* 46, 497-531.

DeShazer, D., Brett, P. J., Carlyon, R. and Woods, D. E. 1997 Mutagenesis of *Burkholderia pseudomallei* with Tn5-OT182: isolation of motility mutants and molecular characterization of the flagellin structural gene. *J. Bacteriol.* 179, 2116-2125.

Douglas, A. E. 1989 Mycetocyte symbiosis in insects. *Biol. Rev.* 64, 409-434.

Douglas, A. E. 1998 Nutritional interactions in insect-microbial symbioses: aphids and their symbiotic bacteria *Buchnera*. *Annu. Rev. Entomol.* 43, 17-37.

Durvasula, R. V., Gumbs, A., Panackal, A., Kruglov, O., Aksoy, S., Merrifield, R. B., Richards, F. F. and Beard, C. B. 1997 Prevention of insect-borne disease: an approach using transgenic symbiotic bacteria. *Proc. Natl. Acad. Sci. USA* 94, 3274-3278.

Edwards, J. L., Lane, M. A. and Nielsen, E. S. 2000 Interoperability of biodiversity databases: biodiversity information on every desktop. *Science* 289, 2312-2314.

Estrada-De, P., Bustillos-Cristales, R. and Caballero-Mellado, J. 2001 *Burkholderia*, a genus rich in plant-associated nitrogen fixers with wide environmental and geographic distribution. *Appl. Environ. Microbiol.* 67, 2790-2798.

Eubanks, M. D., Styrsky, J. D. and Denno, R. F. 2003 The evolution of omnivory in heteropteran insects. *Ecology* 84, 2549-2556.

Favia, G., Ricci, I., Damiani, C., Raddadi, N., Crotti, E., Marzorati, M., Rizzi, A., Urso, R., Brusetti, L., Borin, S., Mora, D., Scuppa, P., Pasqualini, L., Clementi, E., Genchi, M., Corona, S., Negri, I., Grandi, G., Alma, A., Kramer, L., Esposito, F., Bandi, C., Sacchi, L. and Daffonchio, D. 2007 Bacteria of the genus *Asaia* stably associate with *Anopheles stephensi*, an Asian malarial mosquito vector. *Proc. Natl. Acad. Sci. USA* 104, 9047-9051.

Frydman, H. M., Li, J. M., Robson, D. N. and Wieschaus, E. 2006 Somatic stem cell niche tropism in *Wolbachia*. *Nature* 441, 509-512.

Fukatsu, T. and Hosokawa, T. 2002 Capsule-transmitted gut symbiotic bacterium of the Japanese common plataspid stinkbug, *Megacopta punctatissima*. *Appl. Environ. Microbiol.* 68, 389-396.

Glasgow, H. 1914 The gastric caeca and the caecal bacteria of the Heteroptera. *Biol. Bull.* 3, 101-171.

Goto, S., Anbutsu, H. and Fukatsu, T. 2006 Asymmetrical interactions between *Wolbachia* and *Spiroplasma* endosymbionts coexisting in the same insect host. *Appl. Environ. Microbiol.* 72, 4805-4810.

Gualtieri, G. and Bisseling, T. 2000 The evolution of nodulation. *Plant Mol. Biol.* 42, 181-194.

Harrington, J. S. 1960 Studies on *Rhodonius prolixus*: growth and development of normal and sterile bugs, and the symbiotic relationship. *Parasitology* 50, 279-286.

Hill, P., Campbell, J. A. and Petrie, I. A. 1976 *Rhodonius prolixus* and its symbiotic actinomycete: a microbiological physiological and behavioural study. *Proc. R. Soc. B.* 194, 501-525.

Hosokawa, T., Kikuchi, Y., Meng, X. Y. and Fukatsu, T. 2005 The making of symbiont capsule in the plataspid stinkbug *Megacopta punctatissima*. *FEMS Microbiol. Ecol.* 54, 471-477.

Hosokawa, T., Kikuchi, Y., Nikoh, N., Shimada, M. and Fukatsu, T. 2006 Strict host-symbiont cospeciation and reductive genome evolution in insect gut bacteria. *PLoS Biol.* 4, e337.

Hosokawa, T., Kikuchi, Y., Shimada, M. and Fukatsu, T. 2007a How many symbionts are provided by mothers, acquired by offspring, and needed for successful vertical transmission in an obligate insect-bacterium mutualism? *Mol. Ecol.* 16, 5316-5325.

Hosokawa, T., Kikuchi, Y., Shimada, M. and Fukatsu, T. 2007b Obligate symbiont involved in pest status of host insect. *Proc. R. Soc. B.* 274, 1979-1984.

Huber-Schneider, L. 1957 Morphologische und phyaiologische untersuchungen an der wanze *Mesocerus marginatus* L. und Ihren symbionten (Heteroptera). *Z. Morph. Ökol. Tiere* 46, 433-480.

Hunt, T. A., Kooi, C., Sokol, P. A. and Valvano, M. A. 2004 Identification of *Burkholderia cenocepacia* genes required for bacterial survival in vivo. *Infect. Immun.* 72, 4010-4022.

Hypsa, V. and Aksoy, S. 1997 Phylogenetic characterization of two transovarially transmitted endosymbionts of the bedbug *Cimex lectularius* (Heteroptera: Cimicidae). *Insect Mol. Biol.* 6, 301-304.

Jiggins, F. M., Bentley, J. K., Majerus, M. E. and Hurst, G. D. 2001 How many species are infected with *Wolbachia*? Cryptic sex ratio distorters revealed to be common by intensive sampling. *Proc. R. Soc. B.* 268, 1123-1126.

Juchault, P., Rigaud, T. and Mocquard, J. P. 1992 Evolution of sex-determining mechanisms in a wild population of *Armadillidium vulgare* Latr. (Crustacea, Isopoda): competition between two feminizing parasitic sex factors. *Heredity* 69, 382-390.

Kikuchi, Y. and Fukatsu, T. 2003 Diversity of *Wolbachia* endosymbionts in heteropteran bugs. *Appl. Environ. Microbiol.* 69, 6082-6090.

Kikuchi, Y., Hosokawa, T. and Fukatsu, T. 2007 Insect-microbe mutualism without vertical transmission: a stinkbug acquires beneficial gut symbiont from environment every generation *Appl. Environ. Microbiol.* 73, 4308-4316.

Kikuchi, Y., Meng, X. Y. and Fukatsu, T. 2005 Gut symbiotic bacteria of the genus *Burkholderia* in the broad-headed bugs *Riptortus clavatus* and *Leptocorisa chinensis* (Heteroptera: Alydidae). *Appl. Environ. Microbiol.* 71, 4035-4043.

Kittayapong, P., Baisley, K. J., Baimai, V. and O'Neill, S. L. 2000 Distribution and diversity of *Wolbachia* infections in Southeast Asian mosquitoes (Diptera: Culicidae). *J. Med. Entomol.* 37, 340-345.

Lefebre, M. D. and Valvano, M. A. 2002 Construction and evaluation of plasmid vectors optimized for constitutive and regulated gene expression in *Burkholderia cepacia* complex isolates. *Appl. Environ. Microbiol.* 68, 5956-5964.

Lefevre, C., Charles, H., Vallier, A., Delobel, B., Farrell, B. and Heddi, A. 2004 Endosymbiont phylogenesis in the Dryophthoridae weevils: Evidence for bacterial replacement. *Mol. Biol. Evol.* 21, 965-973.

Leslie, M. 2007 Biodiversity: The ultimate life list. *Science* 316, 818.

Magill, A. J. and Reed, S. G. 2000 American trypanosomiasis. In *Hunter's Tropical Medicine and Emerging Infectious Diseases* (ed. G. T. Strickland), pp. 653-664. Philadelphia, P.A.: Saunders.

Margulis, L. and Fester, R. (ed.) 1991 *Symbiosis as a Source of Evolutionary Innovation.* Cambridge, M.A.: MIT Press.

Mazmanian, S. K., Liu, C. H., Tzianabos, A. O. and Kasper, D. L. 2005 An immunomodulatory molecule of symbiotic bacteria directs maturation of the host immune system. *Cell* 122, 107-18.

McGraw, E. A. and O'Neill, S. L. 1999 Evolution of *Wolbachia pipientis* transmission dynamics in insects. *Trends Microbiol.* 7, 297-302.

Mira, A. and Moran, N. A. 2002 Estimating population size and transmission bottlenecks in maternally transmitted endosymbiotic bacteria. *Microb. Ecol.* 44, 137-143.

Miura, T., Braendle, C., Shingleton, A., Sisk, G., Kambhampati, S. and Stern, D. L. 2003 A comparison of parthenogenetic and sexual embryogenesis of the pea aphid *Acyrthosiphon pisum* (Hemiptera: Aphidoidea). *J. Exp. Zool. Part B Mol. Dev. Evol.* 295.

Miyamoto, S. 1961 Comparative morphology of alimentary organs of Heteroptera, with the phylogenetic consideration. *Sieboldia* 2, 197-259.

Moran, N. A. 1996 Accelerated evolution and Muller's rachet in endosymbiotic bacteria. *Proc. Natl. Acad. Sci. USA* 93, 2873-2878.

Moran, N. A. 2007 Symbiosis as an adaptive process and source of phenotypic complexity. *Proc. Natl. Acad. Sci. USA* 104, 8627-8633.

Moran, N. A., Munson, M. A., Baumann, P. and Ishikawa, H. 1993 A molecular clock in endosymbiotic bacteria is calibrated using the insect hosts. *Proc. R. Soc. B.* 253, 167-171.

Moran, N. A., Tran, P. and Gerardo, N. M. 2005 Symbiosis and insect diversification: an ancient symbiont of sap-feeding insects from the bacterial phylum Bacteroidetes. *Appl. Environ. Microbiol.* 71, 8802-8810.

Moulin, L., Munive, A., Dreyfus, B. and Boivin-Masson, C. 2001 Nodulation of legumes by members of the beta-subclass of Proteobacteria. *Nature* 411, 948-950.

Müller, H. J. 1956 Experimentelle studien an der symbiose von *Coptosoma scutellatum* Geoffr (Hem. Heteropt.). *Z. Morph. Ökol. Tiere* 44, 459-482.

Nogge, G. 1976 Sterility in tsetse flies (*Glossina morsitans* Westwood) caused by loss of symbionts *Experimentia* 32, 995-996.

Nyholm, S. V. and McFall-Ngai, M. J. 2004 The winnowing: establishing the squid-*Vibrio* symbiosis. *Nat. Rev. Microbiol.* 2, 632-642.

O'Neill, S. L., Hoffmann, A. A. and Werren, J. H. 1997 *Influential Passenger: Inherited Microorganisms and Arthropod Reproduction.* New York, N.Y.: Oxford University Press.

Oliver, K. M., Moran, N. A. and Hunter, M. S. 2006 Costs and benefits of a superinfection of facultative symbionts in aphids. *Proc. R. Soc. B.* 273, 1273-1280.

Ouvrard, D., Campbell, B. C., Bourgoin, T. and Chan, K. L. 2000 18S rRNA secondary structure and phylogenetic position of Peloridiidae (Insecta, Hemiptera). *Mol. Phylogenet. Evol.* 16, 403-417.

Perret, X., Staehelin, C. and Broughton, W. J. 2000 Molecular basis of symbiotic promiscuity. *Microbiol. Mol. Biol. Rev.* 64, 180-201.

Prado, S. S., Rubinoff, D. and Almeida, R. P. P. 2006 Vertical transmission of a pentatomid caeca-associated symbiont. *Ann. Entomol. Soc. Am.* 99, 577-585.

Puchta, O. 1955 Experimentelle Untersuchungen über die Bedeutung der Symbiose der Kleiderlaus Pediculus vestimenti Burm *Z. Parasitenkd.* 17, 1-40.

Rakoff-Nahoum, S., Paglino, J., Eslami-Varzaneh, F., Edberg, S. and Medzhitov, R. 2004 Recognition of commensal microflora by toll-like receptors is required for intestinal homeostasis. *Cell* 118, 229-241.

Reinhardt, K. and Siva-Jothy, M. T. 2007 Biology of the bed bugs (Cimicidae). *Annu. Rev. Entomol.* 52, 351-374.

Relman, D. A. and Falkow, S. 2001 The meaning and impact of the human genome sequence for microbiology. *Trends Microbiol.* 9, 206-208.

Rio, R. V., Wu, Y. N., Filardo, G. and Aksoy, S. 2006 Dynamics of multiple symbiont density regulation during host development: tsetse fly and its microbial flora. *Proc. R. Soc. B.* 273, 805-814.

Rio, R. V. M., Hu, Y. and Aksoy, S. 2004 Strategies of the home-team: symbioses exploited for vector-borne disease control. *Trends Microbiol.* 12, 325-336.

Rokas, A., Atkinson, R. J., Nieves-Aldrey, J. L., West, S. A. and Stone, G. N. 2002 The incidence and diversity of *Wolbachia* in gallwasps (Hymenoptera; Cynipidae) on oak. *Mol. Ecol.* 11, 1815-1829.

Ruby, E., Henderson, B. and McFall-Ngai, M. 2004 We get by with a little help from our (little) friends. *Science* 303, 1305-1307.

Ryan, N., Peters, B. and Miller, P. 2004 A survey of bedbugs in short-stay lodges. *NSW Public Health Bull.* 15, 215-217.

Sauer, C., Stackebrandt, E., Gadau, J., Holldobler, B. and Gross, R. 2000 Systematic relationships and cospeciation of bacterial endosymbionts and their carpenter ant host species: proposal of the new taxon *Candidatus* Blochmannia gen. nov. *Int. J. Syst. Evol. Microbiol.* 50, 1877-1886.

Schaefer, C. W. and Panizzi, A. R. 2000 *Heteroptera of Economic Importance*. Florida, F.L.: CRC Press.

Schneider, G. 1940 Beiträge zur kenntnis der symbiontischen einrichtungen der heteropteren. *Z. Morph. Ökol. Tiere* 36.

Schorr, H. 1957 Zur verhaltensbiologie und symbiose von *Brachypelta aterrima* Förster (Cydnidae, Heteroptera). *Z. Morph. Ökol. Tiere* 45, 561-602.

Schuh, R. T. 1979 Evolutionary trends in Heteroptera. Pt. 2. Mouthpart-structures and feeding strategies. *Syst. Zool.* 28, 653-656.

Schuh, R. T. and Slater., J. A. 1995 *True bugs of the world (Hemiptera: Heteroptera).* New York, N.Y.: Cornell University Press.

Shigenobu, S., Watanabe, H., Hattori, M., Sakaki, Y. and Ishikawa, H. 2000 Genome sequence of the endocellular bacterial symbiont of aphids *Buchnera* sp. APS. *Nature* 407, 81-86.

Silverman, A. L., Qu, L. H., Blow, J., Zitron, I. M., Gordon, S. C. and Walker, E. D. 2001 Assessment of hepatitis B virus DNA and hepatitis C virus RNA in the common bedbug

(*Cimex lectularius L.*) and kissing bug (*Rodnius prolixus*). *Am. J. Gastroenterol.* 96, 2194-2198.

Stappenbeck, T. S., Hooper, L. V. and Gordon, J. I. 2002 Developmental regulation of intestinal angiogenesis by indigenous microbes via Paneth cells. *Proc. Natl. Acad. Sci. USA* 99, 15451-15455.

Takiya, D. M., Tran, P. L., Dietrich, C. H. and Moran, N. A. 2006 Co-cladogenesis spanning three phyla: leafhoppers (Insecta: Hemiptera: Cicadellidae) and their dual bacterial symbionts. *Mol. Ecol.* 15, 4175-4191.

Thao, M. L., Gullan, P. J. and Baumann, P. 2002 Secondary (□-*Proteobacteria*) endosymbionts infect the primary (□-*Proteobacteria*) endosymbionts of mealybugs multiple times and coevolve with their hosts. *Appl. Environ. Microbiol.* 68, 3190-3197.

Thao, M. L., Moran, N. A., Abbot, P., Brennan, E. B., Burckhardt, D. H. and Baumann, P. 2000 Cospeciation of psyllids and their primary prokaryotic endosymbionts. *Appl. Environ. Microbiol.* 66, 2898-2905.

Tomokuni, M., Yasunaga, T., Takai, M., Yamashita, I., Kawamura, M. and Kawasawa, T. 1993 *A Field Guide to Japanese Bugs.* Tokyo, J.P.: Zenkoku Noson Kyoiku Kyoukai.

Usinger, R. 1966 *Monograph of Cimicidae (Hemiptera, Heteroptera).* College Park, M.D.: Entomol. Soc. Am.

Van Oevelen, S., De Wachter, R., Vandamme, E., Robbrecht, E. and Prinsen, E. 2002 Identification of the bacterial endosymbionts in leaf galls of *Psychotria* (Rubiaceae, angiosperms) and proposal of '*Candidatus* Burkholderia kirkii' sp. nov. *Int. J. Syst. Evol. Microbiol.* 52, 2023-2027.

Van Oevelen, S., De Wachter, R., Vandamme, E., Vandamme, P., Robbrecht, E. and Prinsen, E. 2004 '*Candidatus* Burkholderia calva' and '*Candidatus* Burkholderia nigropunctata' as leaf gall endosymbionts of African *Psychotria*. *Int. J. Syst. Evol. Microbiol.* 54, 2237-2239.

Vessey, J. K. 2003 Plant growth promoting rhizobacteria as biofertilizers. *Plant. Soil.* 225, 571-586.

von Dohlen, C. D. and Moran, N. A. 1995 Molecular phylogeny of the Homoptera: a paraphyletic taxon. *J. Mol. Evol.* 41, 211-223.

Wernegreen, J. J. 2002 Genome evolution in bacterial endosymbionts of insects. *Nat. Rev. Genet.* 3, 850-861.

Wheeler, W. C., Schuh, R. T. and Bang, R. 1993 Cladistic relationships among higher groups of Heteroptera: congruence between morphological and molecular data sets. *Ent. Scand.* 24, 121-137.

Wilson, E. O. 2000 A Global Biodiversity Map. *Science* 289, 2279.

Woods, D. E. and Sokol, P. A. 2000 The genus *Burkholderia*. In *The prokaryotes: an evolving electronic resource for the microbiological community, 3rd ed., release 3.4.* (ed. M. Dworkin, K. H. Schleifer and E. Stackebrandt). New York, N.Y.: Springer-Verlag.

Wu, D., Daugherty, S. C., Van Aken, S. E., Pai, G. H., Watkins, K. L., Khouri, H., Tallon, L. J., Zaborsky, J. M., Dunbar, H. E., Tran, P. L., Moran, N. A. and Eisen, J. A. 2006 Metabolic complementarity and genomics of the dual bacterial symbiosis of sharpshooters. *PLoS Biol.* 4, e188.

In: Microbial Ecology Research Trends
Editor: Thijs Van Dijk, pp. 65- 86

ISBN: 978-1-60456-179-1

Chapter III

Micro-Game Hunting: Predatory Behavior and Defensive Strategies in Ciliates

Giovanna Rosati, Letizia Modeo and Franco Verni[1]
Unità di Protistologia-Zoologia, Dipartimento di Biologia,
Università di Pisa, Italy

Abstract

Using various feeding strategies ciliates consume all sizes of autotrophic cells from bacteria to the larger dinoflagellates or diatoms. In addition they consume other heterotrophic protists. In this way every component of prokaryotic and eukaryotic nanoproducer biomass can be transferred to higher trophic levels. Predator ciliates provide an intermediate level in the microbial loop and to obtain adequate information on their biology is important from an ecological point of view. On the other hand, effort spent in attempting to deepen our understanding of predation among protozoa are completely justified by the basic importance of this process. Protozoa, indeed, were not only the first primary consumers in the primeval Oceans, but the first predators as well. Such a new trophic niche is quite important, due to the two consequences it leads to: 1) it creates new empty spaces for new organisms to settle in; 2) it triggers a sort of evolutionary competition between predators and prey as to their morpho-functional acquisitions. So predation in protozoa offers a wide range of examples of more or less species specific and more or less direct cell interactions leading the organism to specific morphological or behavioral responses. Indeed if predators developed a variety of hunting strategies, the potential prey evolved defensive responses. Nevertheless there are not models exhaustively examined to obtain an overall picture of this phenomenon considered from a more specific sinecological point of view. This chapter deals with predation in ciliates. After an introduction to the various feeding strategies of ciliates and a survey of the ciliate predator-prey systems already studied by other authors, the model *Amphileptus* preying on *Euplotes*, is presented in more detail as a prototype of ciliate

1 Corresponding author.

raptorial feeding behavior. Our attention is successively focused on: a) the ultrastructure of predator toxicysts (the hunting devices); b) the electrophysiological and ultrastructural effects of toxicysts discharge by *Amphileptus* on *Euplotes*; c) the behavioral patterns of the predator (prey recognition from a distance; nearby recognition; engulfment process); d) the peculiar digestion process; e) the defensive strategies of the prey.

Introduction

In spite of their diversity ciliates are easy to recognize; certain unique characters are almost constantly present and set them apart from other protozoa. The most widely recognized characteristic of ciliates is their dimorphic nuclear apparatus consisting of one (or more) large macronucleus accompanied by one (or more) small micronucleus. From a functional viewpoint the macronucleus controls the physiological or general homeostasis of the cell, while the micronucleus maintains the genetic continuity of the species. The second distinctive character of ciliates is the presence of cilia, at least at some stage of their life cycle. Almost all ciliates are heterotroph, the only exceptions being those species that, thanks to photosyntetic symbionts, are mixotroph (i.e. are able of a nutritional pathway which combines heterotrophy and autotrophy). The most remarkable case of photosynthetic endosymbiosis occurs in *Mesodinium rubrum* that depends entirely on the symbionts and has, in fact, become a photosynthetic ciliate (Packard *et al.*, 1978). The animal-like characteristics of protozoa in general, and of ciliates in particular, are nowhere more obvious than in their seeking and ingesting food.

A common feeding mechanism in heterotrophic protozoans is phagocytosis (Radek and Hausmann, 1996). Since the cortex of ciliates is complex, with an alveolar pellicular organization and sets of subpellicular microtubules, phagocytosis can take place only in distinct sites. The presence of a cytostome, that is a sort of cellular mouth where the plasma membrane is deprived of underlying structures, plays a central role in the uptake of food and is another of the distinctive characters present in most ciliates.

In general, two different types of food acquisition are reported in free-living ciliates: filter feeding and predation (Hausmann, 2002; Hausmann and Hülsmann, 1996). A third strategy, namely suctorian feeding, is separately considered by other authors (Verni and Gualtieri, 1997a). This feeding behavior is restricted only to suctorian ciliates, which in the adult stage are generally found attached to the surface of marine or freshwater invertebrates, or occasionally to inanimate substrates, by means of a non-contractile stalk. They have not a typical cytostome but possess one or more tentacles with special extrusomes (ejectable organelles).

Filter feeding in protozoa is food acquisition in which suspended food particles are concentrated through the action of flagella or cilia. Filter feeding ciliates presents a more or less developed oral cavity that leads to the cytostome. This oral cavity is, in most species, surrounded by specialized rows of cilia: on the right side a paroral membrane composed of one or two rows of strictly packed cilia; on the left side the adoral membranelles consisting in a series of parallel ciliary rows perpendicularly inserted to the long axis of the oral cavity. The water current is always generated by adoral membranelles. In Spirotrichea and Colpodea, whose paroral ciliature is often reduced, the membranelles also function as a filter: they pump water out of the oral cavity together with small particles, while particles too large to pass between neighboring membranelles are retained and concentrated at the cytostome. Thus, by this mechanism generally referred to as upstream filtration, food particles are somehow selected according to their size. In other ciliates (Oligohymenophorea) downstream filtration occurs: the water currents generated by the membranelles are driven through the paroral membrane which acts as a sieve allowing retention of smaller particles. If the food particles size would be the only factor diversifying filter feeding ciliates diet, they all should be omnivorous. Hence, as experiences anyone who attempts to grow ciliates in the lab, some of them are bacterivorous, some prefer algae while others choose heterotrophic flagellates and smaller ciliates. In general even grazing upon suspended bacteria is selective. First of all most ciliates graze in general on medium–sized bacteria (Šimek and Chrzanowski, 1992); then other traits of bacteria such as motility, shape, cell surface characteristics, and activity state (growth or division) may influence the selectivity of protistan grazing (Hahn and Höfle, 2001). Selective feeding by algae consuming ciliates has also been reported (Verity, 1991). Moreover even in truly omnivorous filter feeding ciliates food selectivity, independent from the food organism size, has been experimentally demonstrated (Jurgens and Šimek, 2002; Pfister and Arndt, 1998). The diversification of the diet allows ciliates to diversify their ecological niches. On the other hand, the coexistence of ciliate species that compete for the same nutritional resource has been often observed. Three hypothesis to explain this surprising coexistence have been proposed: 1. resource partition; 2. chemically mediated interference competition; 3. differential use of space. Fox and Barreto (2006) conducted a series of experiments to test these hypotheses. Their results show that none of this mechanism is sufficient to explain a long term coexistence and that interspecific relationships in microcosms are complex and still largely not understood.

At variance with filter feeding ciliates, gulper ciliates capture each particle individually, rather than in bulk, following a well-defined behavioral pattern and in general show a greater degree of discrimination. The predator-prey relation can be either non-specific, i.e. the predator feeds on prey encountered by chance as in the case of *Lacrymaria olor* or, more often, specific, i.e. the predator recognizes and selects a specific prey. In the latter case cell-cell recognition mechanisms come into play.

Hausmann (2002) distinguishes vegetarian gulpers from carnivorous gulpers, the latter being the true predators. Vegetarian ciliates feed on filamentous cyanobacteria they capture through a particular structure namely the cyrtos or nasse (Corliss, 1979) consisting of a basket-like microtubular apparatus that surround the oral cavity. Usually the filamentous algae are attached from the tip. But if this is not possible the ciliate will attack somewhere the filament's length. They bend the filament by partially digesting its cell wall and then ingest both sides simultaneously (Hausmann, 1996; Verni and Gualtieri, 1997a). Predatorial feeding is characteristic of those ciliates which obtain their food through the action of special organelles acting as true hunting devices. They are appropriate types of extrusomes which secrete paralytic (and proteolytic) enzymes to immobilize (and cytolyze) the normally fast-swimming prey. The most common and studied are toxicysts. Following Krüger (1936), Hausmann (1978) defined a toxicyst as 'the extrusome in ciliates that is extruded as a tube-like structure and which secretes poisonous material'. Many different types of offensive extrusomes have been described and referred to with different names, however most of them are now considered different forms of a unique extrusome category (Rosati and Modeo, 2003). Predator ciliates feed on other protists, especially other ciliates and even on small metazoa.

The broad spectrum of feeding possibility is an important contributing factor in the ecological role of ciliates (particularly fundamental in marine habitat). Using various feeding strategies they consume all sizes of autotrophic cells from bacteria to the larger dinoflagellates or diatoms. In addition they consume other heterotrophic protists. In this way every component of prokaryotic and eukaryotic nanoproducer biomass can be transferred to higher trophic levels. Predator ciliates provide an intermediate level in the microbial loop (Fenchel, 1987).

On the other hand protozoa, ciliates included, were not only the first primary consumers in the primeval oceans, but the first predators as well (Gould *et al.*, 1977).

In this chapter, after a rapid survey of the ciliate predator-prey systems already studied by other authors and of the most common prey defensive mechanisms, we present in more detail a study on the model *Amphileptus* preying on *Euplotes*, as a prototype of raptorial feeding behavior and of predator-prey interactions in microcosms.

The study includes:

1) A presentation of the main characters of the story, *Euplotes crassus* and *Amphileptus marinus*, comprehensive of a discussion on taxonomic position of *A. marinus*. Indeed a deeper morphological analysis lead us to transfer the predator, previously recognized as *Litonotus lamella* (Ricci and Verni, 1988), to *Amphileptus* genus;
2) the ultrastructure of predator toxicysts;
3) the electrophysiological and ultrastructural analysis of the consequences to *Euplotes* of toxicysts discharge by *Amphileptus*;
4) the behavioral patterns of the predator (prey recognition from a distance; nearby recognition; engulfment process);
5) the peculiar digestion process;
6) the defense system of the prey.

Survey on the Most Studied Carnivorous Predators in Ciliates

Occasional reports on predatory behavior in ciliates can be traced back in the time, but the credit for the first descriptions of prey attacked and immobilized by these protozoa is due to Maupas (1883). This author also noticed the existence of a certain food specificity in predation. A series of subsequent studies are reported by Dragesco (1962) in his extensive review 'Capture et ingestion des proies chez les infusoires ciliè' in which he also describes, on the basis of his own studies, the predatory behavior of 5 haptorian ciliates: *Enchelys mutans, Litonotus lamella, Chaenea vorax, Dididium nasutum*, and *Dileptus anser*. All of them are able to perceive the prey at a distance but only after the first contact, that however is by chance, they attack the prey with specialized, toxic extrusomes (now recognized as toxicysts). There are differences in the specificity of the prey and in the catching methods.

For example *Litonotus lamella* eats prey species of different genera, while *Chaenea vorax* attacks only small ciliates of *Uronema* and *Cyclidium* genera. In the latter ciliate toxicyst discharge is particularly rapid and violent and the ingestion occurs through typical rhythmic movements. Interestingly, however, when the preferred prey is not available *Chaenea* can adapt to eat larger ciliates.

Dragesco's description of *Didinium nasutum* predation has been later extended by Wessenberg and Antipa (1968; 1970) with the aid of electron microscopy. *Didinium nasutum* is the voracious predatory ciliate *par excellence* with a preference for feeding on species of the genus *Paramecium*, although it has been experimentally demonstrated that, following a sort of imprinting, *Didinium* can eat also other ciliates like for example *Colpidium* (Berger, 1979). The series of pictures, obtained by scanning electron microscopy (SEM), which show the successive steps of the engulfment of *Paramecium* by *Didinium* (Wessenberg and Antipa, 1970) are now reported on school text books all over the world. In *Didinium*, the cytoplasm of the cytostome region is elevated to form a 'proboscis'. This structure, which can also be observed in many other haptorians, is usually the first part of the cell to contact the prey during raptorial feeding (Wessenberg and Antipa, 1970; Ricci *et al.*, 1996), and is usually full of toxicysts. When an unfed *Didinium* happens to collide with a *Paramecium* in such a way that its proboscis contacts *Paramecium* there is an instantaneous discharge of extrusive organelles. By electron microscopy Wessenberg and Antipa (1970) recognized two different kinds of offensive extrusomes namely pexicysts and toxicysts. In spite of their structural and functional differences, however, they can be considered two different kinds of toxicysts (Rosati and Modeo, 2003). The ejection of toxicysts takes place only after pexicyst discharge, which can be considered ' a recognition signal set off by accidental contact with a suitable prey'. Discharged toxicysts can penetrate into the prey up to 20 μm like 'harpoons' and cause prey immobilization. Wessenberg and Antipa (1970) assumed the prey immobilization to be due to the toxin carried by the toxicysts because the struck prey soon stops swimming as its somatic ciliature is modified: in *Paramecium* struck by *Didinium*, the plasma and outer alveolar membranes bulge outward and the cilia are detached at the stricken area. Then the cytoplasm undergoes lysis and forms vacuoles with struck *Paramecium* invariably dying in a few minutes. Apparently *Didinium* is much too specialized for predation to fail and not ingested, struck *Paramecium* always die (Wessenberg and Antipa, 1970).

The first extensive study of the predatory mechanism of *Dileptus anser* was by Visscher (1923), then it was redescribed by Dragesco (1962) and Miller (1968). *Dileptus* are large sized ciliates with a long, trunk-like proboscis continuously sweeping the environment. Toxicysts are located on the oral surface of the proboscis and are ejected when the surface contacts a possible prey: their function is not only to kill but also totally or partially cytolyze the prey. Ciliates of the genus *Colpidium* are apparently preferred prey, but *Dileptus* eats also other ciliates, flagellates and small metazoa like rotifers and even *Planaria*e (Brown and Jenkins, 1962). Ingestion of cytolyzed prey occurs several seconds after initial contact. The cytostome opens to a diameter greater than that of non feeding cells and rapidly encloses the 'victim'. After feeding, very often *Dileptus* pull off portions of its proboscis, so that only a proboscis stub remains. Regeneration of the entire proboscis occurs before the next feeding period begins. Indeed *Dileptus* has a daily cycle of feeding and growth behavior. The timing has been precisely described with *Colpidium* as food organism. The most active feeding period is the early morning (4:30 a.m. – 8:45 a. m.), then the regeneration of proboscis occurs and there is no or little feeding until next morning (Miller, 1968).

Another typical carnivorous ciliate is *Lacrymaria olor*. Its binomial name means 'tear of a swan' (from latin *lacrima* = tear; *olor* = swan), which is an excellent and apt description of both its general shape and one of its identifying traits: this ciliate possess the ability to extend its anterior end outwards by means of what looks like an elongating neck. And just as a swan is able to manipulate its neck in many directions, *Lacrymaria* is able to maintain the same type of agility over its extensibility (Tatchell, 1980). An apical dome, trimmed with a ciliary crown, is at the neck apical end. Whether a prey organism come into contact with *Lacrymaria* a number of toxicysts are discharged from the apical dome. Prey organisms are usually paralyzed. At this point the behavior of the neck changes and it becomes rigid. The opening of the cytostome follows by the expansion of the ciliated crown and the collapse of the oral dome into the cavity created. *Lacrymaria* has been observed feeding on a variety of protozoan ciliates, flagellates and amoebe. Some of these prey organisms can be larger than *Lacrymaria* itself. Some of them may be able to escape predation while others (especially if they are exposed to a large number of *Lacrymaria*) dye and a number of predators will participate to their consumption (Tatchell, 1981).

From this rapid *excursus*, far from being exhaustive, it appears that predatory feeding in ciliates is a complex, specialized process that presents some common traits but also great differences among the different species. Common traits are the succession of different steps depending one from the other and the use of hunting devices namely different types of toxicysts. Seshachar *et al.* (1971) reported that *Homalozoon vermiculare* ejects a liquid, toxic substance and did not mention extrusomes. Nevertheless, two types of toxicysts were later described also in this ciliate (Hausmann, 1978; Kuhlmann and Hausmann, 1980; Rosati and Modeo, 2003). Important differences there are as concern the prey specificity, the hunting mechanisms and the prey engulfment.

Defensive Strategies of Prey

The relationship between a predator and its prey is of fundamental ecological interest. Moreover predator-prey interactions are considered an important evolutionary pulling force. In line with these assumptions in ciliates a number of predator induced defensive mechanisms evolved. Since the phenomenon was first reported by Kuhlmann and Heckmann (1985) predator-induced defense responses have been reported in 15 species of ciliates representing three subclasses: Hypotrichia, Stichotrichia, and Hymenostomia. The discovery of this phenomenon in such distantly related groups suggested that it could be more widespread within Ciliophora but evidences of defense response have not been demonstrated in other groups up to now (Fyda *et al.*, 2005). These responses are, in general, induced by chemical signals released by the predator and are expressed only when the organism perceives predation risk. Most studies have primarily dealt with morphological responses. Predator-induced morphological transformation seems widespread among small aquatic organisms. It has been observed in freshwater planktonic cladocerans, insects larvae and rotifers as well as in marine bryozoans, corals, barnacles and other crustacean (Havel, 1987). Indeed, in aquatic ecosystems most organisms, like ciliates, are able to detect the presence of potential predators through chemical signals (for review see Wicklow, 1997 and Snell, 1998). Lass and Spaak (2003) called these kind of substances kairomones and defined them as ‘a class of info-chemicals evoking a behavioral or physiological reaction in the receiver that is adaptively favorable for the receiver but not for the sender’. Even bacteria may express grazing defense mechanisms under continuous grazing pressure (Posch *et al.*, 2001). In ciliates, morphological transformations include change in size and in shape, the development of horns, keels, spines or other structures which render more difficult or even impossible the ingestion by the predator. For example the induced phenotype of *Colpidium kleini* in the presence of the predator *Lembadion bullinum* can be 60% wider and 20% shorter than the typical one and reach an almost spherical shape. A number of ultrastructural differences were also observed between defended and undefended *C. kleini*: an increased localization of mitochondria in the peripheral regions of the cell; a mobilization of mucocysts; and a change in the appearance of plasma membrane (Fyda *et al.*, 2006). The predator needs significantly more time and has to repeat attacks many times to catch and engulf the morphologically transformed prey (Fyda and Wiackoski, 1998). *Sterkiella* cells in response to signals by *Lembadion magnum* develop two keel-like dorsal projections and transform from a typical lanceolate phenotype to an enlarged more ovoid phenotype (Wicklow, 1997). Obviously, induced morphological changes have an energetic cost. However the extent of these responses vary for different predatory species and also depend on predatory density i.e. can be adjusted according to the entity of risk (Kusch, 1995).

Reciprocal phenotypic plasticity in a predator-prey system has also been described: the interaction between an inducible defense and an inducible offense. This is the case of *Euplotes octocarinatus* (the prey) and *Lembadion bullinum* (the predator). When confronted with the predator *E. octocarinatus* develops protective lateral wings, which inhibit the ingestion by the predator. The predator reacts by expressing a plastic increase in cell size and gape size. Anyway this inducible offence does not seem to be an effective counter-adaptation to the inducible defense of the prey (Kopp and Tollrian, 2003). These observations are in accordance with the general assumption that an increase in the defensive ability of prey is less likely to produce evolutionary counter-measures in its partner than is a comparable increase in attack ability of the predator. Finally it is worth noting that also ciliates usually regarded as filter feeding omnivorous rather than predatory, like *Euplotes patella* or *Stylonychia* sp. induce morphological changes in *Colpidium kleini* (Fyda *et al.*, 2005).

Other ciliates use extrusomes as defensive tools. Thus, it was shown, using trichocyst mutants of *Paramecium*, that discharge of trichocysts defends *Paramecium* against the raptorial ciliates *Dileptus margaritifer* (Harumoto, 1994; Harumoto and Miyake, 1991) and *Monodinium balbiani* (Miyake and Harumoto, 1996). *Dileptus* and *Monodinium* are haptorids; although they use different modes of feeding, both use toxicysts to capture their prey. The discharge of the toxicysts appears to evoke trichocyst discharge in *Paramecium*. Sugibayashi and Harumoto (2000) also demonstrated that trichocysts in *P. tetraurelia* function as defensive organelles against the heterotrich ciliate *Climacostomum virens*, which does not use toxicysts; rather it pulls-in ciliates of various size by the adoral ciliation of its well-developed oral apparatus. *Paramecium* shows a rapid backward movement upon explosive extrusion of trichocysts. Wild-type *Paramecium* cells and mutant cells incapable of backward swimming were compared during encounters with *D. margaritifer*. The results show that the primary defense mechanism protecting *Paramecium* is trichocyst discharge *per se*, not the consequent backward swimming. Hence, trichocyst discharge is a powerful defensive device of importance also in the natural habitat where many protozoa feed on *Paramecium*. However, as frequently observed, in the living world there are no absolute rules: trichocysts have proven to be ineffective against the haptorid *Didinium nasutum*, which preys preferentially on *Paramecium*. *Didinium* may have in some way overcome the defensive function of trichocysts over evolutionary time, thus becoming a main consumer of *Paramecium* (Sugibayashi and Harumoto, 2000), yet another example of the endless 'arms race' between prey and predator.

In some cases a defensive function against predation can be ascribed to the presence of symbionts. *Paramecium bursaria* is a ciliate species which is known to harbor zoochlorellae, photosynthetic symbionts that confer mixotrophy to their host. It has been experimentally demonstrated (Berger, 1980) that *Didinium nasutum* choose apochlorotic *P. bursaria* twice as often as normal paramecia; the hypothesis has been presented that the mutualistic zoochlorellae within *P. bursaria* tend to discourage predation by *D. nasutum* by releasing distasteful metabolites that repel them.

Sand dwelling hypotrich ciliates of the genus *Euplotidium* harbor on their dorsal surface peculiar bacterial ectosymbionts, referred to as epixenosomes. The ectosymbiotic association between *Euplotidium* and epixenosomes (Rosati, 1999; Verni and Rosati, 1990) appears constant in nature and is maintained in lab well-fed cultures; but *Euplotidium* stocks tend to lose epixenosomes when moderate starvation slows their cell cycle. The lack of epixenosomes does not modify either the behavior or the fission rate of the ciliate. Thus the symbionts are not vital for their host. This observation led to hypothesize that, in the natural environment, the presence of the ectosymbionts might play a crucial ecological role, such as a defense against predators. This hypothesis was experimentally verified by comparing the food uptake of the ciliate now recognized as *Amphileptus marinus* (whose raptorial feeding behavior will be described below), when preying upon *Euplotes crassus*, *Euplotidium itoi* without epixenosomes and *E. itoi* with epixenosomes. Considering the rate of disappearance of prey it appeared that *Euplotes*, the preferred prey of this predatory species, are eaten completely and rapidly, *Euplotidium* without epixenosomes are eaten, although with minor efficiency, and *Euplotidium* with epixenosomes are never ingested (Rosati *et al.*, 1999). The results obtained strongly support the hypothesis that epixenosomes provide their host with an efficient defense against predation and that its extrusive apparatus is somehow involved in this defense. Even if paralyzed by toxicyst discharge, *Euplotidium* carrying the symbionts can, in a short time, recover their normal swimming behavior. On the contrary, *E. itoi* without epixenosomes are hunted, struck, and engulfed by this predator. Thus in *Euplotidium* ectosymbionts play the defensive role that in other ciliates is played by extrusomes.

Another particular defense methods is performed by the oligotrich ciliates *Strobilidium velox* and *Halteria grandinella*. Both ciliate species initiate jumps after encounters with rotifer predators. They jump at high velocity. *S. velox* during jumps travels a mean distance of 0.37 mm (18 body lengths) at a mean velocity of 2.76 mm/s (131 body lengths/s). This swimming velocity is the highest so far reported for an oligotrich, and equals the highest one reported in any ciliate (*Mesodinium rubrum*). A predation experiment showed that the rotifer

Asplanchna girodi was about twice as, and significantly more, likely to ingest non jumping ciliate species like *Paramecium tetraurelia* as *S. velox* in a mixture of equal number of this ciliates (Gilbert, 1994).

The Model: Amphileptus Marinus Preying on Euplotes Crassus

1.The Prey (Figure 1)

Euplotes crassus (Spirotrichea, Hypotrichia) is a medium-size, ovoid marine ciliate well characterized by morphological and molecular points of view. It has no irregularities in the outline and is ~ 100 x 50 μm in size *in vivo* (SEM sizes: ~ 65 x 35 μm, due to cell shrinkage after fixation). On its dorsal surface 8 longitudinal ridges are present and, between each couple of ridges, 10 kineties consisting in short bristle-like cilia are inserted. The ventral surface is not ornamented and shows 10 frontoventral, 5 transverse, and 5 or 6 caudal cirri. The peristome is long and narrow, covering ~ ⅔ of the cell length. The macronucleus is C-shaped; the roundish micronucleus is usually located laterally to the mid region of the macronucleus and is not detectable *in vivo*.

2. The Predator (Figure 2)

2.1. Description and Systematic Position

Previously recognized as *Litonotus lamella* (Litostomatea, Pleurostomatida) by Ricci and Verni (1988), this species has been here reinvestigated and shifted to the genus *Amphileptus* on the basis of the results of both a deeper morphological analysis (performed with light microscope as well as SEM and transmission electron microscope-TEM), and an original molecular analysis (performed on 18S rRNA gene sequence).

The morphological study was especially focused on the invariable and observer-independent characters which diversify the more similar pleurostomatid genera (namely *Amphileptus*, *Acineria*, *Loxophyllum*, *Siroloxophyllum*, and *Litonotus*) as described by several papers where modern methods of morphological investigation (protargol staining and SEM) have been used (Foissner, 1984; Foissner and Leipe, 1995; Foissner *et al.*, 1995; Lin *et al.*, 2007; Song *et al.*, 2004), i.e. :1) the kind of infraciliature on the ventral side; 2) the presence/absence of a twisted anterior cell end; 3) the shape and the arrangement of extrusomes; 4) the number and position of contractile vacuoles; 5) the presence/absence of spines around the body margin. The comparison of the determined sequence of the gene encoding for 18S rRNA in our predator with those present in data bank revealed that it is more similar to the corresponding sequences from species belonging to *Amphileptus-Pseudoamphileptus* group (unfortunately the latter sequences are unpublished and no morphological data are available for the species they belong to) than to sequences from species within *Acineria-Litonotus-Loxophyllum-Siroloxophyllum* group (I. Andreoli, *pers. commun.*).

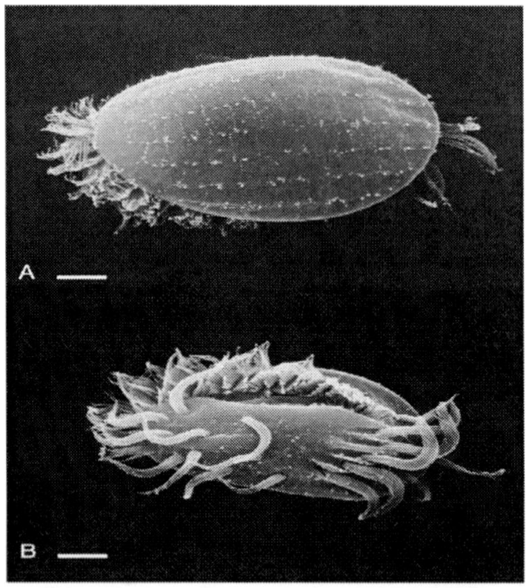

Figure 1. *Euplotes crassus* at SEM. A. Dorsal view. B. Ventral view. Bars = 10 μm.

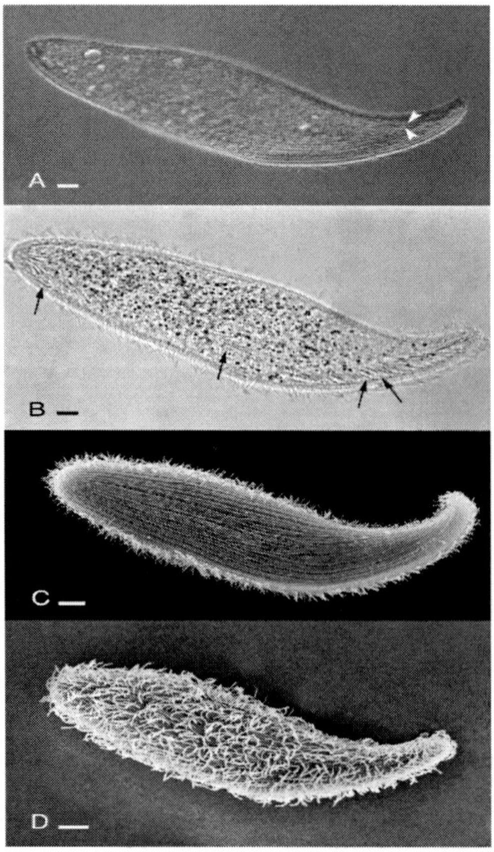

Figure 2. *Amphileptus marinus.* A-B. *In vivo* pictures. A. Superficial view showing the *spica* formed by the somatic kineties (arrowheads). B. A different focus level showing the localization of toxicysts (arrows). C-D. SEM pictures. C. Dorsal view. D. Ventral view. Bars = 10 μm.

This multidisciplinary work lead us to identify our ciliate as the highly variable in size species originally described as *Lionotus (Hemiophrys) marina* by Kahl (1931) and recently redescribed by Song *et al.* (2004) as *Amphileptus marinus*.

Amphileptus marinus (Litostomatea, Pleurostomatida) is a large, conspicuously flexible, and contractile marine predator species, ~ 220 x 70 μm *in vivo* (SEM sizes: ~ 200 x 45 μm; fixed specimens smaller, due to cell contraction and shrinkage: Foissner and Leipe, 1995). It has a slender lanceolate body: the anterior cell end shows a proboscis (or 'neck') reaching ~ ¼ of the cell length, often curving backwards; the posterior cell end is rounded. On the cell left (functionally dorsal) side there are: some sparse somatic ~ 1.5 μm long cilia forming 6-8 somatic kineties; a perioral kinety (called perioral kinety 1); a brush kinety with ~ 2.5 μm long close-set paired cilia. On the cell right (functionally ventral) side a second perioral kinety (called perioral kinety 2) plus 20-22 somatic kineties (consisting in ~ 7 μm long cilia) are visible, the latter forming the specialized ciliary pattern typical of the genus referred to as the *spica* (i. e. an anterior suture in the median area due to the presence of central shortened somatic kineties: Song *et al.*, 2004). Toxicysts are bar-shaped, straight and ~ 10 μm long, densely arranged in a conspicuous, ~ 2.5 μm thick fringe running along the oral bulge (which is the string-like patterned eminence along the oral slit: Foissner and Leipe, 1995) that surrounds the whole ventral side; some are scattered in the cytoplasm, especially at the posterior end and in the dorsal margin of the neck (the latter are less easily detectable *in vivo*). Mucocyst-like extrusomes are as well present under the plasma membrane (TEM observations). The cytostome is slit-like (Corliss, 1979) as typical for pleurostomatids. The oral opening is not permanent, becoming a visible, ~ 80 μm long aperture only during engulfment of the prey: it is actually defined as the oral bulge region accompanied by the portions of the two perioral kineties consisting of paired basal bodies (Foissner and Leipe, 1995), hence this region covers up to ¾ of *A. marinus* body length. Two ellipsoid macronuclear nodules (~ 22 x 13 μm) with a single, not detectable *in vivo* roundish micronucleus in between are present in the mid-region. Near the ventral margin, along the posterior half of the cell, usually from one up to ~ 4 contractile vacuoles are present.

The availability of *A. marinus* 18S rRNA gene sequence will allow to design species-specific fluorescently labelled oligonucleotide probes to always unequivocally detected this species in FISH experiments (Petroni *et al.*, 2003); thus, although we are confident of our identification, possible problems either of misidentification or difficult discrimination between this species and the numerous, morphologically similar, closely related pleurostomatids can be easily solved.

2.2. Toxicyst Structure (Figure 3)

Resting toxicysts of Litostomatea are complex organelles which contain fully developed structures inside a capsule which can be evaginated. The resting toxicysts of *A. marinus* are ~ 10 μm long (*in vivo* size) with a diameter of 0,2-0,5 μm (TEM observations) and consist of a tube lying within a capsule surrounded by a fibro-granular layer. The explosive extrusion takes place in a few milliseconds following a mechanical contact between predator and prey. After the extrusion, the granular layer disappears: very likely this layer plays a role in the tube expulsion. Cytochemical analysis revealed that the capsule and the tube wall are rich in polysaccharides while the granular layer consists of ribosomes embedded in a proteinaceous matrix and contains acid phosphatase (Verni and Ricci, 1986). The presence of this hydrolitic enzyme suggests a possible double function of toxicysts: a first during the paralysis of the prey, and a second during the first phase of the digestion process.

The tube is ejected via evagination, thus the length of ejected toxicysts goes up two times the resting state length (~ 20 μm), with a capsule/tubule ratio 1:2. A similar ratio is reported in the long toxicysts of *Loxophyllum*, while a capsule/tube ratio 1:1 is reported in pexicysts of *Didinium* and in the short toxicysts of *Loxophyllum* and *Dileptus*. A capsule/tube/end tube ratio 1:2:1 is described for the long type toxicysts of *Dileptus* (for a review see Haussman, 1978).

2.3. Locomotory Behavior

The locomotory behavior of *A. marinus* has been analyzed in detail (Ricci and Verni, 1988). Like most interstitial ciliates *A. marinus* performs creeping and swimming. Creeping is the most common movement for this species that creeps on the substrate as much as 70-75% of the time. The general pattern of forward creeping consists of a series of perfect geometrical rightward arcs (long lasting elements) which, because of the particular morphology of the species, are actually bent to the dorsal side of the organism. The creeping pathways is only rarely straight: it is interrupted either by periodic, spontaneous changes of internal conditions or by external stimuli. The general pattern of the changes (short lasting elements) is the same in both cases and does not depend on the kind of stimulus received, i.e. the trajectory changes are stereotyped in their spatial-temporal pattern. Four types of trajectory changes were recognized according to the sharpness of the angle and the variation in velocity (for review see Ricci, 1996). The first type, the continuous trajectory change (CTC), is performed by *A. marinus* rather frequently (26% of all studied cases); the second type, the smooth trajectory change (STC) is only rarely observed (4.6%) while the third type, the rough trajectory change (RTC) is more frequent (10%). The fourth type, the side-stepping reaction (SSR) is the most frequently performed (59%): after stopping suddenly the organism creeps backwards for about 500 μm, then stops again, turns anticlockwise, and eventually starts creeping forward again along a new trajectory to the left of the first track (Ricci and Verni, 1988).

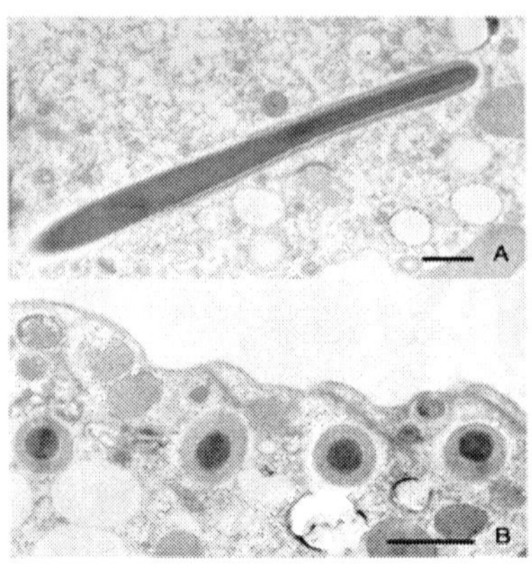

Figure 3. TEM pictures of toxicysts. A. Longitudinal section. B. Cross sections. Bars = 0.5 μm.

Although not as commonly as creeping (25-30% of the studied tracks), swimming is performed by *A. marinus* very smoothly and apparently with great ease. *A. marinus* swims forward along a right-hand helicoid, as gliding clockwise on the lateral surface of a cylinder, and keeping its ciliate surface outwards. As well as in all ciliates each helicoidal tract is periodically interrupted by a stop-and-reorientation reaction (SRR) which precedes a new helicoid.

3. The 'Hunting Game'

3.1. Predatory Behavior

The stages in the *A. marinus* predatory behavior were analyzed step by step. The initial step of the hunt was firstly investigated through a series of experiments in which the kinetics of the spatial distribution of *A. marinus* in the presence or absence of the prey was analyzed. A bidimensional T-maze (Van Houten *et al.*, 1975) was used. This equipment is conventionally divided into three zones: the experimental arm in which the test solutions (either sea water with preys or prey cell free fluid-CFF) were introduced; the neutral arm; and the control arm, in which pure sea water was introduced. The effects of the experiments were measured as the number of *A. marinus* observed at successive times in each arm. The results clearly indicated that *A. marinus* can track *Euplotes* also when it is not in physical contact. This predator can recognize from a distance its prey to intercept them, and upon direct cell-cell recognition, to discharge its toxicysts before to kill (Ricci *et al.*, 1996). These results rule out any random exploration by *A. marinus* while searching for its prey: the species indeed proved to be capable of approaching the prey efficiently. Such a capability shows a predatory strategy which is quite sophisticated and somehow different from that of most predator ciliates. This oriented predatory behavior implies a chemo-orientation. *A. marinus* proved to be able to track *Euplotes* from a distance of at least 3 cm. In the experimental conditions reported above the effects became evident after 20 min and lasted for more than 60 min: within this time lag a gradient forms which is capable of guiding *A. marinus* to its prey. These results indicate that *Euplotes* releases into the medium some chemical(s) which induce in the predator a creeping behavior leading it to the prey. The induction was obtained also with the CFF of *Euplotes*. It appears that the factor released by *Euplotes* and perceived by the predator is proteinaceous (Morelli and Verni, 1996). Successive studies demonstrated that the proteinaceous attractive factors are different molecules lying in the molecular weight ranges 10 kD, 30 kD, and between 30 and 100 kD (Andreoli, 2006). Interestingly, *A. marinus* is capable of finding and killing several *Euplotes* species (*E. vannus*, *E. crassus*, *E. minuta*, and *E. raikovi*) while it can only find (and not kill) *E. charon* and ignores completely *E. magnicirratus* and *E. rariseta* together with other euplotids. Anyway, among the potential preys a difference occurs, at least as far as their capability of eliciting the searching behavior of *A. marinus*; the attracting forces of each species can be ordered according to a decreasing strength, *E. crassus* and *E. vannus* producing the strongest. For this reason the analysis of the successive steps of predation has been performed with *E. crassus* as prey. It is not known whether the difference in the attracting forces depend upon different amounts of attractant released or upon different affinities that different attractant(s) have with the receptors of *A. marinus* (Ricci *et al.*, 1996).

The peculiar fact that *E. charon* can be tracked by *A. marinus* without evoking any discharge of its toxicysts demonstrates that the recognition of the prey actually occurs at two different levels and with the two distinct mechanisms as reported by Ricci and Verni, 1994.

The induced changes in behavior leading *A. marinus* to gather the area where the prey is perceived as abundant are clear-cut: 1) the creeping velocity decreases; 2) the frequency of the behavioral patterns interrupting the track increases; 3) the rightward arcs become more coiled. A new behavioral pattern appears to explore the 'favorable' areas, where the prey is perceived!

A direct contact between predator and prey is necessary for the actual discharge of toxicysts. In general, it seems possible that a basic physiological 'health' of both predator and prey represents a sort of prerequisite for predation to occur efficiently.

The most frequently discharged toxicysts are those located along the oral edge of the neck, even if toxicysts located in the posterior part of the same edge could function effectively in particular when the *A. marinus* has undergone a prolonged starvation. In the predator the discharge of toxicysts was unfailingly followed by a rather slow backward creeping along a segment as 500-1000 μm.

Soon after the backward creeping, *A. marinus* starts searching for the prey by means of three types of behavior which can be used by the same cell, even during the same process of searching without any predetermined sequence. Either several forward and backward anticlockwise jerks along radial arcs separated by sharp angles, or several SSR prolonged in space and time can be performed; more rarely the predator moves according to several successive conical corrections. All reactions end as soon as the prey is contacted again (Ricci and Verni, 1988).

3.2. The Consequence to Euplotes of Toxicysts Discarge by A. marinus

When the toxicysts are discharged, *E. crassus* is immobilized: a specific, dramatic ciliary disarrangement occurs only at the level of the ventral ciliature i.e. the motile ciliature. Both membranelles and cirri lose their compact morphology. As revealed by SEM and TEM, single cilia of both organelles undergo a modification at their terminal portion. At this level, the axoneme becomes folded, and the ciliary membrane enlarges into a conspicuous rounded bulb (Verni, 1985). An electrophysiological analysis showed that the material discharged by *A. marinus* toxicysts causes a series of alterations in the electrical properties of the *Euplotes* membrane. The most immediate effect of the toxin is a depolarization (Morelli *et al.*, 2002). The ultrastructural modifications of ventral ciliary structure could be the direct consequence of the changes in the plasma membrane potential or indirectly derived by fixation procedures upon structures with altered physiological conditions.

3.3. Ingestion (Figure 4)

Ingestion occurs as soon as the paralyzed *Euplotes* is contacted by the oral edge of the searching *A. marinus*. As described above, the predator has no permanent oral funnel. This structure is organized during the prey engulfment below the oral slit. Then the cytostome enlarges and the prey is rapidly engulfed.

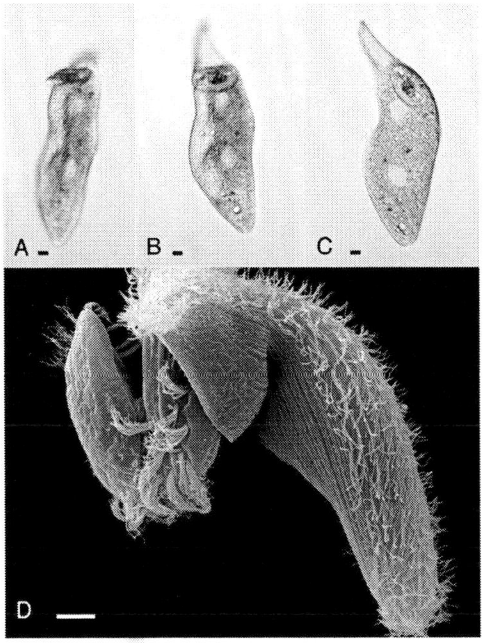

Figure 4. A-C Sequential steps of *Euplotes* ingestion by *A. marinus in vivo* (from Verni and Rosati, 1992). D. SEM picture of *A. marinus* while engulfs a pair of *E. crassus*. Note the enlargement of the cytostome. Bars = 10 μm.

At first it can be seen, by *in vivo* observation at light microscope, perpendicularly oriented in respect to the long axis of the predator, then the latter bends its body and, by a number of peristaltic jerks, turns the still well-recognizable prey parallel to the long axis. This process on the whole takes 1.5-2 min. No more than three preys in succession are ever ingested. Electron microscopy showed that the oral zone of *A. marinus* is delimited only by the plasma membrane. A certain number of undischarged, not oriented toxicysts are present in the neighboring cytoplasm. During the ingestion the oral zone rapidly enlarges and the plasma membrane invaginates around the prey. Toxicysts can enter the forming food vacuole. Once ingestion is completed the oral zone again takes up its normal morphology. Microtubular and microfilamentous structures, play a leading role in the ingestion. The cytostome opening is mediated by a subcortical microtubular basket surrounding the cytopharynx, while peristaltic jerks, which are extended to the whole body, may be caused by the *lamina corticalis*, a fibrous layer beneath the pellicle and marking the ecto-endoplasmic boundary. These jerks may be aimed to orient the predator toward the dying *Euplotes* and to make the entrance of the prey into the cytostome easier. In *A. marinus* the microtubular basket and the *lamina corticalis* cooperate, indeed the destruction of one of them stops the whole process. It has been hypothesized that the basket and the *lamina* may be joined by specific components (Morelli and Verni, 1996)

3.4. Digestion

After the ingestion of the prey the digestion cycle takes place. Although much is known about the digestive cycle of filter feeder ciliates little information is available for raptorial ciliates. The digestion process of *Paramecium*, among filter feeders, is the best understood and has been described in detail: the food vacuoles are rhythmically pinched off from the cytostome and transported through the cell toward the anus (cytopige) on a predetermined course (cyclosis) (for review see Allen and Fok, 2000). Verni and Gualtieri (1997b) illustrated the digestive process of a raptorial ciliate for the first time using *A. marinus marinus* with *Euplotes crassus* as prey as a model. In the first phase acid vesicles, accumulated in the anterior portion of the neck, move from the neck to the forming large food vacuole containing the still apparently intact prey. Then, a great amount of differently sized acid vesicles sizes fuse with the digestive vacuole membrane lowering the internal pH. This phase lasts 7-10 min. Very likely, during this phase the acid phosphatase contained in the toxicysts included in the forming vacuole plays a role in the digestive process. The second phase occurs in many vacuoles, originated by a sort of fragmentation of the large single phagosome, scattered all over the cytoplasm, and containing *Euplotes* fragments. This phase lasts 40-50 minutes; during this period the digestion of the fragmented prey occurs asynchronously and autonomously in each vacuole.

The presence of this asynchronous and independent multi-vacuolar digestion is likely correlated with the size of the prey. A great number of little vacuoles assures a greater surface for acid vesicles attachment, and a smaller volume for the 3-D enzymatic action, in comparison with a single big phagosome. These characteristics reduce the digestion time. At the end of the digestion, i. e. 1h after the capture of the prey, digestive products are released into the cytoplasm and the vacuoles become defecation competent. Then no subcellular digestive compartments can be recognized inside the predator.

It is possible to assume that this digestion pattern is present also in other predator-prey systems, both in the Litostomatea family, where exists a predator-prey specificity, (Ricci and Verni, 1988; Ricci *et al.,* 1996), or in the other raptorial feeder families, where does not exist this specificity (Kuhlmann and Hausmann, 1983; Tatchell, 1981). Baumberg and Hausmann (2007) recently described at the ultrastructural level the digestive cycle in *Homalozoon vermiculare* and, in accordance with the above reported assumption, evidenced that in this predator too numerous smaller vacuoles are formed from the initial single digestive vacuole, These authors also reported a special type of segregation of the food during digestion. Certain prey organelles or cell constituents are segregated into individual digestive vacuoles: nuclei, mitochondria, cilia, extrusomes, zoochlorellae, and complete food vacuoles already present in the prey. So the predator appears to recognize the different prey constituents. How the segregation of these constituents into special vacuoles takes place is unexplained. The ability in separate different kinds of food has been already reported in *Oxytricha bifaria* when this species shifts from the normal bacterivorous to the gigantic carnivorous form that feeds on conspecific organisms. In this form, indeed, some food vacuoles contained only bacteria while others only one ciliate prey. A sort of mechanical explanation, due to the rhythmic formation of food vacuoles, has been tentatively proposed as the digestive pattern in giant *O. bifaria* follows a similar pattern for both kind of food. The transformation of normal *Oxytricha* in giants appears to be an adaptive process that renders this organisms capable of ingesting food particles larger than bacteria without modifying the digestive pattern (Verni and Rosati, 1992). On the contrary *H. vermiculare* is apparently not able to break down bacteria which have been enclosed in food vacuoles of the prey or the symbiotic zoochlorellae that are defecated structurally unchanged via the cytoproct. A possibility, for the moment presented by the authors as a speculation, could be that different enzymes are involved in the degradation of the different food constituents (Baumberg and Hausmann, 2007) or that lysosome of *H. vermiculare* have no ability to overcome the barrier of the membrane food vacuoles of the prey or to digest zoochlorellae, a type of food not suited for them. Once again predatory feeding in ciliates appears a very specialized nutrition way that a more effective digestion and a more complete assimilation of the ingested food, also favored in many cases by the high specialization in the choice of prey, in some way counterbalance the greater cost of food uptake and renders continuous feeding unnecessary.

3.5. Defensive Strategies of the Prey

As in all predator-prey interactions the prey of our system acts a strategy for escape from predators. Actually, *Euplotes* seems to possess two kinds of defensive strategies: a strategy at individual level and a strategy at population level.

The effect of toxicyst discharge by the predator extends around over a sub-circular area ~ 300 μm in diameter. Not stricken *Euplotes crassus* avoid this area by a high frequency of avoiding reactions (Ricci and Verni, 1994). The effect increases up to a maximum for 90 sec. This reaction could represent an individual defensive level. This finding (Ricci and Verni, 1994) strongly suggests that *Amphileptus* releases a factor (or complex of factors) whenever it discharges its toxicysts that might represent the 'stay away' signal for *Euplotes.* The second defensive strategy is more sophisticated. Morelli and Verni (1996) reported that in CFF of *Euplotes* is always present, although in low concentration, a non proteinaceous molecule (5kD), that manifests repulsive activity against the predators. A small number of *Euplotes* in each population may be capable of producing this substance and allowing some organisms to escape and avoid ingestion by the predators more easily. Thus this tonic emission of a repulsive molecule possibly constitutes a population defense.

The defensive morphological changes, that occur only in the presence of the predator (see above), limit metabolic activity and have fitness costs. From this point of view, *Euplotes* defensive strategy is very different. For the prey the metabolic cost is minimal, the fitness is not reduced anyway (structural changes do not occur) and the defensive system is active (the predators run away). Thus, a small number of producer organisms are sufficient to facilitate the survival of the population.

Conclusion

Amphileptus–Euplotes interaction, as far as we know, is the most deeply investigated among predatory interaction in ciliates, and perhaps in protozoa. There are still many aspects to be clarified; for example, both the attractant and repellent substances produced by *Euplotes* and that emanated by predator following toxicyst discharge have not yet been precisely identified. Anyway, the knowledge acquired through the study of this model well illustrates that predation among protozoa involves a variety of general biological processes. Among them: mechanisms of cell-cell recognition from a distance and following direct contact; chemical communication; membrane receptors; excretion through specific organelles (extrusomes); phagocytosis; cellular digestion; induction of morphological and behavioral changes. With regard to ciliates this could render the subject of interest for scientists of various branches of biological learning. Protozoa in general and ciliates in particular are ubiquitous, easy to be collected and a number of them cheaply cultivable in the lab. Thus, they could represent a suitable tool in the study of each of the above mentioned branches.

Over the past 30 years it has become increasingly obvious that phagotrophic protists (and ciliates are all phagotrophic) play important ecological roles in aquatic food webs. Studies based on laboratory analysis of cultured species that grow well at prey concentrations order of magnitude higher than typical of most natural systems (like those reported in this chapter), are certainly useful from an ecological point of view to have an idea of the great potentiality and variety existing among ciliates. These studies clearly evidenced that the great differentiation in ciliate feeding mechanisms and in their diet allows them to occupy a variety of ecological niches, many of which neglected by other organisms. Studies of feeding and growth rates under natural condition are now required to have the actual evaluation of the ecological role of predacious ciliates.

References

Allen, R. D. and Fok, A. K. (2000). Membrane trafficking and processing in *Paramecium*. *Int. Rev. Cytol, 198*, 277-318.

Andreoli, I. (2006). Relazioni interspecifiche fra protozoi ciliati: un approccio molecolare. Tesi di Dottorato di Ricerca in Biologia Evoluzionistica (IV ciclo NS). Biblioteca Nazionale Centrale, Roma.

Baumberg, D. and Hausmann, K. (2007). Vesiculation of food vacuoles and segregation of ingesta: peculiarities of the digestive cycle in the raptorial ciliate *Homalozoon vermiculare*. *Acta Protozool., 46*, 1-8.

Berger, J. (1979). The feeding behavior of *Didinium nasutum* on an atypical prey ciliate (*Colpidium campylum*). *Trans. Am. Micros. Soc., 98*: 487-494.

Berger, J. (1980). The feeding behavior of *Didinium nasutum* on *Paramecium bursaria* with normal or apochlorotic zoochlorellae. *J. Gen. Microbiol., 118*, 497-404.

Brown, H. P. and Jenkins, M. M. (1962). A protozoan (*Dileptus:* ciliate) predatory upon metazoa. *Science, 136*, 710.

Corliss, J. O. (1979). The Ciliated Protozoa: Characterization, Classification, and Guide to the Literature (2nd edition). London, England: Pergamon Press.

Dragesco, J. (1962). Capture et ingestion des proies chez les infusoires cilié. *Bull. Biol. Belg., 96*, 123-167.

Fenchel, T. (1987). Ecology of Protozoa: the Biology of free-Living Phagotrophic Protists. Berlin, Germany: Springer.

Fyda, J. and Wiąckowski, K. (1998). Benefits and costs of predator-induced morphological changes in the ciliate *Colpidium kleini* (Protozoa, Ciliophora) *Europ. J. Protistol., 34*, 118-123.

Fyda, J., Warren, A. and Wolińska, J. (2005). An investigation of predator-induced defence responses in ciliated protozoa. *J. Nat. Hist., 39*, 1431-1442.

Fyda, J., Kennaway, G., Adamus, K. and Warren, A. (2006). Ultrastructural events in the predator-induced defence response of *Colpidium kleini* (Ciliophora: Hymenostomatida) *Acta Protozool., 45*, 461-464.

Foissner, W. (1984). Taxonomie und Ökologie einiger Ciliaten (Protozoa, Ciliophora) des Saprobiensystems. I: Genera *Litonotus, Amphileptus, Opisthodon. Hydrobiologia, 119*, 193-208.

Foissner, W. and Leipe, D. (1995). Morphology and ecology of *Siroloxophyllum utriculariae* (Penard, 1922) n. g., n. comb. (Ciliophora, Pleurostomatida) and an improved classification of pleurostomatid ciliates. *J. Eukaryot. Microbiol., 42*, 476-490.

Foissner, W., Berger, H., Blatterer, H. and Kohmann, F. (1995). Taxonomische und ökologische Revision der Cliaten des Saprobiensystems. Band IV: Gymnostomatea, Loxodes, Suctoria. Munich, Germany: Informationsberichte des Bayer. Landesamtes für Wasserwirtschaft; Heft 1/95; p 1-540.

Fox, J. W. and Barreto, C. (2006). Surprising competitive coexistence in a classic model system. *Community Ecol.*, *7*, 143-154.

Gilbert, J. J. (1994). Jumping behavior in the oligotrich ciliates *Strobilidium velox* and *Halteria grandinella*, and significance as a defense against rotifer predators. *Micr. Ecol., 27*, 189-200.

Gould, S. J., Raup, D. M., Sepkoski, J. J. Jr, Schopf, T. J. M. and Simberloff, D. S. (1977). The shape of evolution. A comparison of real and random clades. *Paleobiology, 3*, 173-181.

Hahn, W. M. and Höfle, M. G. (2001). Grazing of protozoa and its effect on populations of aquatic bacteria. *FEMS Microbiol. Ecol., 35*, 113-121.

Harumoto, T. (1994). The role of trichocyst discharge and backward swimming in escaping behavior of *Paramecium* from *Dileptus margaritifer*. *J. Eukaryot. Microbiol., 41*, 560-564.

Harumoto, T. and Miyake, A. (1991). Defensive function of trichocysts in *Paramecium. J. Exp. Zool., 260*, 84-92.

Hausmann, K. (1978). Extrusive organelles in Protists. *Int. Rev. Cytol., 52*:197-276.

Hausmann, K. (2002). Food acquisition, food ingestion and food digestion by protists. *Jpn. J. Protozool. 35,* 85-95.

Hausmann, K. and Hülsmann, N. (1996). Protozoology (2nd edition). Stuttgart, Germany: Georg Thieme Verlag.

Havel J. E. Predator-induced defenses: a review. In: W. C. Kerfoot and Sih A. (eds.). *Predation: Direct and indirect impacts on aquatic community.* Hanover: University Press of New England; 1987; p 263-278.

Jurgens, K. and Šimek, K. (2000). Functional response and particle size selection of *Halteria cfr. grandinella*, a common freshwater oligotrichous ciliate. *Aquat. Microb. Ecol., 22*, 57-68.

Kahl, A. (1931). Urtiere oder Protozoa I: Wimpertiere oder Ciliata (Infusoria) 2. *Tierwelt Dtl., 21*, 181-398.

Kopp, M. and Tollrian, R. (2003). Reciprocal phenotypic plasticity in a predator-prey system: inducible offences against inducible defences? *Ecol. Letters, 6*, 742-748.

Krüger, F. (1936). Die Trichocysten der Ciliaten im Dunkelfeldbild. *Zoologica, 34*:1-83.

Kuhlmann, S. and Hausmann, K. (1980). Untersuchungen zu Nahrungserwerb und Nahrungsaufahme bei *Homalozoon vermiculare*, Stokes 1887. 2. Mucocysten, Konocysten, Toxicysten. *Protistologica, 16,* 125-134.

Kuhlmann, S. and Hausmann, K. (1983). Ultrastrukturelle Untersuchungen zur Nahrungsaufnahme bei *Homalozoon vermiculare*, Stokes 1887. *Protistologica, 19*, 111-122.

Kuhlmann, H. W. and Heckmann, K. (1985). Interspecific morphogens regulate prey-predator relations in protozoa. *Science, 227*, 1347-1349.

Kusch, J. (1995). Adaptation of inducible defense in *Euplotes daidaleos* (Ciliophora) to predation risks by various predators. *Microb. Ecol., 30*, 79-88.

Lass, S. and Spaak, P. (2003). Chemically induced anti-predator defences in plankton. *Hydrobiologia, 491*, 221-239.

Lin, X., Song, W. and Li, J. (2007). *Amphileptus aeschtae* nov. spec. and *Amphileptus eigneri* nov. spec. (Ciliophora, Pleurostomatida), two new marine pleurostomatid ciliates from China. *Europ. J. Protistol.*, *43,* 77-86.

Maupas, E. (1883). Contribution à l'étude morphologique et anatomique des Infusories Ciliés. *Arch. Zool. Exper.,1*, 427-664.

Miller, S. (1968). The predatory behavior of *Dileptus anser. J. Protozool., 15*, 313-319.

Miyake, A. and Harumoto, T. (1996). Defensive function of trichocysts in *Paramecium* against the predatory ciliate *Monodinium balbiani. Europ. J. Protistol., 32,* 128-133.

Morelli, A. and Verni, F. (1996). Predator-prey interactions: effects of the substances released by the prey *Euplotes crassus* (Protozoa, Ciliata) on the predator *Litonotus lamella* (Protozoa, Ciliata). *Microbios, 87*, 185-198.

Morelli, A., Ricci, N. and Verni, F. (2002). Action of *Litonotus* (predator) toxicysts on the electric properties of *Euplotes* (prey) cell membrane. *Ital. J. Zool., 69*, 103-107.

Packard, T. T., Blasco, D. and Barber, R. T. *Mesodinium rubrum* in Baja California upwelling system. In: R. Boje and M. Tomczak (eds). *Upwelling systems*. Berlin: Springer-Verlag; 1978, p. 73-89.

Petroni, G., Rosati, G., Vannini, C., Modeo, L., Dini, F. and Verni, F. (2003). *In situ* identification by fluorescently labeled oligonucleotide probes of morphologically similar, closely related ciliate species. *Microb. Ecol., 45*, 156-162.

Pfister, G. and Arndt, H. (1998). Food selectivity and feeding behavior in omnivorous filter-feeding ciliates: a case study for *Stylonychia. Europ. J. Protistol., 34*, 446-457.

Posch, T., Jezbera, J., Vrba, J., Šimek, K., Pernthaler, J., Andreatta, S. and Sonntag, B. (2001). Size selective feeding in *Cyclidium glaucoma* (Ciliophora, scuticociliatida) and its effects on bacterial community structure: a study from continuous cultivation system. *Microb. Ecol., 42*, 217-227.

Radek, R. and Hausmann, K. Phagotrophy of Ciliates. In: K. Hausmann. and P. C. Bradbury (eds). *Ciliates cells as organisms*. Stuttgart, Germany: Fischer. 1996; p 197-219.

Ricci, N. Ethology of Ciliates. In: K. Hausmann. and P. C. Bradbury (eds). *Ciliates cells as organisms*. Stuttgart, Germany: Fischer. 1996; p 403-416.

Ricci, N. and Verni, F. (1988). Motor and predatory behavior of *Litonotus lamella. Can. J. Zool., 66*, 1973-1981.

Ricci, N. and Verni F. (1994). Experimental perturbations of the *Litonotus-Euplotes* system. *Zool. Sci., 11*, 399-406.

Ricci, N., Morelli, A. and Verni, F. (1996). The predation of *Litonotus* on *Euplotes,* a two step cell-cell recognition process. *Acta Protozool., 35*, 234-268.

Rosati, G. (1999). Epixenosomes: symbionts of the hypotrich ciliate E*uplotidium itoi. Symbiosis, 26*: 1-23.

Rosati, G. and Modeo, L. (2003). Extrusomes in ciliates: diversification, distribution, and phylogenetic implications. *J. Eukaryot. Microbiol., 50*: 383-402.

Rosati, G., Petroni, G., Quochi, S., Modeo, L. and Verni, F. (1999). Epixenosomes, peculiar epibionts of the hypotrich ciliate *Euplotidium itoi*, defend their host against predators. *J. Eukaryot. Microbiol., 46*:278-282.

Seshachar, B. R., Saxena, K. N. and Girgla, H. (1971). Some factors governing feeding behavior of *Homalozoon vermiculare* (Ciliophora, Holotricha). *J. Protistol., 18*, 90-95.

Šimek, K. and Chrzanowski, T. H. (1992). Direct and indirect evidence of size-selective grazing on pelagic bacteria by freshwater nanoflagellates. *Appl. Environ. Microbiol., 58*, 3715-3720.

Snell, T. W. (1998). Chemical ecology of rotifers. *Hydrobiologia, 338*, 267-276.

Song, W., Wilbert, N. and Hu, X. (2004). Redescription and neotypication of *Amphileptus marinus* (Kahl, 1931) nov. comb. (Ciliophora, Pleurostomatida), and reactivation of *A. parafusidens* Song and Wilbert, 1989. *Europ. J. Protistol.*, *40*, 1-11.

Sugibayashi, R. and Harumoto, T. (2000). Defensive function of trichocysts in *Paramecium tetraurelia* against heterotrich ciliate *Climacostomum virens*. *Europ. J. Protistol.*, *36*, 415-422.

Tatchell, E. C. (1980). An ultrastructural study of extension and contraction in *Lacrymaria olor* (O.F. Müller). *Protistologica, 16*, 167-175.

Tatchell, E.C. (1981). An ultrastructural study of prey capture and ingestion in *Lacrymaria olor* (O.F.M. 1786). *Protistologica, 17*, 59-66.

Van Houten, J. L., Hansma, H. and Kung, C. J. (1975). Two quantitative assays for chemotaxis in *Paramecium*. *J. Comp. Physiol.*, *104*, 211-223.

Verity, P. G. (1991). Feeding in planktonic protozoans: evidence for non-random acquisition of prey. *Mar. Microb. Food Webs, 5*, 69-76.

Verni, F. (1985). *Litonotus-Euplotes* (predator-prey) interaction: ciliary structure modifications of the prey caused by toxicysts of the predator (Protozoa, Ciliata). *Zoomorphology, 105*, 333-335.

Verni, F. and Gualtieri, P. (1997a). Feeding behavior in ciliated protists. *Micron, 28*, 487-504.

Verni, F and Gualtieri, P (1997b) Digestive process of the raptorial feeder ciliate *Litonotus lamella* (Rabdophora, Litostomatea) visualized by fluorescence microscopy. *Micron 28*, 447-451.

Verni, F. and Ricci, N. (1986). *Litonotus-Euplotes*, an interesting model of predator-prey interaction. *Simposia Biologica Hungarica, 33*, 213-216.

Verni F. and Rosati, G. (1990). Peculiar epibionts in *Euplotidium itoi* (Ciliata, Hypotrichida). *J. Protozool.*, 37, 337-343.

Verni, F. and Rosati, G. (1992). A comparative study of digestion in a raptorial ciliate and in the facultative carnivorous form of a filter-feeding ciliate. *Tissue Cell, 24*, 443-453.

Visscher, J.P. (1923). Feeding reactions in the ciliate *Dileptus gigas* with special reference to the function of the trichocysts. *Biol. Bull., 45*, 113-143.

Wessenberg, H. and Antipa, G. (1968). Studies on *Didinium nasutum*. I. Structure and ultrastructure. *Protistologica, 4*, 427-447.

Wessenberg, H. and Antipa, G. (1970). Capture and ingestion of *Paramecium* by *Didinium nasutum*. *J. Protozool.*, *17*, 240-270.

Wicklow, B. J. (1997). Signal-induced defensive phenotypic changes in ciliated protists: morphological and ecological implications for predator and prey. *J. Eukaryot. Microbiol., 44*, 176-188.

In: Microbial Ecology Research Trends
Editor: Thijs Van Dijk, pp. 87-114
ISBN: 978-1-60456-179-1

Chapter IV

Laboratory Model Systems for the Study of Microbial Communities

Olga Sánchez[*], *Isabel Ferrera*[1*] *and Jordi Mas*
Departament de Genètica i Microbiologia, Universitat Autònoma de Barcelona, 08193 Bellaterra, Spain

Abstract

Laboratory cultures have been extensively used in the past as a tool to study the behavior of model organisms subject to a variety of environmental conditions and hence to extrapolate the type of behavior that could be expected from similar microorganisms living in nature. The information obtained using this approach has been crucial for our current understanding of the functional role of certain microbial groups in the environment, but falls short when facing the complexity of interactions operating in real microbial communities.

The main drawback associated with the use of laboratory cultures lays on the fact that conditions in such cultures are usually homogeneous and strongly selective. Batch cultures contain initial concentrations of growth substrates much higher than the growth-limiting threshold thus selecting for organisms with very high specific growth rates. Continuous cultures overcome this limitation by forcing the organisms to grow at low substrate concentrations, well within the growth-limiting range. As a result, organisms with a high affinity for the limiting substrate are selected. None of these cases allows the establishment of a complex microbial community.

The development of complexity in laboratory reactors requires the existence of a certain degree of heterogeneity, either in time or in space, which allows organisms with different requirements to find their niche and avoid competitive exclusion.

In an attempt to provide such heterogeneity we have developed a type of bioreactor where solid surfaces within the reactor allow the structuring of the microbial community in space. Analysis of microbial diversity within the reactor using culture-independent molecular techniques reveals the existence of complex assemblages of bacteria diverse

[*] These authors contributed equally to this work.
[1] Present address: Department of Biology, Portland State University, PO Box 751, Portland, OR, 97297, USA.

both taxonomically and functionally. While the taxonomic composition of these assemblages seems to vary depending on the origin of the inoculum, the functional composition remains quite constant. The results indicate that the use of relatively simple laboratory reactors incorporating a small degree of heterogeneity allows the development of a significant level of biological complexity and can constitute a valuable tool to study the factors determining microbial diversity in natural environments.

Introduction

Microbial ecology, understood as the set of biotic and abiotic interactions found in microbial communities is a relatively new discipline.

Early studies that could fit into this category were carried out during the first half of the 20th century by microbiologists like Beijerink, Winogradskii or even Van Niel. The studies performed during that period used enrichment cultures and eventually, pure culture isolates to describe the diversity of microorganisms and functional groups occurring in natural processes.

This approach proved to be very successful and allowed to identify most microbial activities relevant to microbial communities while at the same time providing the basis for our current understanding of biogeochemical cycling. Beyond this descriptive contribution, however, this approach failed to elucidate the intricate nuances of the interactions between the members of the community.

Development of microbial ecology as we currently know it, probably begins in the late fifties and early sixties with a new breed of microbiologists more concerned with the interactions occurring within microbial communities than with the overall processes.

The analysis of microbial interactions required the study of complex microbial communities and pure culture and enrichment techniques were ill suited for this purpose. In his 1987 assay "The study of microorganisms in situ: progress and problems" [3] TD Brock tried to assess the state of Microbial Ecology 20 years after the publication of his pioneer book "Principles of Microbial Ecology" [4]. One of his main complains about the state of the art at that time was methodological and related to the fact that 41% of the papers published in the General Microbial Ecology section of Applied an Environmental Microbiology used *pure cultures* under highly controlled conditions "in the rich culture media used by physiologists", and 12% were *mixed culture* studies carried out using complex samples incubated in the laboratory under poorly defined conditions. By critizising this approaches Brock questioned the methodologies that had constituted the cornerstone of microbial ecology work during the previous 50 years and summarized very well the type of concerns prevailing in the field at that time.

While being extremely critical with these aspects, he also mentioned a number of "meritorious" approaches such as direct microscopy, the use of model systems, biomass/activity studies, microenvironments or integrated field/laboratory studies.

Since 1987, several of the considerations put forward in this paper have found their way into mainstream microbial ecology.

The use of direct techniques for the study of microbial communities constitutes one of the revolutionary breakthroughs that have allowed the unbiassed characterization of the different microbial populations found in natural environments. While Brock envisioned direct techniques based in the use of direct microscopy, the actual breakthrough has come by the hand of molecular biology and the use of culture independent techniques based on the analysis of 16S rRNA or functional genes. The use of these techniques combined with a variety of methods to assess microbial or enzymatic activity has proved to be an excellent tool to describe the taxonomic composition and the different biological processes involved in the microbial community [18, 29].

The use of laboratory models as analytical tools to simulate the behaviour of simplified versions of microbial ecosystems, however, has not been as successful.

Attempts to study microbial communities in their natural environment often faces two major inconvenients. First, the number of environmental variables shaping the community is large, many of these variables are unknown and most of them are beyond our control. As a consequence, many of the dynamic changes observed in nature can be difficult to explain according to the set of variables measured. Second, once we develop a working hypothesis about how a certain microbial community is regulated, experimental testing of this hypothesis is virtually impossible unless we have a stringent control over all the factors involved. This type of total control is difficult to achieve in nature and often requires the use of laboratory model systems developing under strictly well defined conditions.

In principle most model laboratory models are essentially some type of conventional culture systems in which environmental conditions are carefully crafted to mimic those occurring in a particular environment. The ideal model should provide a scaled down version of the microbial community, perhaps with a reduced set of components but still retaining enough complexity to simulate the main set of interactions occurring within the system.

Open Systems Versus Closed Systems

Laboratory model systems can be classified as either open or closed systems depending on whether they exchange matter with the external environment or not. Conventional batch cultures are closed systems in which a chemical environment is initially defined (carbon source, energy source, nutrients, pH,...) and inoculated with a certain amount of a pure culture or a complex natural sample. In either case the system evolves allowing growth of biomass while at the same time depleting the resources initially present in the culture. As a consequence these systems evolve continuously during their lifespan until resources are exhausted, microbial growth slows down and the system enters a new phase in which survival and recycling of resources constitute the main driving forces. During the period where nutrients are in excess, these systems behave essentially as enrichment cultures in which a strong selection occurs favoring organisms with a higher μmax (maximum specific growth rate). Once nutrients are depleted survival to starvation becomes the main adaptive factor. Later on, when microbial populations begin to die due to starvation, the ability to scavenge the nutrients released in this process provides an additional selective force. Rather different incarnations of closed model systems that retain the same conceptual principles can be found in microcosms. Microcosms are built by isolating a certain amount of sample (usually soil or sediment) and incubating it in the laboratory under controlled conditions. Although attempts might be made to provide constant controlled conditions, the internal succession of events triggered by resource depletion resembles very closely the situation described above for batch cultures.

To overcome part of the sequence of events mentioned above as a result of nutrient depletion, microbiologists have often resorted to the use of continuous cultures. Continuous cultures are open systems in which fresh medium is continuously pumped into the culture vessel and microbes are continuously removed at a well defined rate. These cultures eventually reach a spontaneous point of equilibrium (steady state) where microorganisms grow at exactly the same rate at which they are removed [20]. Depending on how they are operated a distinction is made between chemostats and auxostats [14]. In chemostats the dilution rate of the system is held constant for the length of the experiment while the organisms in the culture vessel progressively adapt their specific growth rate to equal the dilution rate of the system. In auxostats the dilution rate is variable and with the aid of appropriated control equipment, it is adjusted to match the metabolic rate of the organism. Examples of auxostats that differ in the type of control variable used can be found all over the literature. Thus, turbidostats, pH auxostats, oxygen auxostats, or redoxstats attempt to match dilution rate and microbial activity using turbidity (microbial biomass), pH, oxygen or redox potential as the control variables [27, 28, 37, 39, 44, 47, 48, 59, 62]. In all cases, these types of reactors have been successfully used to perform physiological studies using pure cultures.

Attempts to grow two or more microorganisms simultaneously in a continuous culture usually fail as a consequence of the existence of a strong selective pressure caused by competition for limiting resources. When a chemostat is inoculated using a complex sample, most organisms in the inoculum are outcompeted by the organism with a higher affinity for substrate limiting growth [56, 57]. Any deviations from this behavior are usually accounted for by the fact that some organisms are able to attach to surfaces and escape washout, being also at an advantage in this type of system. Thus, although open systems avoid the problems associated with the mid-term evolution of the system due to nutrient depletion, they still constitute far from ideal systems to simulate complex microbial communities.

Heterogeneity and the Shaping of Microbial Communities

It seems, then, that most experimental setups based in the utilization of conventional stirred reactors running in an open or in a closed configuration will be selective enough as to yield pure or highly enriched cultures after a relative short period of operation. What elements are missing to allow the establishment of a certain degree of microbial diversity? What are the factors explaining the stable coexistence of different organisms in natural samples? From 1959 to 1961 Hutchinson wrote two seminal papers addressing these very same questions, "Homage to Santa Rosalia" and "The paradox of the plankton" [21, 22]. The answers have long since then appeared published in many different scientific papers, but in general, there is agreement about the fact that species richness and coexistence leading to high diversity require a certain degree of heterogeneity in the environment. Isotropic systems with no space structure and constant conditions will behave like conventional laboratory cultures. As we introduce space heterogeneity, temporal fluctuations or chemical heterogeneity, the ability of the system to allow the simultaneous presence of thriving populations of different microorganisms will increase. The effect of space heterogeneities in coexistence was demonstrated long time ago in the particular case of predator prey interations by Gause [16], using *Paramecium* and *Didinium*, by Van den Ende [54] using *Tetrahymena* and *Klebsiella*, and by Roper and Marshall [43] using *Vexillifera* and *Escherichia*. In all cases, the presence of space heterogeneity due to the presence of clay particles, or to surface growth in the form of biofilms, makes possible for one organism to avoid the pressure exerted by predators allowing coexistence of both. Examples of time heterogeneity allowing coexistence of organisms competing for the same substrate can be found in the classical experiments of Hans van Gemerden [55] using *Chromatium vinosum* and *Chromatium weissei* competing for hydrogen sulfide. In these experiments, both organisms coexisted when growing under light/dark cycles. The steady-state abundance of each microorganism was related to the length ratio between light and dark periods. Stable coexistence of the diatoms *Cyclotella pseudostelligera* and *Fragillaria crotonensis* was also achieved under fluctuating temperature in the experiments of Descamps-Julien and González [8].

Finally, chemical heterogeneity provides also a mechanism to promote coexistence and the development of complexity within microbial communities. This was thoroughly addressed by David Tilman in a number of papers [51, 52] discussing the role of nutrient fluctuations in the control of the species composition and diversity of phytoplankton.

Introducing Complexity in Laboratory Model Systems

From all the above it is quite obvious that any type of model system designed to develop microbial communities with a certain degree of complexity must consider the existence of some type of heterogeneity. These heterogeneities can be introduced in the systems in different ways. Chemical heterogeneity can be easily achieved with media of defined composition containing multiple carbon sources (or N, P, etc). Space heterogeneity requires the introduction of some type of attachment surface. Surfaces are actually present in any reactor but usually the surface to volume ratio is quite low. Increasing this ratio can be achieved by using panels, tubing or columns with diverse types of fillings. Temporal heterogeneity can be introduced as periodic fluctuations such as dial cycles of one ore more environmental variables (usually light, but temperature or other variables are also possible), as random fluctuations (usually in nutrient supply), or indirectly as a consequence of the existence of space heterogeneity. In model systems in which contents are stirred or circulated through, microorganisms are transported across gradients and boundaries established in the spatial structure of the reactor. As a result, the environmental conditions experienced by the organism change with time following a pattern closely modulated by the architecture of the system.

In order to explore these principles, we have developed a bioreactor specially designed to allow the introduction of space heterogeneity through an increase in the ratio of exposed surface to reactor volume. The increase in exposed surface has been achieved through the addition of a packed column to a stirred reactor, and circulation of the reactor contents through the packing material. Chemical and temporal variability have been kept at a minimum to better evaluate the effect of the spatial heterogeneity introduced.

Our results indicate that the introduction of a low degree of heterogeneity provides a considerable increase in the complexity of the community being developed.

Bioreactor Design and Operation

The reactor we used for this work was built around a glass column (15.6 mm inner diameter, 275 mm length, 52.5 mL total volume) surrounded by a double wall through which water circulated at 25 °C. The column was packed with glass rings that provided a large surface for biofilm development and was attached at the basis to a stirred reactor vessel containing the pH and redox sensors, sampling ports, effluent collector and the inlets for the synthetic medium and the solution for pH adjustment (Figure 1) [9, 11]. The contents of the reactor circulated in an upward flow through the column in a closed loop powered by a peristaltic pump (Watson Marlow 313S) at a constant flow rate of 0.1 $L.min^{-1}$ and were continuously diluted by a second pump (Watson Marlow 501U) externally controlled to provide a constant concentration of hydrogen sulfide. While the column was illuminated by two oposed light sources, the body of the stirred reactor was kept in the dark.

The reactor was operated as an auxostat using redox potential as the control variable [44]. Redox potential was mainly determined by the concentration of hydrogen sulfide and ultimately by biological activity. When sulfide oxidation was slow the pump supplying fresh medium to the culture operated at a very low speed. The speed of this pump increased automatically whenever sulfide oxidation rates increased, thus keeping the concentration of hydrogen sulfide at a constant level.

A synthetic mineral medium for anaerobic phototrophic bacteria [32] was used throughout the experiments. This medium contains carbonate as the only carbon source and hydrogen sulfide as electron donor, and it was prepared as two separate solutions. One of them was alkaline (pH 12) and contained carbonate and sodium sulfide. The second solution was acidic (pH 5) and contained the rest of the components; for marine samples, it was supplemented with 3.7% sodium chloride. The two solutions were pumped at equal rates into the reactor resulting in a slightly alkaline medium. Final pH adjustment was achieved with an independent tritator (Cole Parmer pH/ORP controller 5652-10) connected to a pump (Masterflex 7554-60), which added a 1 M HCl solution at the rate required to maintain a constant pH of 6.8-7.2.

Two separate sets of experiments were run using natural samples from two different sulfurous environments as inocula. In one case, samples were taken from the sediment of Lake Estanya, a karstic sulfurous lake located in Huesca (Spain) where blooms of phototrophic sulfur bacteria have often been described [19]. In the second case, the inoculum was taken from the microbial mats developing on the sand flats found in the delta of the Ebro river, in northeast Spain. These mats are stratified and contain different layers of oxygenic and anoxygenic phototrophs, colourless sulfur bacteria as well as sulfate reducers [34].

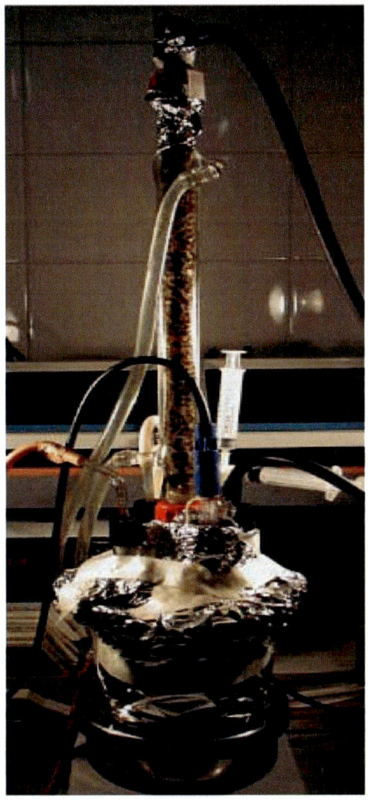

Figure 1. The bioreactor consisted of a stirred vessel and a column packed with glass rings to provide a large surface for biofilm development. The vessel contained the pH and redox sensors, sampling ports, effluent collector and inlets for synthetic medium and pH adjustment.

At start up, the reactor was inoculated and the medium was recycled through the column in a closed loop. As a sulfide-oxidizing biofilm developed, the concentration of hydrogen sulfide in the reactor progressively decreased. When this decrease reached a preset value (in our case 50 μM), pumping of the medium into the reactor started at a rate controlled by the redox titrator (Crison 52-61) which was proportional to the rate of sulfide oxidation of the biofilm.

Development Kinetics of the Microbial Communities. Activity/Biomass Ratios in Experimental Model Reactors

After inoculation, the reactors were allowed to operate for up to 2 - 3 weeks. During this period microbial growth occurred at the surface of the column filling where a thick biofilm developed. As a result of biofilm formation, the aspect of the reactor changed. In the experiments inoculated with microbial mats from the Ebro delta sand flats the column was colonized slowly. Complete coverage of the column surface by microbial biofilms was only observed after 10 days of continuous operation (Figure 2).

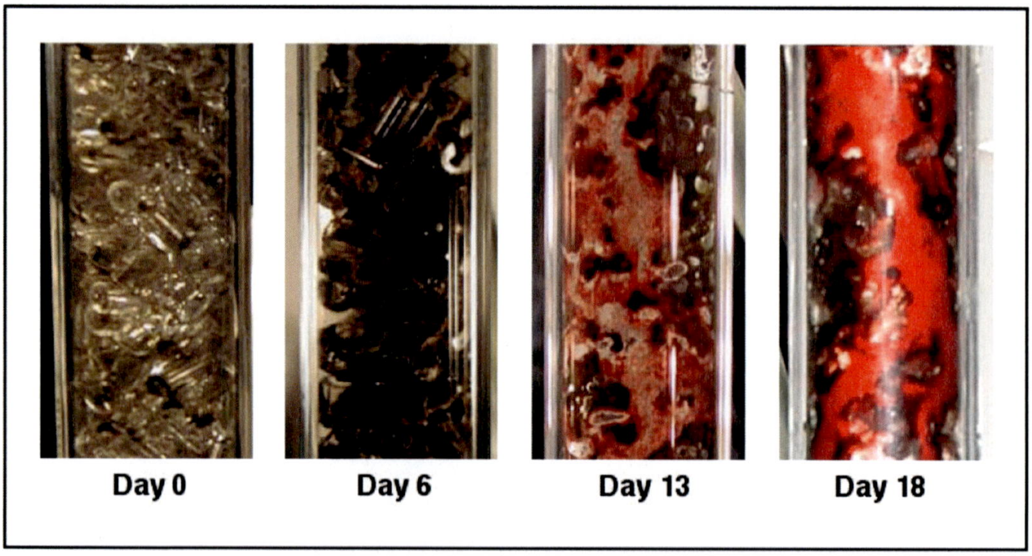

Figure 2. Sequence of different images showing the development of the biofilm from the microbial mat inoculum.

During the first stages the column was covered by a thick black layer, presumably sulfide depositions that were replaced after 10 days by a continuous pink-whitish layer. This was later on substituted by a purple layer that stayed until the end of the experiment.

In the experiment inoculated with sediment from lake Estanya, after only eight days of operation the column developed a purple green layer completely covering the exposed surfaces of the column filling (Figure 3).

In addition to macroscopic observations, biofilm development was also monitored indirectly through the levels of microbial activity. Since the culture medium contained no organic carbon and sulfide was provided as an external electron donor, sulfide oxidation rates were measured as an index of activity. The results are shown in Figure 4. In the reactor inoculated with microbial mats (Figure 4A), sulfide oxidation rates were virtually zero during the first days and increased sharply after 7 days of operation coinciding with the establishment of a visible biofilm. Maximum activity was achieved after 11 days and remained fairly constant until the end of the experiment. In the reactor inoculated with lake sediment (Figure 4B), biological activity was established much faster with oxidation rates increasing right from the beginning and reaching maximum stable values after approximately 1 week. As in the previous case, activity levels remained high until the end of the experiment. After several days of stable activity values we considered that the reactors had reached steady state. At that point, the columns were emptied and the biofilms were recovered and analyzed. Biofilm recovery required sonication of the column filling in a known volume of saline solution using an ultrasonic water bath (Bransonic 5, Branson) for a period of three minutes. After sonication, the volume was brought up to 1000 mL, stirred to homogeneity and split into aliquots that were centrifuged 20 min at 8000 g (Beckman J2-21). Part of the pellets were frozen and kept at -80 °C for DNA extraction, the remaining were used for protein quantification.

Day 8

Figure 3. Image of the column with the biofilm developed from the lake sediment inoculum.

Microbial biomass was determined as protein, using the method of Lowry [30] in samples previously extracted with methanol to remove elemental sulfur that interferes with this assay [33].

Comparison of attached microbial biomass and sulfide oxidation activities between the reactors shows that both were very similar (Table 1). While the reactor inoculated with lake sediment had somewhat higher biomass (627 vs 431 mg protein) and activity (216 vs 158 μmol $S^{2-}\cdot h^{-1}$), specific sulfide oxidation rates (ie. activity normalized by biomass) were almost identical (0.35 vs 0.37 μmol $S^{2-}\cdot$mg protein$^{-1}\cdot h^{-1}$). For the sake of discussion we also compared data from these biofilm reactors to data from a previous work [45] using a stirred reactor that contained a pure culture of an anoxygenic phototroph operating under similar conditions. Interestingly enough, although sulfide oxidation rates of the biofilm communities (158 and 216 μmol $S^{2-}\cdot h^{-1}$) fall well in the range of the rates described for the stirred reactor (84.85 and 649.64 μmol $S^{2-}\cdot h^{-1}$), the specific activities are much lower (0.35 - 0.37 vs 2.60 - 12.43 μmol $S^{2-}\cdot$mg protein$^{-1}\cdot h^{-1}$). While the biomass is homogeneously illuminated and active in stirred reactors, biomass is much higher in biofilm reactors, due to the attachment to surfaces and the low impact of washout removal. Slow-growing organisms that would be removed by dilution from a stirred reactor, have a chance to thrive when attached to surfaces.

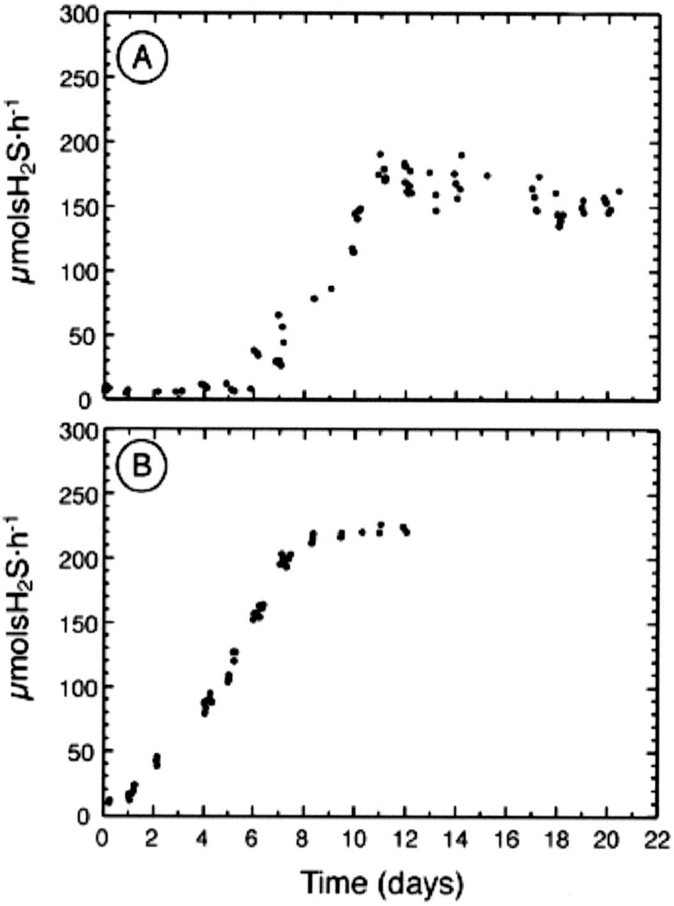

Figure 4. Sulfide oxidation rates as a function of time for the microbial mat (A) and lake Estanya biofilms (B).

Besides, as it will be seen later on in this chapter, many of the organisms detected are heterotrophs. Although these organisms do not contribute to the oxidation of sulfide, they presumably feed on the organic material of the biofilm, contributing to its recycling.

Because biomass is immobilized in biofilms and therefore not subject to washout, high concentrations of slow-growing microorganisms can be achieved [46]. As a consequence, the dynamic behaviour of the system is slower but more stable during sudden environmental changes.

Characterization of the Microbial Communities

The microbial composition for both experiments was determined by using a molecular approach, particularly 16S rRNA cloning and sequencing. DNA from both biofilms was extracted and amplified following the procedure described by Ferrera et al. [10]: 0.1 g of glass beads (150 µm diameter) was added to a solution containing 50 mg of biofilm in 500 µl of lysis buffer (40 mM EDTA, 50 mM Tris-HCl, 0.75 M sucrose).

Table 1. Protein content and oxidation rates corresponding to microbial mat and lake Estanya biofilms, as well as to stirred reactors of *Allochromatium vinosum*. Data have been expressed as a "per reactor" basis

	Microbial mat	Lake sediment	Stirred reactor*
Total activity (μmol $S^{2-}\cdot h^{-1}$)	158	216	84.85 - 649.64
Protein content (mg protein)	431	627	32.62 – 52.30
Specific activity (μmol $S^{2-}\cdot$mg protein$^{-1}\cdot h^{-1}$)	0.37	0.35	2.60 – 12.43

*Data from Sánchez et al. (1998) corresponding to an equivalent volume of 1.8 L

The sample was bead-beated for three cycles of 80 s and was kept for 30 s in an ice-bath between cycles. Then, the lysate was collected by centrifugation (13000 rpm, 1 min) and the supernatant transferred to a new tube. Lysozyme (1 mg·ml^{-1} final concentration) was added and samples were incubated at 37°C for 45 min. Then, sodium dodecyl sulfate (1% final concentration) and proteinase K (0.2 mg·mL^{-1} final concentration) were added and samples were incubated at 55°C for 60 min. Nucleic acids were extracted at least twice with phenol-chloroform-isoamyl alcohol (25:24:1, by vol.) and the residual phenol was removed once with chloroform-isoamyl alcohol (24:1, v/v). Nucleic acids were purified, desalted and concentrated with a Centricon-100 concentrator (Millipore). DNA integrity was checked by agarose gel electrophoresis.

PCR mixtures contained 10 ng of template DNA, each deoxynucleoside triphosphate at a concentration of 200 μM, 1.5 mM $MgCl_2$, each primer at a concentration of 0.3 μM, 2.5 units of *Taq* DNA polymerase (Invitrogen) and PCR buffer supplied by the manufacturer. The PCR cycle consisted of an initial denaturation step at 94°C for 5 min, followed by 30 cycles of 1 min at 94°C, 1 min at 55°C and 2 min at 72°C, with a final extension step of 10 min at 72°C. 16S rRNA genes were amplified between positions 27 and 1492 (*Escherichia coli* 16S rRNA gene sequencing numbering) using the primers 27F (5'-AGA GTT TGA TCM TGG CTC AG-3') and 1492R (5'- GGT TAC CTT GTT ACG ACT T-3'). The PCR product was cloned with the TOPO TA cloning kit (Invitrogen) according to the manufacturer's instructions. Ninety-six (Lake Estanya biofilm) and one hundred and seventy (microbial mat biofilm) putative positive colonies were picked, transferred to a multi-well plate containing Luria-Bertani medium and 7% glycerol, and stored at –80°C. Recombinant plasmids were extracted using the QIAprep spin miniprep kit (QIAgen) and screened by RFLP (restriction fragment length polymorphism) analysis using *Hae*III (Invitrogen). Forty-two (lake biofilm) and thirty-seven (mat biofilm) clones with different band patterns were chosen for partial sequencing. Clone library coverage was calculated according to the following equation: $C=1-(n/N)$, where n is the number of unique clones and N is the total number of clones examined [40]. The coverage of the libraries was 73% and 86.5% for lake and mat biofilms respectively.

Sequencing reactions were performed by Sistemas Genómicos (Spain) [http://www.sistemasgenomicos.com] and QIAgen DNA Sequencing Services (Germany) with primer 27F and run in an ABI PRISM 377 XL lane sequencer. Sequences were subjected to a BLAST search [2] to get a first indication of the phylogenetic aphiliation and to the CHECK-CHIMERA program from RDP [31] to determine potential chimeric artifacts, and they were submitted to the EMBL database (http://www.ebi.ac.uk/embl), receiving the following accession numbers: AJ548890 to AJ548931 (Lake Estanya biofilm) and AJ627979 to AJ628015 (microbial mat biofilm).

Table 2. Phylogenetic affiliation of clones to the closest match and to the closest cultured strain in GenBank (n=sequence length)

Name	N° clones	n	Closest match (Acc. number)	% similarity	Cultured closest match (Acc. number)	% similarity
α-Proteobacteria						
BIOEST-7	1	898	*Rhodobacter* sp. (AB251412)	99.4	The same	99.4
BIOEST-14	16	860	*Rhodobacter* sp. (AB251412)	99.9	The same	99.9
BIOEST-19	3	877	*Rhodobacter* sp. (AB251412)	99.6	The same	99.6
BIOEST-15	1	427	*Rhodobacter* sp. (AB251412)	99.5	The same	99.5
BIOEST-26	1	631	*Rhodobacter* sp. (AB251412)	99.8	The same	99.8
BIOEST-23	2	913	*Rhodobacter* sp. (AB251412)	99.0	The same	99.0
BIOEST-16	2	838	*Rhodobacter capsulatus* (DQ342320)	99.2	The same	99.2
BIOEST-24	1	718	Uncultured α-proteobacterium (AY701453)	98.3	*Rhodobacter sphaeroides* (CP000663)	97.5
BIOEST-27	1	832	Uncultured bacterium (DQ444142)	97.8	*Rhodobacter sphaeroides* (CP000663)	97.0
BIOEST-36	3	834	*Rhodopseudomonas palustris* (AF416663)	99.8	The same	99.8
BIOEST-20	1	770	*Caulobacter fusiformis* (AJ227759)	97.4	The same	97.4
ß-Proteobacteria						
BIOEST-17	1	807	Uncultured bacterium (DQ415745)	97.3	*Thiobacillus baregensis* (Y09280)	94.4
BIOEST-18	1	653	Unidentified eubacterium (Y12371)	99.5	*Rhodoferax antarticus* (AY609198)	99.2

Name	N° clones	n	Closest match (Acc. number)	% similarity	Cultured closest match (Acc. number)	% similarity
BIOEST-28	1	789	Uncultured bacterium (DQ211500)	100	*Ferribacter polymixa* (EF133508)	92.2
BIOEST-37	1	634	Unidentified eubacterium (Y12371)	99.4	*Rhodoferax antarticus* (AY609198)	99,1
γ-Proteobacteria						
BIOEST-40	2	730	Uncultured bacterium (AB247478)	99.9	*Thiothrix unzii* (L79961)	99.5
BIOEST-41	2	694	Uncultured bacterium (AB247478)	99.9	*Thiothrix unzii* (AB166732)	95.5
BIOEST-42	3	813	Uncultured bacterium (AB247478	99.8	*Thiothrix unzii* (L79961)	99.4
BIOEST-34	1	731	Uncultured bacterium (AB247478)	99.7	*Thiothrix unzii* (L79961)	99.3
BIOEST-38	1	758	Uncultured bacterium (AJ247478)	99.6	*Thiothrix unzii* (L79961)	99.2
BIOEST-30	1	858	Uncultured bacterium (AB247476)	98.0	*Thiothrix* sp. (AF148516)	97.9
BIOEST-39	1	661	Uncultured bacterium (AB247471)	97.0	*Thiothrix* sp. (AB166732)	96.8
BIOEST-32	11	623	*Thiocapsa roseopersicina* (AF112998)	97.6	The same	97.6
BIOEST-33	1	790	*Thiocapsa roseopersicina* (AF112998)	96.5	The same	96.5
BIOEST-31	1	894	*Pseudomonas stutzeri* (AJ312176)	99.8	The same	99.8
δ-Proteobacteria						
BIOEST-35	1	663	Uncultured bacterium (DQ133908)	98.2	*Desulfocapsa thiozymogenes* (X95181)	97.9
CFB Group						
BIOEST-9	3	869	Uncultured bacterium (EF121348)	98.8	Cf. *Bergeyella* sp. (AJ575430)	93.9
BIOEST-21	1	884	Cf. *Bergeyella* sp. (AJ575430)	98.8	The same	93,8
BIOEST-10	1	840	Cf. *Bergeyella* sp. (AJ575430)	93.9	The same	93.9

Table 2. (Continued)

Name	N° clones	n	Closest match (Acc. number)	% similarity	Cultured closest match (Acc. number)	% similarity
BIOEST-29	2	886	Uncultured bacterium (DQ066985)	97.7	*Flexibacter canadensis* (AB078046)	85.4
BIOEST-11	2	818	Uncultured bacterium (DQ066989)	97.8	*Flexibacter canadensis* (AB078046)	84.8
BIOEST-8	2	775	Uncultered bacterium (DQ376571)	97.2	*Leadbetterella byssophila* (AY854022)	92.0
BIOEST-12	5	848	Uncultured bacterium (AB247474)	95.9	*Cytophaga* sp. (AJ431254)	86.6
BIOEST-13	2	833	Uncultured bacterium (AB247474)	94.0	*Cytophaga* sp. (AJ431254)	85.4
BIOEST-2	1	821	Rhizosphere soil bacterium (AJ252591)	97.9	*Flavobacterium ferrugineum* (M62798)	92,9
BIOEST-3	1	849	Uncultured eubacterium (AF050545)	90.5	*Flexibacter canadiensis* (M62793)	87,6
Cyanobacteria/Plasts						
BIOEST-6	10	860	*Scenedesmus obliquus* chloroplast (AF394206)	98.9	The same	98.9
BIOEST-5	1	892	*Scenedesmus obliquus* strain UTEX 393 cloroplast (DQ396875)	95.0	The same	95.0
BIOEST-4	1	867	Uncultured cyanobacterium (EF072789)	96.7	*Scenedesmus obliquus* strain UTEX 393 cloroplast (DQ396875)	92.4
BIOEST-1	1	571	Uncultured cyanobacterium (DQ512831)	97.2	*Synechocistys* PCC6308 (AB039001)	94.4
Low G+C Gram-positive						
BIOEST-22	1	864	Uncultured bacterium (AB237721)	99.6	*Catabacter hongkongensis* (AY574991)	88.4
BIOEST-25	1	802	Uncultured bacterium (AB237721	99.1	*Catabacter hongkongensis* (AY574991)	85.8

Table 3. Phylogenetic affiliation of clones to the closest match and to the closest cultured strain in GenBank (n=sequence length)

Name	Nº clones	n	Closest match (Acc. number)	% similarity	Cultured closest match (Acc. number)	% similarity
α-Proteobacteria						
2B4	1	692	*Rhodovulum adriaticum* (D16418)	100	The same	100
1E5	1	674	*Erythrobacter litoralis* (AF465836)	99.6	The same	99.6
1G1	1	669	*Roseospirillum parvum* (AJ011919)	99.1	The same	99.1
2E12	2	681	*Roseospirillum parvum* (AJ011919)	99.1	The same	99.1
2H6	1	686	*Roseospirillum parvum* (AJ011919)	98.9	The same	98.9
1B9	8	693	*Roseospirillum parvum* (AJ011919)	98.8	The same	98.8
γ-Proteobacteria						
1F4	1	648	*Marichromatium purpuratum* (AF294031)	99.7	The same	99.7
3A9	1	700	*Marichromatium indicum* (AM179449)	99.0	The same	99.0
2G6	3	715	*Marichromatium indicum* (AM179449)	99.0	The same	99.0
1A10	3	621	*Marichromatium purpuratum* (AF294030)	99.5	The same	99.5
2G3	1	709	*Marichromatium purpuratum* (AF294031)	99,4	The same	99,4
1B2	69	695	*Marichromatium purpuratum* (AF294031)	98.6	The same	98.6
1G4	1	690	*Thiorhodococcus drewsii* (AF525306)	99.4	The same	99.4
2C9	2	697	*Thiomicrospira* sp. (AF013975)	99.6	The same	99.6

Table 3. Continued)

Name	N° clones	n	Closest match (Acc. number)	% similarity	Cultured closest match (Acc. number)	% similarity
2E11	1	675	Uncultured *Thiomicrospira* (DQ218327)	98.5	*Thiomicrospira* sp. (AF013978)	95.4
1C3	1	667	Uncultured gamma proteobacterium (AY580854)	99.3	*Vibrio* sp. (DQ530293)	99.0
2E10	2	651	Uncultured bacterium (EF040532)	98.9	*Vibrio parahemolyticus* (AF388386)	98,9
1D3	1	673	Uncultured proteobacterium (DQ330798)	95.7	Gamma proteobacterium NEP17 (AB212804)	95.5
1F10	1	656	Uncultured proteobacterium (DQ330798)	95.7	Gamma proteobacterium NEP17 (AB212804)	95.6
1G5	1	611	Uncultured gamma proteobacterium (DQ811846)	95.4	Sulfur-oxidizing bacterium NDII1.1 (AF170424)	91.8
ε-Proteobacteria						
1B6	1	654	Uncultured *Sulfospirillum* (AF513952)	97.2	*Sulfurospirillum* sp. (AF357199)	92.2
1C6	26	639	Uncultured epsilon proteobacterium (AJ441205)	96.9	*Sulfurovum* sp. (AB197160)	94.2
1F5	1	656	Uncultured epsilon proteo- bacterium (AB294904)	97.3	Proteobacterium BH I60-14 (AJ431221)	94.4
1B8	1	630	Uncultured epsilon proteobacterium (AB247901)	95.6	*Sulfurimonas paralvinella* (AB252048)	95.3
2A6	5	642	Uncultured epsilon proteobacterium (AB247901)	95.5	*Sulfurimonas paralvinella* (AB252048)	95.2
1D10	4	660	Uncultured epsilon proteobacterium (AY280393)	93.0	*Sulfurimonas paralvinella* (AB252048)	92.0
CFB Group						
1A6	1	648	*Marinilabilia salmonicolor* (M62423)	96.0	The same	96.0

Name	N° clones	n	Closest match (Acc. number)	% similarity	Cultured closest match (Acc. number)	% similarity
1G6	1	688	*Marinilabilia salmonicolor* (M62423)	94.8	The same	94.8
1A5	1	672	*Marinilabilia salmonicolor* (M62423)	94.5	The same	94.5
1A4	2	645	Uncultured Bacteroidetes (AJ347762)	94.3	*Cytophaga* sp. (AJ431253)	84.4
1D5	6	690	Uncultured bacterium (DQ662503)	92.0	*Cytophaga* sp. (AJ431253)	83.9
2A10	1	643	Sphingobacteriales bacterium (AB260041)	92.5	The same	92.5
2F2	1	616	Sphingobacteriales bacterium (AB260041)	93.0	The same	93.0
Chlorobiales						
1B5	8	650	*Prosthecochloris* sp. (AB301726)	99.1	The same	99.1
Cyanobacteria/Plasts						
1D12	7	697	Uncultured cyanobacterium (AY580403)	98.9	*Phaeodactylum tricornutum* chloroplast (EF067920)	98.3
ND-Bacteria						
2G8	1	663	Uncutured bacterium (DQ513083)	97.7	*Sphaerochaeta* sp. (DQ833403)	85.2
1F8	1	676	Uncultured bacterium (AY568882)	84.8	*Dehalococcoides* sp. (AJ431247)	80.1

The retrieved sequences, their frequency in the library, and their closest relatives are listed in Tables 2 and 3.

In the lake biofilm, Alpha-Proteobacteria constituted the dominant group in the clone library (33% of the total clones), with a high representation of *Rhodobacter*-like sequences (up to 29%), although *Rhodopseudomonas* and *Caulobacter* spp. were also present. Gamma-Proteobacteria accounted for 25% of the total clones (11% belonging to the *Thiothrix*-like aerobic sulfur oxidizers, 13% anaerobic sulfur oxidizers of the family Chromatiaceae and 1% to the *Pseudomonas* group), while the CFB (Cytophaga-Flavobacterium-Bacteroides) group was represented by 21% of the total clones. Other groups, such as chloroplasts from algae, and Cyanobacteria, constituted up to 14% in the clone library, whereas few clones belonged to Beta-, Delta-Proteobacteria and to low G+C Gram-positive bacteria (Firmicutes).

It is remarkable to state that several of the retrieved sequences showed a considerable degree of microdiversity [15], particularly for the *Rhodobacter* sp. and *Thiothrix* sp. sequences. Six *Rhodobacter* sp.-like sequences were on average 99.6% similar to each other, while for the *Thiothrix*-like sequences an average of 99.8% similarity was found among them. Microdiversity was also observed for clones BIOEST-18 and BIOEST-37 (Beta-Proteobacteria), BIOEST-22 and BIOEST-25 (Firmicutes group), and BIOEST-10 and BIOEST-21 (CFB group). This microheterogeneity in the 16S rRNA gene has been previously reported in clone libraries of Bacteria, Archaea and eukaryotic microorganisms from several environments [6], and might be explained by *Taq* polymerase errors during PCR amplification [13] or by rRNA multioperons belonging to a single population [36]. However, it has been demonstrated that, in many cases, microdiversity allows niche specialization and therefore the coexistence of ecotypes with different substrate affinities or different light-dependent phisiology [6, 60]. Nevertheless, the possible role of microdiversity in our reactor remains to be established, but due to the heterogeneity of the biofilm and the microenvironments developed, niche specialization seems a reasonable explanation.

Different phylogenetic groups were also identified in the biofilm developed from the microbial mat sample, such as Gamma-Proteobacteria (52% of total clones), which dominated the library mostly with representatives of *Marichromatium* and relatives. Epsilon-Proteobacteria constituted another well-represented group in the library (22% of the total clones), while the remaining clones belonged to Alpha-Proteobacteria, Cyanobacteria, Chlorobi, and the CFB group. One percent of the clones was not affiliated to a significant extent with any phylogenetic group. Microdiversity was also observed for different phylogenetic groups, such as the Alpha- Gamma and Epsilon subclasses of Proteobacteria, as well as for the CFB group.

In both experiments, highly diverse biofilms, covering a wide range of phylogenetic and physiological microbial groups, developed in the bioreactor [10, 12]. Both biofilms presented high species richness and many different groups were represented. However, differences were observed among the most abundant phylogenetic groups presented in both cases (Table 4). For the Proteobacteria, clones affiliated to the Alpha and Gamma subdivisions were retrieved in both biofilms. Conversely, minor members belonging to Beta and Delta Proteobacteria were present only in the freshwater biofilm, whereas many clones belonging to the Epsilon-Proteobacteria were recovered only from the marine biofilm. Members of the Cytophaga-Flavobacterium-Bacteroides group were abundant in both biofilms, most of them belonging to uncultured species. Chloroplasts from algae appeared in the lake biofilm, while Cyanobacteria were recovered in both libraries. Clones affiliated to the low G+C Gram-positive bacteria (Firmicutes) were specific from the marine biofilm, while Chlorobi-affiliated clones were specific from the marine biofilm. However, the development of Chlorobiaceae in the freshwater biofilm would have been limited by the presence of oxygen produced by algae, which were very abundant. In addition, we cannot discard that minor components could appear after analyzing a larger number of clones.

Although some phylogenetic groups were present in both biofilms, we found totally different species composition. Such differences agree with the fact that each inoculum came from a different origin (i.e., freshwater and marine), and we used the medium supplemented with 3.7% of sodium chloride in the microbial mat experiment.

Some of the differences are in accordance with what is usually found in microbial assemblages from natural environments. For instance, Glöckner et al. [17] compared bacterioplankton from many lakes and oceans and found that the Beta subclass of Proteobacteria was abundant in freshwater environments, while they were essentially absent in the marine samples analyzed. On the contrary, most of environmental Epsilon-Proteobacteria have been isolated from marine sediments, in particular from deep-sea hydrothermal vents [1, 42, 50], but also from deep subsurface sediments, oil fields, activated sludge, marine snow, coastal sediments, and in the oxic/anoxic chemocline from the Black Sea [26, 58, 61]. These organisms are considered to play a key role in the cycling of carbon, nitrogen, and sulfur in sulfur-rich environments [5], and although their importance, only a limited number of species have been isolated. There is evidence indicating their use of the reverse TCA cycle for chemolithoautotrophy [5, 49], and autotrophic growth on sulfide, elemental sulfur, thiosulfate, and hydrogen as electron donors under anoxic or microaerophilic conditions has been described [5, 49].

Cytophaga-Flavobacterium-Bacteroides-related organisms also developed in both biofilms. These bacteria are abundant in many organic-material-rich habitats and play a major role in the turnover of organic matter in such environments [25]. Within this group, aerobic, microaerophilic, and anaerobic species are found, and most cultured CFB are chemoorganotrophs. In the past, they could not be appropriately investigated due to the lack of suitable techniques for their identification, but recent studies have revealed their importance in sediments [35]. Most 16S rRNA gene sequences from the CFB cluster included in databases belong to uncultured organisms. Some of the sequences detected in the mat biofilm were related to the marine species *Marinilabilia salmonicolor* (synonym *Cytophaga salmonicolor*), which like most Cytophaga species is able to grow under microaerophilic conditions [41].

In our fixed-biomass reactor, biofilms with high cell densities and high genetic and physiological diversity developed after inoculation with different types of communities. Since we used inorganic mineral medium with carbonate as the only carbon source, hydrogen sulfide as electron donor, a continuous light regime and anaerobic conditions, the development of phototrophic sulfur bacteria could be expected. Surprisingly, oxygenic phototrophs and colorless sulfur bacteria were also present, together with purple sulfur bacteria. Thus, aerobic and anaerobic sulfur oxidizers coexisted in the system, and oxygen was internally supplied by oxygenic photosynthesis.

The species of purple sulfur bacteria found in marine and freshwater environments are usually different [24], as it occurred in our biofilms, in which *Marichromatium purpuratum*-like dominated the marine biofilm while *Thiocapsa roseopersicina*-like was the species present in the freshwater biofilm. The first species is found in association with marine sponges [23] rather than in suspension, and this ability may have facilitated its profuse attached growth; its optimal growth is at 5% NaCl, while *Tca. Roseopersicina* has not salt requirements for growth [24].

Table 4. Phylogenetic groups recovered from the complex biofilms developed. The closest match in Genbank for the clones recovered are listed. Numbers indicate the percentage of clones retrieved in the libraries

Phylogenetic groups	Freshwater biofilm	Marine biofilm
Alpha-Proteobacteria	33% Rhodobacter sp. Rhodobacter capsulatus Rhodopseudomonas palustris Caulobacter fusiformis Uncultured Alpha-Proteobacteria	8% Rhodovulum adriaticum Erythrobacter litoralis Roseospirillum parvum
Beta-Proteobacteria	4% Uncultured bacteria	-
Gamma-Proteobacteria	25% Thiocapsa roseopersicina Pseudomonas stutzeri Uncultured bacteria	52% Marichromatium purpuratum Marichromatium indicum Thiorhodococcus drewsii Thiomicrospira sp. Uncultured Gamma-Proteobacteria
Delta-Proteobacteria	1% Uncultured bacteria	-
Epsilon-Proteobacteria	-	22% Uncultured Epsilon-Proteobacteria
Chlorobi	-	5% Prosthecochloris sp.
CFB group	21% Uncultured CFB Bergeyella sp.	8% Uncultured CFB Marinilabilia salmonicolor
Cyanobacteria/Plasts	14% Scenedesmus obliquus Uncultured cyanobacterium	4% Uncultured cyanobacterium
Low G+C Gram-positive	2% Uncultured bacteria	-
Others	-	1% Uncultured bacteria

Green sulfur bacteria highly related to *Prosthecochloris* sp. were present in the mat biofilm but, as occurs in nature, at a lower level than purple sulfur bacteria, very likely because the former are strict anaerobes whereas the latter tolerate low oxygen concentrations, and oxygen producers were also present in the biofilm.

Macroscopically both biofilms looked also different. The predominant color in the freshwater biofilm was green, in agreement with algae and Cyanobacteria represented by an important percentage of clones in the library. Oxygen production by those microorganisms was likely to occur and would allow the development of aerobic oxidizers while, probably, inhibited the development of strict anaerobic sulfur oxidizers (green sulfur bacteria). On the contrary, in the marine biofilm the predominant color was red in accordance with the dominance of anaerobic purple sulfur bacteria recovered in the library (46% of clones); aerobic sulfur oxidizers represented a low percentage of the library. Probably, due to the low oxygen production, green sulfur bacteria could develop in this biofilm.

Despite differences found in the species composition and in the macroscopic aspect, biofilms were very similar from a functional perspective. The same physiologic groups developed in both biofilms, represented by different species in each case (Table 5). Aerobic, anaerobic, phototrophic, chemotrophic and heterotrophic organisms were recovered in both experiments.

Table 5. Physiologic groups recovered from the complex biofilms developed. The closest match in Genbank for the clones recovered are listed. Numbers indicate the percentage of clones retrieved in the libraries

Microbial Guild	Freshwater biofilm	Marine biofilm
Oxygenic phototrophs	14% Scenedesmus obliquus Uncultured cyanobacterium	4% Uncultured cyanobacterium
Phototrophic sulfur bacteria Purple sulfur bacteria Green sulfur bacteria	13% Thiocapsa roseopersicina	51% Marichromatium purpuratum Marichromatium indicum Thiorhodococcus drewsii Prosthecochloris sp.
Aerobic sulfur oxidizers	12% Uncultured bacteria	2% Thiomicrospira sp.
Phototrophic non-sulfur bacteria	34% Rhodobacter sp. Rhodobacter capsulatus Rhodopseudomonas palustris Uncultured Proteobacteria	8% Roseospirillum parvum Rhodovulum adriaticum
Heterotrophic	26% CFB members Uncultured Delta-Proteobacterium Pseudomonas stutzeri Caulobacter fusiformis Firmicutes members	10% CFB members Erythrobacter litoralis Uncultured Gamma-Proteobacteria
Others	1% Uncultured Beta-Proteobacterium	25% Uncultured Epsilon-Proteobacteria Uncultured Gamma-Proteobacteria Uncultured bacteria

Anaerobic sulfur oxidizers were represented by members of the purple sulfur bacteria in both biofilms and also, by green sulfur bacteria in the marine sample. These organisms carry out anaerobic light-driven oxidation of sulfide. The relative abundance of this functional group was higher in the marine case, probably because of the lower oxygen production. Aerobic sulfur oxidizers (also known as colorless sulfur bacteria), represented by

Thiobacillus-like and *Thiothrix*-like species in the freshwater biofilm, were abundant while only a few clones related to the genus *Thiomicrospira* were found in the marine biofilm. Again, the higher algae contribution would have produced higher oxygen concentration that would have favored aerobic oxidizers in the freshwater biofilm.

Oxidation of sulfide can be also performed by purple nonsulfur bacteria (some phototrophic species of the Alpha and Beta-Proteobacteria). This group was present in both cases, but was Cspecially abundant in the freshwater biofilm. Heterotrophs were represented mainly by members of the Cytophaga-Flavobacterium-Bacteroides group, but also by pseudomonads, Caulobacteriales and Firmicutes.

Functional Model of the Microbial Communities

The presence of organisms belonging to several well-known functional groups in both experiments suggests a complex set of interactions (Figure 5). Purple sulfur bacteria (*Thiocapsa*-like) and purple nonsulfur bacteria (*Rhodobacter*-like, *Rhodopseudomonas*-like) can perform anaerobic light-driven oxidation of sulfide.

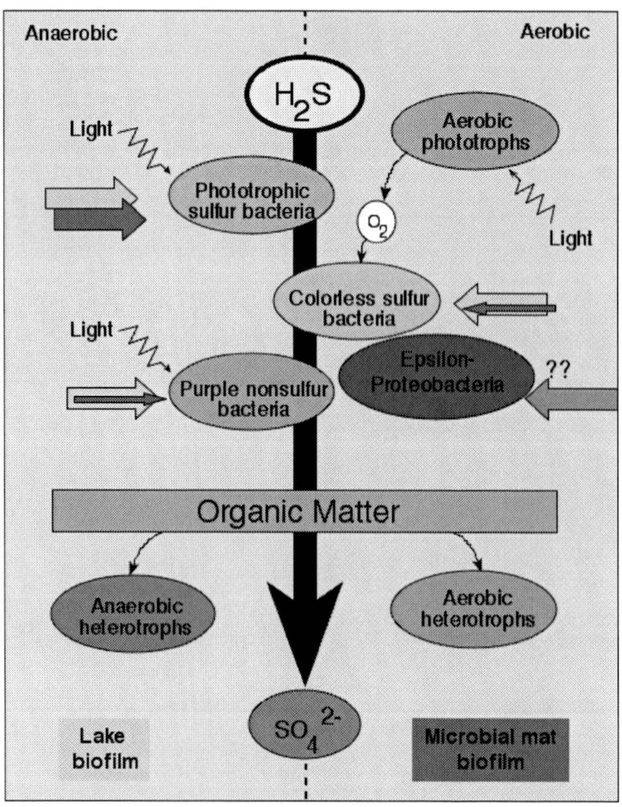

Figure 5. Hypothetical microbial activities among members of the microbial communities. Sulfide would be aerobically and anaerobically oxidized. Autotrophic members would produce organic matter that would be subsequently used by heterotrophs.

Also, sulfide can be oxidized aerobically by colorless sulfur bacteria (*Thiobacillus*-like, *Thiothrix*-like) and by some purple sulfur bacteria (*Thiocapsa*-like; [7]) using the oxygen produced by oxygenic phototrophs such as algae and Cyanobacteria. The organic matter generated by primary producers can be subsequently utilized by heterotrophs which comprise CFB, Pseudomonadales, Caulobacteriales, Firmicutes and purple nonsulfur bacteria, in metabolic processes including fermentation and aerobic oxidation of organic matter.

Although only indirect experimental evidence is presented here and these statements are merely speculation, the processes outlined above seem plausible for this kind of reactor. Techniques combining genetic identity and function, such as microautoradiography-FISH [29, 38] and stable isotope probing of nucleic acids [18], would help revealing the real contribution of each active population.

Conclusion

Biofilms consisting of complex microbial assemblages could be successfully developed using a packed-column reactor inoculated with a freshwater sediment and a microbial mat sample. A high genetic diversity was found, together with the coexistence of metabolically diverse groups, which can be explained by the formation of different microenvironments through the large and heterogeneous attachment surface. Clusters of closely related sequences below the "species-level" were observed, but in addition, new 16S rRNA genes were retrieved, substantially different from any previously known sequence. Many sequences were affiliated to uncultured organisms, especially for the CFB group and for the Epsilon-Proteobacteria. These results indicate that the bioreactor provides a growth environment different enough from the conditions traditionally used for isolating new organisms. Over the past years, investigation on bacterial physiology has been carried out in the laboratory using pure cultures in liquid suspensions or agar plates, and under fairly uniform conditions. Most of the microorganisms retrieved were not those abundant in nature, and therefore, these methods have weakly contributed to our understanding of the ecology of microorganisms. Now, the challenge for microbial ecophysiologists is to mimic in the laboratory the microbial scenarios as they occur in nature, and to use methods and tools that allow detailed studies of relevant features under controlled laboratory conditions [53]. Bioreactor systems and DNA-based molecular tools open a new perspective for microbial ecophysiology studies in the XXI century.

References

[1] Alain, K., Olagnon, M., Desbruyères, D., pagé, A., Barbier, G., Juniper, S. K., Quérellou, J. and Cambon-Bonavita, M. A. (2002). Phylogenetic characterization of the bacterial assemblage associated with mucous secretions of the hydrothermal vent polychaete *Paralvinella palmiformi*. *FEMS Microbiology Ecology, 42,* 463-476.

[2] Altschul, S. F., Madden, T. L., Schäffer, A. A., Zhang, J., Zhang, Z., Miller W. and Lipman, D. J. (1997). Gapped BLAST and PSI-BLAST: a new generation of protein database search programs. *Nucleic Acids Research, 25,* 3389-3402.

[3] Brock, T. D. (1987). The study of microorganisms in situ: progress and problems. In M. Fletcher, T. R. G. Gray and J. G. Jones (Eds), *Ecology of microbial communities* (pp. 1-17). Cambridge, England, Society for General Microbiology: Cambridge University Press.

[4] Brock, T.D. (1966). *Principles of microbial ecology.* Englewood Cliffs, NJ, USA, Prentice-Hall Biological Science Series: Prentice-Hall.

[5] Campbell, B. J., Engel, A. S., Porter, M. L. and Takai, K. (2006). The versatile epsilon-proteobacteria: key players in sulphidic habitats. *Nature Reviews Microbiology, 4,* 458-468.

[6] Casamayor, E. O., Pedrós-Alió, C., Muyzer, G. and Amann, R. (2002). Microheterogeneity in 16S ribosomal DNA-defined bacterial populations from a stratified planktonic environment is related to temporal changes ant to ecological adaptations. *Applied and Environmental Microbiology, 66,* 1706-1714.

[7] De Wit, R. and Van Gemerden, H. (1990). Growth and metabolism of the purple sulfur bacterium Thiocapsa roseopersicina under combined light/dark and oxic/anoxic regimens. *Archives of Microbiology, 154,* 459-464.

[8] Descamps-Julien, B. and Gonzalez, A. (2005). Stable coexistence in a fluctuating environment: an experimental demonstration. *Ecology, 86,* 2815-2824.

[9] Ferrera, I. (2004). *Development and characterization of sulfide-oxidizing biofilms.* Thesis, Universitat Autònoma de Barcelona.

[10] Ferrera, I., Massana, R., Casamayor, E. O., Balagué, V., Sánchez, O., Pedrós-Alió, C. and Mas, J. (2004). High diversity biofilm for the oxidation of sulfide-containing effluents. *Applied Microbiology and Biotechnology, 64,* 726-734.

[11] Ferrera, I., Sánchez, O. and Mas, J. (2004). A new non-aerated illuminated packed column reactor for the development of sulfide-oxidizing biofilms. *Applied Microbiology and Biotechnology, 64,* 659-664.

[12] Ferrera, I., Sánchez, O. and Mas, J. (2007). Characterization of a sulfide-oxidizing biofilm developed in apacked-column reactor. *International Microbiology, 10,* 29-37.

[13] Field, K. G., Gordon, D. Wright, T., Rappe, M., Urbach, E., Vergin, K. and Giovannoni, S. J. (1997). Diversity and depth-specific distribution of SAR11 cluster rRNA genes from marine planktonic bacteria. *Applied and Environmental Microbiology, 63,* 63-70.

[14] Fraleigh, S.P., Bungay, H. R. and Clesceri, L. S. (1989). Continuous culture, feedback control and auxostats. *Trends in Biotechnology, 7,* 159-164.

[15] Fuhrman, J. A. and Campbell, L. (1998). Microbial microdiversity. *Nature, 393,* 410-411.

[16] Gause, G. F. (1934). *The struggle for existence.* Baltimore: Williams and Wilkins.

[17] Glöckner, F. O., Fuchs, B. M. and Amann, R. (1999). Bacterioplankton compositions of lakes and oceans: a first comparison based on fluorescence in situ hybridization. *Applied and Environmental Microbiology, 65,* 3721-3726.

[18] Gray, N. D. and Head, I. M. (2001). Linking genetic identity and fuction in communites of uncultured bacteria. *Environmental Microbiology, 3,* 481-492.

[19] Guerrero, R., Pedrós-Alió, C., Esteve, I. and Mas, J. (1987). Communities of phototrophic sulfur bacteria in lakes of the Spanish Mediterranean region. In T. Lindholm (Ed.), *Ecology of Photosynthetic Prokaryotes* (pp. 125-151). Åbo Academy Press.

[20] Herbert, D., Elsworth, R. and Telling, R. C. (1956). The continuous culture of bacteria; a thoretical and experimental study. *Journal of General Microbiology, 14,* 601-622.

[21] Hutchinson, G. E. (1959). Homage to Santa Rosalia; or, Why are there so many kinds of animals? *The American Naturalist, 93,* 145–159.

[22] Hutchinson, G. E. (1961). The paradox of the plankton. *The American Naturalist, 95,* 137–145.

[23] Imhoff, J. F. and Trüper, H. G. (1976). Marine sponges as habitats of anaerobic phototrophic bacteria. *Microbial Ecology, 3,* 1-9.

[24] Imhoff, J. F. (2006). The Chromatiaceae. In M. Dworkin, S. Falkow, E. Rosenberg, K. H. Schleifer and E. Stackebrandt (Eds), *The prokaryotes* (3rd edition, Vol 7, part 3, pp. 846-873). New York: Springer.

[25] Kirchman, D. L. (2002). The ecology of *Cytophaga-Flavobacteria* in aquatic environments. *FEMS Microbiology Ecology , 39,* 91-100.

[26] Kodama, Y. and Watanabe, K. (2003). Isolation and characterization of a sulfur-oxidizing chemolithotroph growing on crude oil under anaerobic conditions. *Applied and Environmental Microbiology, 69,* 107-112.

[27] Konstantinov, K., Kishimoto, M., Seki, T. and Yoshida, T. (1990). A balanced DO-stat and its applications to the control of acetic acid excretion by recombinant *Escherichia coli. Biotechnology and Bioengeneering, 36,* 750-758.

[28] Larson, G., Enfors, S. O. and Pham, H. (1990). The pH-auxostat as a tool for studying microbial dynamics in continuous fermentation. *Biotechnology and Bioengeneering, 36,* 224-232.

[29] Lee, N., Nielsen, P. H., Andreasen, K.H., Juretschko, S., Nielsen J. L., Schleifer, K. H. and Wagner, M. (1999). Combination of fluorescent in situ hybridization and microautoradiography – a new tool for structure-function analyses in microbial ecology. *Applied and Environmental Microbiology, 65,* 1289-97.

[30] Lowry, O. H., Rosebrough, N. J., Farr, A. L. and Randall, R. J. (1951). Protein measurement with the Folin phenol reagent. *Journal of Biological Chemistry, 193,* 265-275.

[31] Maidak, B. L., Cole, J. R., Lilburn, T. G., Parker, C. T. Jr., Saxman, P. R., Stredwick, J. M., Garrity, G. M., Li, B., Olsen, G. J., Pramanik, S., Schmidt, T. M. and Tiedje, J. M. (2000). The RDP (Ribosomal Database Project) continues. *Nucleic Acids Research, 28,* 73-174.

[32] Mas, J. and Van Gemerden, H. (1987). Influence of sulfur accumulation and composition of the sulfur globule on cell volume and buoyant density of *Chromatium vinosum. Archives of Microbiology, 146,* 362-369.

[33] Mas, J. and Van Gemerden, H. (1992). Phosphate-limited growth of *Chromatium vinosum* in continuous culture. *Archives of Microbiology, 157,* 135-140.

[34] Mir, J., Martínez-Alonso, M., Esteve, I. and Guerrero, R. (1991). Vertical stratification and microbial assemblage of a microbial mat in the Ebro Delta (Spain). *FEMS Microbiology Ecology, 86,* 59-68.

[35] Mouné, S., Caumette, P., Matheron, R. and Willison, J. C. (2003). Molecular sequence analysis of prokaryotic diversity in the anoxic sediments underlying cyanobacterial mats of two hypersaline ponds in Mediterranean salterns. *FEMS Microbiology Ecology, 44,* 117-130.

[36] Nübel, U., Engelen, B., Felske, A., Snaid, J., Wieshuber, A., Amann, R. I., Ludwig, W. and Backhaus, H. (1996). Sequence heterogeneities of genes encoding 16S rRNAs in *Paenobacillus polymyxa* detected by temperature gradient gel electrophoresis. *Journal of Bacteriology, 178,* 5636-5643.

[37] Oltmann, L. F., Schoenmaker, G. S., Reijnders, W. N. and Stouthamer, A. H. (1978). Modification of the pH-auxostat culture method for the mass cultivation of bacteria. *Biotechnology and bioengeneering, 20,* 921-925.

[38] Ouverney, C. C. and Fuhrman, J. A. (2000). Marine planktonic archaea take up aminoacids. *Applied and Environmental Microbiology, 66,* 4829-33.

[39] Pooley, D. T., Larsson, J., Jones, G., Rayner-Brandes, M. H., Lloyd, D., Gibson, C. and Stewart, W. R. (2004). Continuous culture of photobacterium. *Biosensors and Bioelectronics, 19,* 1457-63.

[40] Ravenschlag, K, Sahm, K., Pernthaler, J. and Amann, R. (1999). High bacterial diversity in permanently cold marine sediments. *Applied and Environmental Microbiology, 65,* 3982-3989.

[41] Reichenbach, H. (2006). The Order Cytophagales. In M. Dworkin, S. Falkow, E. Rosenberg, K. H. Schleifer and E. Stackebrandt (Eds), *The prokaryotes* (3rd edition, Vol 7, part 6, pp. 549-590). New York: Springer.

[42] Reysenbach, A. L., Longnecker, K. and Kirshtein, J. (2000). Novel bacterial and archaeal lineages from an in situ growth chamber deployed at a Mid-Atlantic Ridge hydrothermal vent. *Applied and Environmental Microbiology, 66,* 3798-3806.

[43] Roper, M. M. and Marshall, K. C. (1978). Effects of a clay mineral on microbial predation and parasitism of *Escherichia coli. Microbial Ecology, 4,* 279-289.

[44] Sánchez, O., Van Gemerden, H. and Mas, J. (1996). Description of a redox-controlled sulfidostat for the growth of sulfide-oxidizing phototrophs. *Applied and Environmental Microbiology, 62,* 3640-3645.

[45] Sánchez, O., Van Gemerden, H. and Mas, J. (1998). Utilization of reducing power in light-limited cultures of *Chromatium vinosum. Archives of Microbiology, 170,* 411-417.

[46] Schramm, A. and Amann, R. (1998). In situ structure and function analysis of biofilms. In H. Märkl and R. Stegmann (Eds), Technik anaerober Prozesse (pp. 45-54). Frankfurt am Main, Germany: Dechema.

[47] Skipnes, O., Eide, I. and Jensen, A. (1980). Cage culture turbidostat: a device for rapid determination of algal growth rate. *Applied and Environmental Microbiology, 40,* 318-325.

[48] Sorgeloos, P., Van Outryve, E., Persoone, G. and Cattoir- Reynaerts, A. (1976). New type of turbidostat with intermittent determination of cell density outside the culture vessel. *Applied and Environmental Microbiology, 31,* 327-331.

[49] Takai, K., Campbell, B. J., Cary, S. C., Suzuki, M., Oida, H., Nunoura, T., Hirayama, H., Nakagawa, S., Suzuki, Y., Inagaki, F. and Horikoshi, K. (2005). Enzymatic and genetic characterization of carbon and energy metabolisms by deep-sea hydrothermal chemolithoautotrophic isolates of Epsilonproteobacteria. *Applied and Environmental Microbiology, 71,* 7310-7320.

[50] Takai, K., Suzuki, M., Nakagawa, S., Miyazaki, M., Suzuki, Y., Inagaki, F. and Horikoshi, K. (2006). *Sulfurimonas paralvinellae* sp. nov., a novel mesophilic, hydrogen- and sulfur-oxidizing chemolithoatotroph within the Epsilonproteobacteria isolated from a deep-sea hydrotermal vent polychaete nest, reclassification of *Thiomicrospira denitrificans* and *Sulfurimonas denitrificans* comb. nov. and emended description of the genus *Sulfurimonas*. *International Journal of Systematic and Evolutionary Microbiology, 56,* 1725-1733.

[51] Tilman, D. (1976). Ecological competition between algae: Experimental confirmation of resource-based competition theory. *Science, 192*, 463-465.

[52] Tilman, D., Kilham, S. and Kilham, P. (1982). Phytoplankton community ecology: the role of limiting nutrients. *Annual Reviews of Ecology and Systematics, 13,* 349-372.

[53] Tolker-Nielsen, T. and Molin, S. (2000). Spatial organization of microbial biofilm communities. *Microbial Ecology, 40,* 75-84.

[54] Van den Ende, P. (1973). Predatory-prey interactions in continuous culture. *Science, 181,* 562-564.

[55] Van Gemerden, H. (1974). Coexistence of organisms competing for the same substrate: An example among the purple sulfur bacteria. *Microbial Ecology, 1,* 104-119.

[56] Veldkamp, H. (1976). Mixed culture studies with the chemostat. In A. C. R. Dean, D. C. Ellwood, C. G. T. Evans and J. Melling (Eds), *Continuous culture 6: Applications and new fields* (pp. 315-326). Chichester, United Kingdom: Ellis Horwood, Publisher, Ltd.

[57] Veldkamp, H. (1977). Ecological studies with the chemostat. In M. Alexander (Ed), *Advances in microbial ecology* (Vol. 1, pp. 59-89). London: Plenum Publishing Co, Ltd.

[58] Vetriani, C., Tran, H. V. and Kerkhof, L. J. (2003). Fingerprinting microbial assemblage from the oxic/anoxic chemocline on the Black Sea. *Applied and Environmental Microbiology, 69,* 6481-6488.

[59] Watson, T. G. (1969). Steady state operation of a continuous culture at maximum rate by control of carbon dioxide production. *Journal of General Microbiology, 59,* 83-89.

[60] West, N. J. and Scanlan, D. J. (1999). Niche-partitioning of *Prochlorococcus* populations in a stratified water column in the eastern North Atlantic Ocean. *Applied and Environmental Microbiology, 65,* 2585-2591.

[61] Wirsen, C. O., Sievert, S. M., Cavanaugh, C. M., Molyneaux, S. J., Ahmad, A., Taylor, L. T., DeLong, E. F. and Taylor, C. D. (2002). Characterization of an autotrophic sulfide-oxidizing marine *Arcobacter* sp. that produces filamentous sulfur. *Applied and Environmental Microbiology, 68,* 316-325.

[62] Yamada, S., Wada, M. and Chibata, I. (1979). Oxygen transfer as a parameter of automatic control of the continuous cultivation for the conversion of sorbitol to sorbose by *Acetobacter suboxydans*. *Journal of Fermentation Technology, 57,* 210-214.

In: Microbial Ecology Research Trends
Editor: Thijs Van Dijk, pp. 115- 137
ISBN: 978-1-60456-179-1

Chapter V

Ecological Significance and Possible Risks of Nonculturable Intestinal Bacteria in Water Systems

I. Arana and I. Barcina
Department of Immunology, Microbiology and Parasitology,
Faculty of Science and Technology, Basque Country University,
Apdo. 644, E-48080 Bilbao, Spain

Abstract

The detection, enumeration and fate of *Escherichia coli* in the aquatic systems are on the priority objectives for researches as well as public and health administrations. The role of this bacterium as universal indicator of faecal pollution and of pathogen presence in waters justifies the interest in its study. The method directed to detection and enumeration of this indicator microorganism are gathered and standardized in the prevailing regulations concerning water quality (International Standard Office, American Protection Agency, etc) and are based on their ability to grow in culture. Out of the intestine, in a hostile environment such the aquatic systems, *E. coli* adopts the viable but nonculturable (VBNC) state. In this state, *E. coli* loses its ability to grow on culture media but remains metabolically active. Apparently, this fact nullifies the regulations for its detection justifying the renewed interest in the biological significance of VBNC phenomena and its relation with the survival of microorganisms in the aquatic systems. It is proposed that VBNC state is a strategy for the survival which implies the recovery of growth ability after reversion of the adverse conditions. Another theory hints this state to be a previous step for cellular death. In this review it is presented a global vision of works studying the proposed theses and it is analyzed the ability to resuscitate or to resuming culturability after the entry into the VBNC state, the participation of molecular signals inducing the entry into and the exit from this state, as well as, the genes implied in the overall process through proteomic studies. Moreover, studies analyzing relationship with protozooplankton are too reviewed, due to the VBNC phenomena can alter the ecological relationship between *E. coli* and predator microorganisms from aquatic systems. This approach provides an ecological vision, very close to reality for repercussions of entry into the VBNC state during *E. coli* survival in aquatic systems.

Why is it Important to Study the Behaviour and Fate of *Escherichia Coli* in a Water Environment?

Water is a limited resource which is exhaustively used, especially in what are known as the developed countries. The massive use of water, together with the relation many of its uses have with public health (the drinks, food and pharmaceutical industries etc.), means that a strict control and regulation of water quality is essential. Allochthonous bacteria and most enteric and pathogenic microorganisms are released directly or through wastewater into rivers and coastal sea water. *Escherichia coli* is the intestinal bacterial species which is universally accepted as an indicator of faecal contamination for which reason its detection, quantification, behaviour and fate in water environments is constantly studied by both the public administration and the scientific community.

Several studies and reviews have presented data on the process of disappearance, known as "die off", of this bacterium in water environments, in which its survival depends on both biotic and abiotic environmental factors (McCambridge and McMeekin 1980, 1981, Borrego *et al.* 1983, Barcina *et al.* 1986a, 1986b, 1997, Crane and Moore 1986, Davies *et al.* 1995, Burkhardt III *et al.* 2000, Menon *et al.* 2003, Noble *et al.* 2004, Easton *et al.* 2005, John and Rose 2005).

Among the abiotic factors which negatively influence survival, the following are of prime importance: a lack of nutrients (Sjϕgren and Gibson 1981, Sinclair and Alexander 1984, Henis and Alexander 1990, Baleux *et al.* 1998, Ramos *et al.* 2001), salinity (Barcina *et al.* 1990, García-Lara *et al.* 1991, Davies *et al.* 1995, Joux *et al.* 1997, Baleux *et al.* 1998, Marco-Noales *et al.* 1999, Trousellier *et al.* 1999, Rozen and Belkin 2001) and solar radiation (McCambridge and McMeekin 1981, Fujioka and Narikawa, 1982, Baleux *et al.* 1998, Muela *et al.* 2000, Sinton *et al.* 2002, Yukselen *et al.* 2003, Whitman *et al.* 2004, Chandran and Hatha 2005). Moreover, these factors interact with each other, having a negative, sometimes synergic effect on survival. One of the most widely known examples of this interaction is that which takes place between salinity and solar radiation (Grigsby and Calkins 1980, Fujioka *et al.* 1981, Bailey *et al.* 1983, Baleux *et al.* 1998, Bordalo *et al.* 2002, Sinton *et al.* 2002, Zaafrane *et al.* 2004).

Among the biotic factors which most affect the survival of *E. coli* is its competitive relationship with bacterioplankton (Scheuerman *et al.* 1988, Henis and Alexander 1990, Marco-Noales *et al.* 2004, Long *et al.* 2005), its parasitic relationship with virus and lytic bacteria (McCambridge and McMeekin 1980, Guerrero *et al.* 1986, Flint 1987, Chandran and Hatha 2005), and the fact that it is the prey of protozooplankton (Barcina *et al.* 1986a, 1986b, 1991, 1997, González *et al.* 1990a, 1992, Iriberri *et al.* 1994a, 1994b, Davies *et al.* 1995, Muela *et al.* 1998, Rozen and Belkin 2001, Arana *et al.* 2003); this last factor being that which most effectively contributes to the elimination of both *E. coli* and other allochthonous bacteria, including pathogenic bacteria.

The Viable but Nonculturable State (VBNC)

Under adverse conditions, some bacteria are able to maintain detectable metabolic functions, but they are unculturable by available methods. Xu *et al.* (1982) defined this phenomenon as the "viable but nonculturable (VBNC) state" and the number of species known to enter VBNC constantly increases (for a review see Oliver 2005a). The list not only includes non-pathogens, but also a large number of human pathogens and the universal indicator, *E. coli.* This raises serious questions about, for example, the interpretation of routine water or food testing in which it is assumed that infecting bacteria can be detected by their ability to grow on culture media.

A typical VBNC response is shown in Figure 1. The abiotic factors present in sea water (lack of nutrients, non-optimum temperature, salinity and solar radiation) provoke a clear decrease in culturable cells (CFU counts). Nevertheless, the cells maintain their integrity ([TDC] total direct counts, constants). The active cell population descends more slowly than culturable cells, and therefore provides a reference for the estimation of the VBNC population (comparing active and culturable cell counts) and the number of non-viable (NV) cells (comparing total and active cell counts).

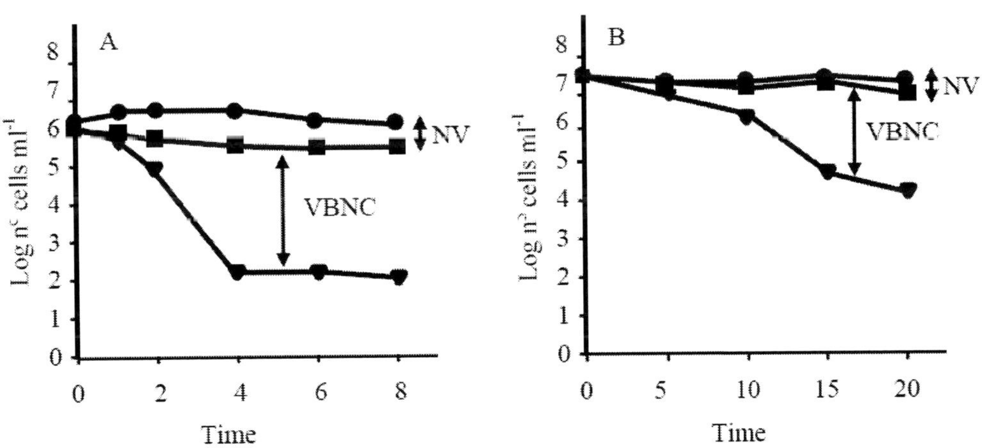

Figure 1. A typical response of *E. coli* maintained in illuminated freshwater (A) and non-illuminated seawater (B) and formation of VBNC and NV subpopulations. Total number of bacteria (●); number of bacteria with intact cytoplasmic membranes (■), number of culturable bacteria (▼). Modified from Arana *et al.* 2003, 2007a.

Therefore, it is possible to differentiate three distinct subpopulations; culturable, VBNC and non-viable. The relative percentages of the different subpopulations formed and the VBNC state induction period are both species and stress-dependent (Oliver *et al.* 1991, Barcina *et al.* 1997, Colwell and Gray 2000, Wong and Wang 2004, Oliver 2005a, Arana *et al.* 2007a). Readers can find information about the different methods described for the detection of cellular activity and integrity in Hobbie *et al.* (1977), Kogure *et al.* (1979), Porter and Feig (1980), Rodríguez *et al.* (1992), Barcina *et al.* (1995), McFeters *et al.* (1995), Zweifel and Hagström (1995), Barer and Harwood (1999), Muela *et al.* (1999), Breeuwer and

Abee (2000), Joux and Lebaron (2000), Lleó *et al.* (2000), Shapiro (2000) or Yaron and Matthews (2002).

From this perspective, many studies have been carried out in which the effect of environmental factors on the survival of *E. coli* and many other pathogenic and non-pathogenic bacteria has been reviewed. Using this new approach, the conclusion has been reached that abiotic factors, previously considered as the principal factors responsible for the disappearance of these bacteria from water systems, do, in many cases, cause them to enter the VBNC state (Barcina *et al.* 1989, 1990, 1997, Wang and Doyle 1998, Muela *et al.* 1999, 2000, Arana *et al.* 2003, 2004, García-Armisen and Servais 2004). Moreover, it has been demonstrated that some treatments aimed at reducing the number of microorganisms, such as the pasteurisation of milk (Gunasekera *et al.* 2002) and other disinfection treatments, also induce the VBNC state (Singh *et al.* 1990, Tryland *et al.* 1998, Arana *et al.* 1999, Bjergbæk and Roslev 2005, Oliver, 2005b).

The Biological Significance of VBNC

Both the morphological characteristics and the physiological properties of cells in the VBNC state have been widely studied, and it has been reported that VBNC cells exhibit morphological changes and a reduction in nutrient transport, respiration rates and macromolecular synthesis (Barcina *et al.* 1989, 1990, Oliver 1993, González-Escalona *et al.* 2006, Vattakaven *et al.* 2006, Coutard *et al.* 2007). However, novel stress proteins (Morton and Oliver 1994, Srinivasan *et al.* 1998) are formed and cytoplasmic membrane fatty acid composition (Day and Oliver 2004, Wong *et al.* 2004a) and cell wall peptidoglycan (Signoretto *et al.* 2000, 2002) is modified. Moreover, gene expression has also been demonstrated (Lleó *et al.* 2000, Smith and Oliver 2006).

Consequently, we know the factors which cause cells to enter this state and the characteristics of cells when in this state, but what is the biological significance of this? What role does it play in nature?

Since this state was described, the interpretation of its significance has provoked a great deal of controversy within the scientific community. Some authors have hypothesized that cells exposed to hostile environments become nonculturable due to cellular deterioration, which can be a stochastic (Bogosian and Bourneuf 2001) or genetically programmed (Nyström 2003, Aertsen and Michiels 2004) phenomenon. This fact might be an adaptive strategy to ensure persistence of population. Cells with reduced or nil colony-forming capacity could provide nutrients for remaining culturable cells and allow the presence of persistors under adverse conditions (Lewis 2000, Arana *et al.* 2004, Cuny *et al.* 2005, Coutard *et al.* 2007).

Another theory states that nonculturability is the culmination of an adaptive pathway generating dormant survival forms, similar to spore formation in differentiating bacteria (McDougald *et al.* 1999, Kell and Young 2000). The VBNC state could be a significant means of survival if VBNC cells were able to again become culturable, that is to say, if cells underwent a process of resuscitation (McDougald *et al.* 1999, Oliver 2005a; Keep *et al.* 2006a, Panutdaportn *et al.* 2006, Coutard *et al.* 2007). If the latter theory proved to be correct, its repercussions in the field of medicine, bioremediation, the use of bacteria as faecal indicators, disinfection processes and indeed, in most microbiological studies where culturability is employed as the indicator of viability, would not only be evident, but would also provide great cause for concern.

Demonstrating resuscitation is not always a simple matter. A major problem is the difficulty in showing conclusively that the "resuscitated" population comes from a true resuscitation, and is not just the result of the growth of some culturable cells hidden in the nonculturable population (Bogosian *et al.* 1998, Kell *et al.* 1998, Bogosian and Bourneuf 2001). Wai *et al.* (2000) and Oliver (2005a) recommend the dilution method in order to preclude the possibility of any culturable cells being present and growing. The addition of some bacterial inhibitors is another option. Jiang and Chai (1996), Ohtomo and Saito (2001) and Wong *et al.* (2004b) report that the inhibition of the DNA synthesis (i.e. the addition of nalidixic acid) prevents the multiplication of culturable cells, but not the resuscitation process.

An apparently simple way to deal with this problem involves the physical separation of the previously mentioned subpopulations (CFU, VBNC and NV) with the aim of obtaining a fraction strictly comprising only of nonculturable cells.

The Separation of Culturable and Nonculturable Cells

Density gradient centrifugation is a common method for the separation and isolation of cells and sub-cellular particles. On the basis of their buoyant density, homogeneous fractions can be obtained from heterogeneous mixtures of microorganisms (Weiss *et al.* 1988), subpopulations of bacteria (Hanson *et al.* 1981, Whiteley *et al.* 2000) and so on. Various mechanisms have been cited to explain buoyant density variation during growth phase transition, such as changes in cell volume and shape, compaction of the nucleoid, alteration in cell-wall composition and accumulation of storage compounds (Kolter *et al.* 1993, Huisman *et al.* 1996, Makinoshima *et al.* 2003), and many of these changes have been detected in VBNC cells (see before). However, the most critical factor responsible for the variation in buoyant density remains unknown.

Percoll and Urografin (an iodated contrast medium marketed also as Renografin-76 or Radioselectan) have been used to separate, on the basis of their buoyant density, cellular fractions throughout growth, following long periods of starvation, and also culturable from nonculturable subpopulations. Pertoft (2000) described advantages and disadvantages of the use of Percoll density gradients to separate cells or particles. Percoll has been used to separate subpopulations of cells during the growth of *E. coli* and *Vibrio parahaemolitycus* (Makinoshima *et al.* 2002, 2003, Nishino *et al.* 2003). Makinoshima *et al.* (2002, 2003) indicates that *E. coli* cultures can be fractioned into many cell subpopulations, more than 20, and attributed the cell separation to the increase in buoyant density from the exponential to stationary phase of growth. Arana *et al.* (2007b, in press), when trying to separate different subpopulations of *E. coli* from growth curves or following stress exposure, found that the bands were heterogeneous and comprised of both culturable and nonculturable cells.

Whiteley *et al.* (2000) combined the tetrazolium salt method and Percoll density centrifugation in order to separate active from nonactive bacteria from natural communities. The intracellular accumulation of insoluble formazan increases the buoyant density of respiring bacteria, thus enabling them to be separated.

Siegele *et al.* (1993) and Desnues *et al.* (2003) report the use of Urografin to separate culturable from nonculturable *E. coli* cells according to their buoyant densities. They assign greater density to nonculturable cells, which are thus understood to concentrate in the high-density bands (HD). Arana *et al.* (2007b, in press) examined physiologically heterogeneous *E. coli* populations in different phases of growth, and populations subjected to stress, and their results showed a common pattern (see Figure 2). High-density (HD) bands were composed mainly of nonculturable cells and, in contrast, low-density (LD) bands had a very heterogeneous composition, being mixtures of culturable and nonculturable cells. Cuny *et al.* (2005) obtained two clearly separate bands (LD and HD) from a culturable population of *E. coli*, but both bands consisted exclusively of culturable cells. They concluded that the high-density culturable cells were in transition to the VBNC state.

Percoll is regarded as non-toxic to cells (Pertoft 2000, Nayak *et al.* 2005) but Urografin has a bactericidal and bacteriostatic toxic effect upon cells and, moreover, a negative effect upon culturability (Hanson *et al.* 1981, Kim and Lachman 1982, Tisa *et al.* 1982, Dory and Wautelet 1985, Rockne *et al.* 2003). Apparently, the experimental procedure itself, fundamentally, contact time with the contrast medium and temperature, principally determines the level of toxicity.

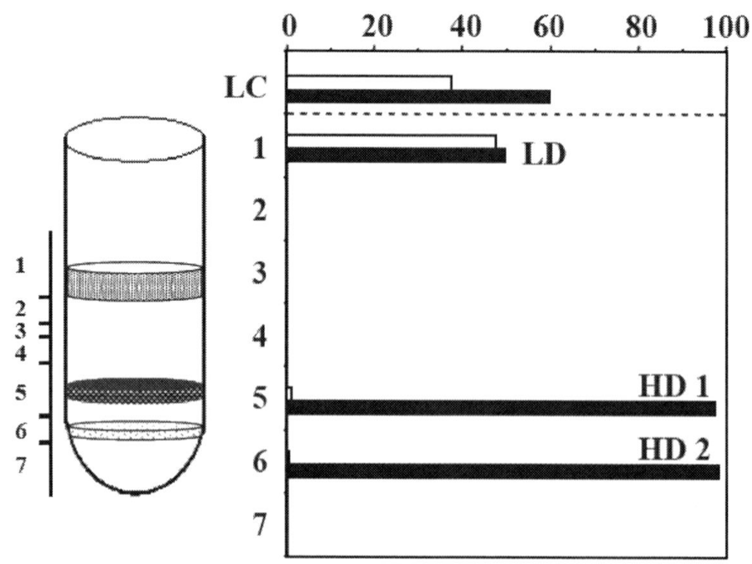

Figure 2. Tube shows the distribution, into arbitrary zones, of the bands generated after centrifugation of a mixed suspension of *E. coli*(aged cells and culture from stationary growth phase) in Urografindensity gradient. In the graph, percentages of culturable (☐) and nonculturablecells (■) obtained from layered cells (LC) and from generated bands. The percentages of cells in each band with respect to the total layered cells were for LD band 10-50%, HD 1 band 50-90%, and HD 2 band <10%. Modified from Arana *et al.* 2007 b.

Percoll is regarded as non-toxic to cells (Pertoft 2000, Nayak *et al.* 2005) but Urografin has a bactericidal and bacteriostatic toxic effect upon cells and, moreover, a negative effect upon culturability (Hanson *et al.* 1981, Kim and Lachman 1982, Tisa *et al.* 1982, Dory and Wautelet 1985, Rockne *et al.* 2003). Apparently, the experimental procedure itself, fundamentally, contact time with the contrast medium and temperature, principally determines the level of toxicity.

In accordance with that previously stated, it is evident that the data in the literature related to this question are not homogeneous. A deeper knowledge of the factors which are responsible for the fractioning of bacteria populations is required and, moreover, the possibility that factors other than buoyant density may intervene in fractioning by means of Percoll and Urografin gradients cannot be ruled out.

Resuscitation Process

The recovery of the capacity to grow by cells which have entered the VBNC state is the point which generates the most conflict within the controversy surrounding this newly described biological state. In addition to the methodological problems explained above, there have also been several other factors which have contributed to the disparity among the published results and the varying opinions of the scientific community. Among these factors, the following should be highlighted: the lack of stringent controls in many studies (Kell *et al.* 1998), the time spent by the cells in the VBNC state, the differences among the experimental designs used, the use of cell populations subjected to very different adverse conditions which provoke a loss of culturability due to different lesions, and, obviously, the complexity of the biological process itself.

However, studies have demonstrated that true resuscitation does occur, at least in some bacteria. The removal of environmental stress by temperature up-shift from 5°C to 20°C results in the resuscitation of *Vibrio vulnificus* (Whitesides and Oliver 1997). The same strategy, i.e. the removal of environmental stress, led to the resuscitation of *Salmonella* (Gupte *et al.* 2003), *Enterococcus* (Lleó *et al.* 1998, 2001), *Pasteurella* (Magariños *et al.* 1997), diverse species of *Vibrio* (Ramaiah *et al.* 2002, Coutard *et al.* 2007), *E. coli* (Ohtomoto and Saito 2001) and *Aeromonas* (Wai *et al.* 2000), etc.

A second category of treatments to induce resuscitation includes the activation of a replicative function. Comparing with the germination following sporulation, stimulation of VBNC cells could be necessary, as occurs with heat shock (Gupte *et al.* 2003) and ammonium salt treatment (Wai *et al.* 1996). Peptidoglycan hydrolysis alters the mechanical properties of the cell wall to facilitate cell division (Signoretto *et al.* 2002) or releases lysis products acting as resuscitation signals (Keep *et al.* 2006b).

The team of Kaprelyants, Kell and Mukamolova (1993, 1994, 1998, 1999, 2000, 2002, 2006) identified a protein secreted by dividing *Micrococcus luteus*, a rapidly growing *Actinobacteria*, which increased the numbers of colonies that could be formed by dormant cultures by several orders of magnitude, even at picomolar concentrations. They called this the protein resuscitation promotion factor (Rpf) and demonstrated that *M. luteus* Rpf protein cleaves to peptidoglycan and has strong structural similarities to other lysozymes. An Rpf-like protein secreted by growing cells of *Salmonella typhimurium* has been described (Panutdaporn *et al.* 2006). This protein induces resuscitation of VBNC cells of the food pathogen *Salmonella enterica*, but not of *E. coli.*

The case of the Rpf protein of *Micrococcus luteus* is not unique. Other authors (Weichart and Kell 2001, Shleeva *et al.* 2004, Downing *et al.* 2005) have found compounds secreted by different organisms during growth or starvation periods. Srinivasan *et al.* (1998) demonstrated the effect of acylated homoserine lactones secreted by *Vibrio* sp. strain S14 in the expression of proteins crucial to the development of starvation and stress-resistant phenotypes. The resuscitation of *Legionella pneumophila* could only be demonstrated following co-incubation of the VBNC cells with the amoeba *Acantamoeba castellani* (Steiner *et al.* 1997). However, Arana *et al.* (2007a, in press) do not find that the free supernatants of cells from actively growing populations have any effect upon the resuscitation of *E. coli.*

MacDougald *et al.* (1999) reported that the identification of specific genes that are important for the development of the VBNC response will allow us to address whether the VBNC response is a true life cycle event or a default response that occurs in the absence of maintenance of culturability. Hengge-Aronis (1993, 2002) described and studied a set of dependent σ^S genes which codify the proteins known as stress proteins. In a culture, these genes are expressed at the end of the exponential phase, prior to entering the stationary phase in which growth ceases. Stress proteins help cells to resist adverse conditions and delay their loss of culturability under conditions of environmental stress. Apparently, they delay entry into the VBNC state. *E. coli* cells in the exponential phase are less resistant to stress and enter the VBNC state more quickly than cells in the stationary phase in which these proteins are present (Munro *et al.* 1995, Dukan and Touati 1996, Gourmelon *et al.* 1997, Diez *et al.* 2000, Richard and Foster 2003, Dantur and Pizarro 2004). Aertsen and Michiels (2004) report that although the VBNC phenomenon is thought of as a protective response, comparable with sporulation (*Bacillus*) or dormancy (*Micrococcus luteus*), so far no mutations or genes have been found as evidence for an underlying pathway triggered in response to stress.

In bacteria with a sequenced genome, the proteomic comparative analysis between culturable populations and VBNC populations may provide information about possible genes which are involved in the entry into this state. Protein expression patterns between the stationary phase and the VBNC state of *Enterococcus faecalis* differ markedly, indicating a distinct physiological state (Heim *et al.* 2002). Wang *et al.* (2006) compared the alkaline proteomes of *E. coli* populations between the period of majority live bacteria (18-h culture), the period of similar amount of live and dead bacteria (30-h culture) and the period of majority dead bacteria (48-h culture) and they indicated that *E. coli* expressed differentially six alkaline proteins. Muela *et al.* (data unpublished) found that differences in the proteome of the external membrane of *E. coli* in the stationary phase and the VBNC state bear no relation to this cellular state, but to the adverse environmental conditions which induce it.

Cell Death: Individual or Populational, Aleatory or Altruistic?

Cell death can either be interpreted as an individual event, restricted to the cellular environment, or as a populational phenomenon. Aertsen and Michiels (2004) have published an excellent, extensive review of this topic.

Joux *et al.* (1997), in a study using flow cytometry and staining procedures, report that adverse conditions in a marine environment cause a progressive deterioration which leads to cell death in *Salmonella typhimurium* and in *E. coli*. They describe a sequence of the process through the loss of measurable physiological capacities. A similar chain of events culminating in the degeneration of the nucleoid was also described by Muela *et al.* (1999) for *E. coli* populations maintained in river water (see Figure 3). Nyströn (2001) and Desnues *et al.* (2003) highlight the importance of irreversible oxidative damage which affects different cell compartments.

A large number of studies, in which the fate of intestinal bacteria in water environments has been analysed, have shown that a "remaining" population exists whose density oscillates between 10^3 and 10^4 cells ml^{-1} and which continues to be culturable despite the deterioration of the majority of the population (Flint 1987, Barcina *et al.* 1989, 1990, Muela *et al.* 1999, Turner *et al.* 2000, Arana *et al.* 2003, 2004, Ritchie *et al.* 2003, Craig *et al.* 2004).

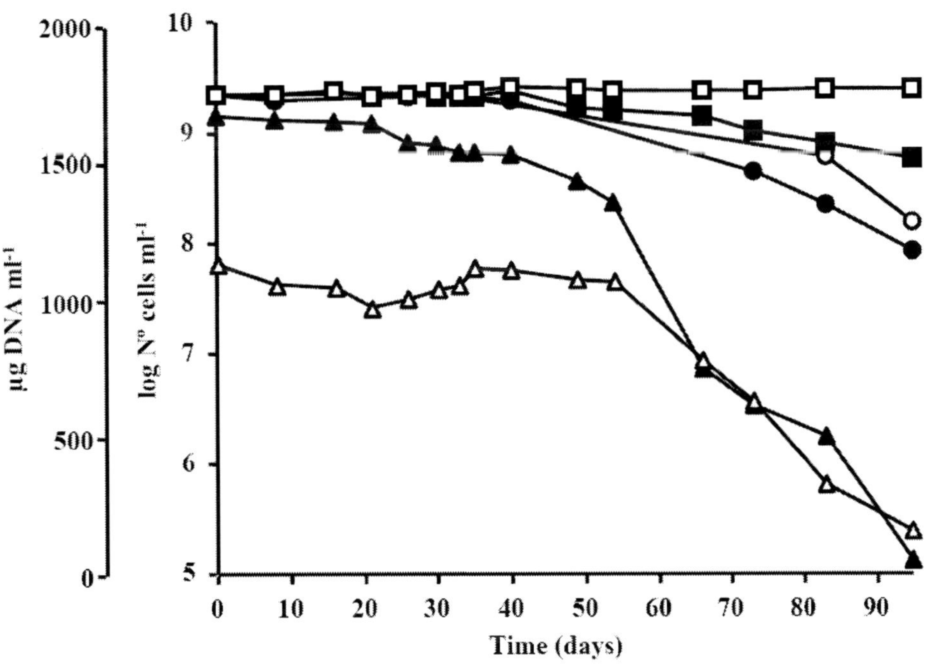

Figure 3. Total number of cells(□), number of nucleotid-containing cells(■) number of cells with active electron transport system(○), number of cells with active cellular synthesis (●) number of culturable cells(△) and chromosomal DNA content(▲) during starvation-survival of *E. coli* strain J62 in Butrón river water. Taken from Muela *et al.* 1999.

This fact is attributed to the intrinsic heterogeneity of bacteria populations and may be an extra mechanism to assure the perdurability of the species. Basing their study on buoyant density as determined by Percoll gradient centrifugation (see "The separation of culturable and nonculturable cells" Section), Makinoshina *et al.* (2002) reported the existence of between 5 and 10 different fractions of *E. coli* in both the exponential and the stationary phase, in apparently homogeneous cultures. Following cell division, the distribution of mRNA proteins among the daughter cells is not symmetric, which means that there is proteomic heterogeneity among the cells which form a population. These differences may result in different morphological and physiological phenotypes, especially if this unequal protein share involves minority proteins (Aertsen and Michiels 2004).

An interpretation of cell death within a populational environment is clearly illustrated in the studies of Zambrano *et al.* (1993), Zambrano and Kolter (1996) and Zinser and Kolter (1999, 2000). From batch cultures de *E. coli* subjected to long periods of starvation, these authors isolated mutants whose intake and metabolism of sustrates were highly efficient, and they named them GASP (growth advantage in stationary phase). These mutants use the energy obtained from the few nutrients which remain in the environment or from the organic material released by dead cells. In the latter case, the death of part of a population may be seen as an altruistic event oriented towards providing the GASP mutants with nutrients. Arana *et al.* (2004) observed that under adverse conditions and when in the process of entering the VBNC state, *E. coli* released into its surroundings organic molecules, amino-acids, proteins and carbohydrates in sufficient amounts so as to permit growth. Lewis (2000) and Cuny *et al.* (2005) proposed that in the case of serious damage by stressful factors, cells will donate their nutrients to other members of the population to assure species survival until better times come.

What is the Situation in Water Environments?

When intestinal bacteria arrive in water systems after being transported there in waste water, they must face the adverse abiotic conditions present in the water system which receives them. The result is deterioration of the cells and the consequent entry into the VBNC state. Moreover, they must also face the biotic factors, the natural microbial population.

Within the published literature on this theme we can find studies containing a description of the establishment of competitive relationships for the available nutrients between intestinal bacteria and heterotrophic bacteria (Marco-Noales *et al.* 2004, Long *et al.* 2005), of parasitism with the phages and the lytic bacteria like *Bdellovibrio* (Guerrero *et al.* 1986, Flint 1987, Chandran and Hatha 2005) and of predation with the protozoa (González *et al.* 1990a, 1992, Rozen and Belkin 2001, Arana *et al.* 2003).

These relationships of parasitism and predation eliminate intestinal bacteria in water environments and incorporate their biomass in the food chain. From a quantitative point of view, predation by the flagellate protozoa and ciliate protozoa is the process which most efficiently contributes to the elimination of allochthonous bacteria (Barcina *et al.* 1986a, 1986b, 1991, Iriberri *et al.* 1994b). As individuals, the ciliate protozoa are more efficient predators than the flagellate protozoa (González *et al.* 1990b), but given the greater density of the latter, especially in sea water, their predation is the main biological process affecting allochthonous bacteria density (Barcina *et al.* 1991, Weinbauer and Höffe 1998).

Despite the large amount of published research about the VBNC state, it is surprising that so few works have been aimed at studying the efficiency of predation on VBNC cells (Arana *et al.* 2003). Perhaps this lack of information is due to the necessity of having a live biological model which is distinguishable from the natural bacteria and in which it is possible to distinguish the viable but nonculturable forms. With the aim of distinguishing *E. coli* from bacterioplankton, fluorescent-labelled bacteria (FLB) are used as tools. FLB are bacteria inactivated by heat and marked with 5-(4-6-dicholotriazin-2-yl) aminofluoriscein (Sherr *et al.* 1987, González *et al.* 1992, Iriberri *et al.* 1994b). The use of FLBs does not permit the study of the predation on VBNC cells because these cells are dead. González *et al.* (1992) describe the advantages of using Rhodamine stain bacteria (RSB, Landry *et al.* 1991) which are live. However, the staining protocol to obtain RSB cells is very tedious and deleterious for the cells.

The gene *gfp* found naturally in the jellyfish *Aequorea victoria* encodes the green fluorescent protein (Tsien 1998). The utility of this gene to act as a marker with which to detect and visualise bacteria in environmental samples has been demonstrated in several studies (Leff and Leff 1996, Ammendola *et al.* 1998, Cho and Kim 1999, Lowder *et al.* 2000, Arana *et al.* 2003). The green fluorescent protein (GFP) maintains its fluorescence under adverse conditions, even when exposed to light radiation (Arana *et al.* 2003), which allows one to track, in a natural environment and over a period of time, the bacteria which are carriers of the *gfp* gene (Leff and Leff 1996, Lowder *et al.* 2000, Arana *et al.* 2003, Na *et al.* 2006) and study the way in which VBNC cells are formed. Cho and Kim (1999) and Lowder *et al.* (2000) discuss the advantages and disadvantages of using this gene in environmental studies in comparison with the use of the *lux, luxAB* and *lacZY* genes, respectively. Arana *et al.* (2003) studied the behaviour of *E. coli* DH5α pGreenTIR in the water of the Butrón river, with a special interest in the dynamics of the formation of VBNC cells in a situation in which the predator-prey relationship with the protozooplankton of the river is of priority (Barcina *et al.* 1986a, 1986b, González *et al.* 1990a).

Arana *et al.* (2003) demonstrated that the river bacteria behaved as predation-escaping prey and maintained their population density, while *E. coli* DH5α pGreenTIR, behaved as predation-non-escaping prey, and were eliminated from the system. It was also demonstrated that flagellate protozoa, which were mainly responsible for the elimination of *E coli* DH5α pGreenTIR, do not discriminate between live and dead cells. The most interesting result of this study was that it showed that the elimination of *E coli,* through predation, takes place before the abiotic factors cause cell deterioration and subsequent entry into the VBNC.

In a balanced water environment with active protozooplankton, the previously mentioned results would seem to remove importance from the VBNC state of allochthonous bacteria. However, there is no reason to believe that these results are possible to extrapolate to the behaviour of intestinal Gram positive bacteria such as *Enterococcus faecalis.* The efficiency of predation on *Enterococcus faecalis* is inferior to that on *E. coli* and, therefore, it is eliminated from water systems more slowly (González *et al.* 1990b). The application of a tool similar to that described for *E. coli* which permits us to know the interrelationships among the fecal enterococci, the protozooplankton and the formation of VBNCs is required.

Conclusion

While the behaviour of *E. coli* should not necessarily be a standard for all bacteria, it does provide a standard for Gram negative intestinal bacteria, including pathogenic bacteria. Alternation of favourable and unfavourable situations is typical in aquatic systems, i.e. light and dark, feast and famine, cycles of optimal and suboptimal temperatures etc. For the autochthonous bacterial species such as *Vibrio vulnificus*, a life-cycle involving the entry into the VBNC state and a subsequent resuscitation process in response to environmental conditions may be a strategy which contributes to assuring the perpetuation of the species (Oliver 2005a). *E. coli* lives, in high density populations in a stable environment, in the intestine. This situation is clearly different from that described for *Vibrio vulnificus* and autochthonous species. As a consequence of this, we believe that the genetic endowment of intestinal bacteria probably does not include a programme in which resuscitation constitutes a strategy for the perpetuation of the species. The altruistic death strategy previously commented on is difficult to extrapolate to aquatic environments in which bacteria densities are low and there is fierce competition for the available nutrients. Moreover, if we also take into account the efficiency of the predation carried out by the protozooplankton, the presence of intestinal bacteria in the VBNC state does not seem to be of importance from a quantitative point of view, and it is very unlikely that its presence may imply a sanitary risk.

At this stage it is necessary to ask the following question: what role do the VBNC forms have in aquatic systems? We are only able to find one reason to justify the existence in water systems of VBNC forms or even non-viable cells, but which retain their morphological integrity intact, and, obviously this reason is totally speculative. In our opinion it may be to preserve the organic material contained within the cell, thereby facilitating, or even ensuring, that it passes to a higher level of the food chain, fundamentally through predation.

The hypotheses have been put forward and the unanswered questions asked about all the aspects reviewed in this manuscript, all of which leads us to embark upon new lines of research which will enable us, in the near future, to discover the causes and reasons for the existence of this biological phenomenon whose application is unquestionable.

References

Aertsen, A. and Michiels, C. W. (2004). Stress and how bacteria cope with death and survival. Critical Reviews in Microbiology 30, 263-273.

Ammendola, A., Geisenberger, O., Andersen, J. B., Givskov, M., Schleifer, K. H. and Eberl, L. (1998). *Serratia liquefaciens* swarm cells exhibit enhanced resistance to predation by *Tetrahymena* sp. FEMS Microbiology Letters 164, 69-75.

Arana, I., Santorum, P., Muela, A. and Barcina, I. (1999). Chlorination and ozonation of waste-water: comparative analysis of efficacy through the effect on *Escherichia coli* membranes. *Journal of Applied Microbiology* 86, 838-888.

Arana, I., Irizar, A., Seco, C., Muela, A., Fernández-Astorga, A. and Barcina, I. (2003). *gfp*-tagged cells as a useful tool to study the survival of *Escherichia coli* in the presence of the river microbial community. *Microbial Ecology* 45, 29-38.

Arana, I., Seco, C., Epelde, K., Muela, A., Fernández-Astorga, A. and Barcina, I. (2004). Relationship between *Escherichia coli* cells and the surrounding medium during survival processes. Antonie van Leeuwenhoek 86, 189-199.

Arana, I., Orruño, M., Pérez-Pascual, D., Seco, C., Muela, A. and Barcina, I. (2007a). Inability of *Escherichia coli* to resuscitate from the viable but nonculturable state. FEMS *Microbial Ecology*. (In press).

Arana, I., Orruño, M., Seco, C., Muela, A. and Barcina, I. (2007b). Ability of Urografin density gradients to separate nonculturable subpopulations of *Escherichia coli*. Antonie van Leeuwenhowk. (In press).

Bailey, C. A., Neihof, R. A. and Tabor, P. S. (1983). Inhibitory effect of solar radiation on amino acid uptake in Chesapeake Bay bacteria. *Applied and Environmental Microbiology* 46, 44-49.

Baleux, B., Caro, A., Lesne, J., Got, P., Binard, S. and Delpeuch B. (1998). Survie et maintenance de la virulence de *Salmonella typhimurium* VNC exposée simultanément à trois facteurs stressants expérimentaux. *Oceanologica Acta* 21, 939-950.

Barcina, I., Arana, I., Iriberri, J. and Egea L. (1986a). Factors affecting the survival of *Escherichia coli* in a river. *Hydrobiologia* 141, 249-253.

Barcina, I., Arana, I., Iriberri, J. and Egea, L. (1986b). Influence of the light and natural microbiota of the Butron river on *Escherichia coli* survival. Antonie van Leeuwenhoek 52, 555- 566.

Barcina, I., González, J. M., Iriberri, J. and Egea, L. (1989). Effect of visible light on progressive dormancy of *Escherichia coli* cells during the survival process in natural fresh water. *Applied and Environmental Microbiology* 55, 246-251.

Barcina, I., González, J. M., Iriberri, J. and Egea, L. (1990). Survival strategy of *Escherichia coli* and *Enterococcus faecalis* in illuminated fresh and marine systems. Journal of Applied Bacteriology 2, 189-198.

Barcina, I., González, J. M., Iriberri, J. and Egea, L. (1991). Role of protozoa in the regulation of enteric bacteria populations in seawater. *Marine Microbial Food Webs* 5, 179-187.

Barcina, I., Arana, I., Santorum, P., Iriberri, J. and Egea, L. (1995). Direct viable count of gram-positive and gram-negative bacteria using ciprofloxacin as inhibitor of cellular division. *Journal of Microbiological Methods* 22, 139-150.

Barcina, I., Lebaron, P. and Vives-Rego, J. (1997). MiniReview: Survival of allochtonous bacteria in aquatic systems: a biological approach. *FEMS Microbiology Ecology* 23, 1-9.

Barer, M. R. and Harwood, C. R. (1999). Bacterial viability and culturability. Advances in *Microbial Physiology* 41, 93-137.

Bjergbæk, I. A. and Roslev, P. (2005). Formation of nonculturable *Escherichia coli* in drinking water. *Journal of Applied Microbiology* 99, 1090-1098.

Bogosian, G., Morris, P. J. L. and O'Neil, J. P. (1998). A mixed culture recovery method indicates that enteric bacteria do not enter the viable but nonculturable state. Applied and Environmental Microbiology 64, 1736-1742.

Bogosian, G. and Bourneuf, E. V. (2001). A matter of bacterial life and death. EMBO Reports 2, 770-774.

Bordalo, A. A., Onrassami, R. and Dechsakulwatana, C. (2002). Survival of faecal indicator bacteria in tropical estuarine waters (Bangpakong River, Thailand). *Journal of Applied Microbiology* 93, 864-871.

Borrego, J. J., Arrabal, F., de Vicente, A., Gómez, L. F. and Romero, P. (1983). Study of microbial inactivation in the marine environment. *Journal of Water Pollution Control Federation* 55, 297-302.

Breeuwer, P. and Abee, T. (2000). Assessment of viability of microorganisms employing fluorescence techniques. International *Journal of Food Microbiology* 55, 193-200.

Burkhardt III, W., Calci, K. R., Watlins, W. D., Rippey, S. R. and Chirtel, S. J. (2000). Inactivation of indicator microorganisms in estuarine waters. *Water Research* 34, 2207-2214.

Chandran, A. and Hatha, A. A. M. (2005). Relative survival of *Escherichia coli* and *Salmonella typhimurium* in a tropical estuary. *Water Research* 39, 1397–1403.

Cho, J. C. and Kim, S. J. (1999). Viable, but non-culturable, state of a green fluorescence protein-tagged environmental isolate of *Salmonella typhi* in groundwater and pond water. *FEMS Microbiology Letters* 170, 257-264.

Colwell, R. R. and Gray, D. J. (2000). Nonculturable microorganisms in the environment. Washington D. C.: American Society for Microbiology Press.

Coutard, F., Crassouss, P., Droguet, M., Gobin, E., Colwell, R. R., Pommepuy, M. and Hervio-Heath, D. (2007). Recovery in culture of viable but nonculturable *Vibrio parahaemolyticus* regrowth or resuscitation? *The ISME Journal* 1, 111-120.

Craig, D. L., Fallowfield, H. J. and Cromar, N. J. (2004). Use of microcosms to determine persistence of *Escherichia coli* in recreational coastal water and sediment and validation with in situ measurements. *Journal of Applied Microbiology* 96, 922-930.

Crane, S. R. and Moore, J. A. (1986). Modelling enteric bacterial die-off: A review. *Water, Air, and Soil Pollution* 27, 411-439.

Cuny, C., Dukan, L., Fraysse, L., Ballesteros, M. and Dukan, S. (2005). Investigation of the first events leading to loss of culturability during *Escherichia coli* starvation: future nonculturable bacteria form a subpopulation. *Journal of Bacteriology* 187, 2244-2248.

Dantur, K. I. and Pizarro, R. A. (2004). Effect of growth phase on the *Escherichia coli* response to ultraviolet-A radiation: influence of conditioned media, hydrogen peroxide and acetate. *Journal of Photochemistry and Photobiology B: Biology* 75, 33-39.

Davies, C. M., Long, J. A. H., Donald, M. and Ashbolt, N. J. (1995). Survival of fecal microorganisms in marine and freshwater sediments. *Applied and Environmental Microbiology* 61, 1888-1896.

Day, A. P. and Oliver, J. D. (2004). Changes in membrane fatty acid composition during entry of *Vibrio vulnificus* into the viable but nonculturable state. *Journal of Microbiology* 42, 69-73.

Desnues, B., Cuny, C., Grégori, G., Dukan, S., Aguilaniu, H. and Nyström, T. (2003). Differential oxidative damage and expression of stress defence regulons in culturable and non-culturable *Escherichia coli* cells. *EMBO Reports* 4, 400-404.

Diez, A., Gustavsson, N. and Nyström, T. (2000). The universal stress protein A of *Escherichia coli* is required for resistance to DNA damaging agents and is regulated by a RecA/FtsK-dependent regulatory pathway. *Molecular Microbiology* 36, 1494-1503.

Dory, M. A. and Wautelet, M. J. (1985). Arthroscopy in septic arthritis. Licodaine- and iodine-containing contrast media are bacteriostatic. *Arthritis and Rheumatism* 28, 198-203.

Downing, K. J., Mischenko, V. V., Shleeva, M. O., Young, D. I., Young, M., Kaprelyants, A. S., Apt, A. S. and Mizrahi, V. (2005). Mutants of *Mycobacterium tuberculosis* lacking three of the five *rpf-like* genes are defective for growth in vivo and for resuscitation in vitro. *Infection and Immunity* 73, 3038-3043.

Dukan, S. and Touati, D. (1996). Hypochlorous acid stress in *Escherichia coli*: resistance, DNA damage, and comparison with hydrogen peroxide stress. *Journal of Bacteriology* 178, 6145-6150.

Easton, J. H., Gauthier, J. J., Lalor, M. M. and Pitt R. E. (2005). Die-off of pathogenic *E. coli* O157:H7 in sewage contaminated waters. *Journal of the American Water Resources Association* 41, 1187-1193.

Flint, K. P. (1987). The long-term survival of *Escherichia coli* in river water. *Journal of Applied Bacteriology* 63, 261-270.

Fujioka, R. S., Hashimoto, H. H., Siwak, E. B. and Young, R. H. T. (1981). Effect of sunlight on survival of indicator bacteria in seawater. *Applied and Environmental Microbiology* 41, 690-696.

Fujioka, R. S. and Narikawa, O. T. (1982). Effect of sunlight on enumeration of indicator bacteria under field conditions. *Applied and Environmental Microbiology* 44, 395-401.

Garcia-Armisen, T. and Servais, P. (2004). Enumeration of viable *E. coli* in rivers and wastewaters by fluorescent in situ hybridization. *Journal of Microbiological Methods* 58, 269-279.

García-Lara, J., Menon, P., Servais, P. and Billen, G. (1991). Mortality of fecal bacteria in seawater. *Applied and Environmental Microbiology* 57, 885-888.

González, J. M., Iriberri, J., Egea, L. and Barcina, I. (1990a). Differential rates of digestion of bacteria by freshwater and marine phagotrophic protozoa. *Applied and Environmental Microbiology* 56, 1821-1837.

González, J. M., Sherr, E. B. and Sherr, B. F. (1990b). Size-selective grazing on bacteria by natural assemblages of estuarine flagellates and ciliates. *Applied and Environmental Microbiology* 56, 583-589.

González, J. M., Iriberri, J., Egea, L. and Barcina, I. (1992). Characterization of culturability, protistan grazing, and death of enteric bacteria in aquatic ecosystems. *Applied and Environmental Microbiology* 58, 998-1004.

González-Escalona, N., Fey, A., Hofle, M. G., Espejo, R. T. andGuzman, C. A. (2006). Quantitative reverse transcription polymerase chain reaction analysis of *Vibrio cholerae* cells entering the viable but non-culturable state and starvation in response to cold shock. *Environmental Microbiology* 8, 658-666.

Gourmelon, M., Touati, D., Pommepuy, M. and Cormier, M. (1997). Survival of *Escherichia coli* exposed to visible light in seawater: analysis of *rpoS*-dependent effects. Canadian *Journal of Microbiology* 43, 1036-1043.

Grigsby, P. and Calkins, J. (1980). The inactivation of a natural populatin of coliform bacteria by sunlight. *Photochemistry and Photobiology* 31, 291-294.

Guerrero, R., Pedrós-Alió, C., Esteve, I., Más, I., Chase, D. and Margulis, L. (1986). Predatory prokaryotes: predation and primary consumption evolved in bacteria. Proceeding of National Academy of Sciences 83, 2138-2142.

Gunasekera, T. S., Sϕrensen, A., Attfield, P. V., Sϕrensen, S. J. and Veal, D. A. (2002). Inducible gene expression by nonculturable bacteria in milk after pasteurization. *Applied and Environmental Microbiology* 68, 1988-1993.

Gupte, A. R., de Rezende, L. E. and Joseph, S. W. (2003). Induction and resuscitation of viable but nonculturable *Salmonella enterica* serovar typhimurium DT104. *Applied and Environmental Microbiology* 69, 6669-6675.

Hanson, B. A., Wisseman Jr., C. L., Waddell, A. and Silverman, D. J. (1981). Some characteristics of heavy and light bands of *Rickettsia prowazekii* on Renografin gradients. *Infection and Immunity* 34, 596-604.

Heim, S., Lleó, M. M., Bonato, B., Guzman, C. A. and Canepari, P. (2002). The viable but nonculturable state and starvation are different stress responses of *Enterococcus faecalis*, as determined by proteome analysis. *Journal of Bacteriology* 184, 6739-6745.

Henis, Y. and Alexander, M. (1990). Effect of organic amendments on bacterial multiplication in lake water. Antonie van Leeuwenhoek 57, 9-20.

Hengge-Aronis, R. (1993). Survival of hunger and stress: the role of *rpoS* in early stationary phase gene regulation in *E. coli*. *Cell* 72, 165-168.

Hengge-Aronis, R. (2002). Recent insights into the general stress response regulatory network in *Escherichia coli*. Journal of Molecular Microbiology and Biotechnology 4, 341-346.

Hobbie, J. E., Caley, R. J. and Jasper, S. (1977). Use of Nucleopore filters for counting bacteria by epifluorescence microscopy. *Applied and Environmental Microbiology* 33, 1225-1228.

Huisman, G. W., Siegele, D. A., Zambrano, M. M. and Kolter, R. (1996). Morphological and pysiological changes during stationary phase. In F. C. Neidhart, R. Curtiss III, J. L. Ingraham, E. C. C. Lin, K. B. Low, B. Magasanik, W. S. Reznikoff, M. Riley, M. Schaechter and H. E. Umbarger (Eds.) *Escherichia coli* and *Salmonella*: Cellular and Molecular Biology (2nd ed., pp 1672-1682), Washington, D. C.: ASM Press.

Iriberri, J., Azua, I., Labirua-Iturburu, A., Artolozaga, I. and Barcina, I. (1994a). Differential elimination of enteric bacteria by protist in a freshwater system. *Journal of Applied Bacteriology* 77, 476- 483.

Iriberri, J., Ayo, B., Artolozaga, I., Barcina, I. and Egea, L. (1994b). Grazing on allochthonous *vs* autochthonous bacteria in river water. Letters in Applied Microbiology 18, 12-14.

Jiang, X. and Chai, T. J. (1996). Survival of *Vibrio parahaemolyticus* at low temperatures under starvation conditions and subsequent resuscitation of viable, nonculturable cells. *Applied and Environmental Microbiology* 62, 1300-1305.

John, D. E. and Rose, J. B. (2005). Review of factors affecting microbial survival in groundwater. *Environmental Science and Technology* 39, 7345-7356.

Joux, F., Lebaron, P. and Troussellier, M. (1997). Succession of cellular states in *Salmonella typhimurium* population during starvation in artificial seawater microcosms. *FEMS Microbiology Ecology* 22, 65-76.

Joux, F. and Lebaron, P. (2000) Use of fluorescent probes to assess physiological functions of bacteria at single-cell level. *Microbiology and Infection* 2, 1523-1535.

Kaprelyants, A. S., Gottschal, J. C. and Kell, D. B. (1993). Dormancy in non-sporulating bacteria. *FEMS Microbiology Review* 104, 271-286.

Kaprelyants, A. S., Mukamolova, G. V. and Kell, D. B. (1994). Estimation of dormant *Micrococcus luteus* cells by penicillin lysis and by resuscitation in cell-free spent culture medium at high dilution. *FEMS Microbiology Letters* 115, 347-352.

Keep, N. H., Ward, J. M., Robertson, G., Cohen-Gonsaud, M. and Henderson, B. (2006a) Bacterial resuscitation factors: revival of viable but non-culturable bacteria. *Cellular and Molecular Life Sciences* 63, 2555-2559.

Keep, N. H., Ward, J. M., Cohen-Gonsaud, M. and Henderson, B. (2006b). Wake up! Peptidoglycan lysis and bacterial non-growth states. *TRENDS in Microbiology* 14, 271-276.

Kell, D. B., Kaprelyants, A. S., Weichart, D. H., Harwood, C. R. and Barer, M. R. (1998). Viability and activity in readily culturable bacteria: a review and discussion of the practical issues. Antonie van Leeuwenhoek 73, 169-187.

Kell, D. B. and Young, M. (2000). Bacterial dormancy and culturability: the role of autocrine growth factors. *Current Opinion in Microbiology* 3, 238–243.

Kim, K. S. and Lachman, R. (1982). In vitro effects of iodinated contrats media on the growth of staphylococci. *Investigative Radiology* 17, 305-309.

Kogure, K., Simidu, U. and Taga, N. (1979). A tentative direct microscopic method for counting living marine bacteria. *Candian Journal of Microbiology* 25, 415-420.

Kolter, R., Siégele, D. A. and Tormo, A. (1993). The stationary phase of the bacterial life cycle. *Annual Review of Microbiology* 47, 855-874.

Landry, M. R., Lehner-Fournier, J. M., Sundstrom, J. A., Fagerness, K. E. and Selph, K. E. (1991). Discrimination between living and heat-killed prey by a marine zooflagellate, *Paraphysomonas vestita* (Stokes). *Journal of Experimental Marine Biology and Ecology* 146, 139-151.

Leff, L. G. and Leff, A. A. (1996). Use of green fluorescent protein to monitor survival of genetically engineered bacteria in aquatic environments. *Applied and Environmental Microbiology* 62, 3486-3488.

Lewis, K. (2000). Programmed death in bacteria. *Microbiology and Molecular Biology Review* 64, 503-514.

Lleó, M. M., Tafi, M. C. and Canepari, P. (1998). Nonculturable Enterococcus faecalis cells are metabolically active and capable of resuming active growth. Systematic and Applied Microbiology 21, 333-339.

Lleó, M. M., Pierobon, S., Tafi, M. C., Signoreto, C. and Canepari, P. (2000). mRNA detection by reverse transcription-PCR for monitoring viability over time in *Enterococcus faecalis* viable but nonculturable population maintained in laboratory microcosms. *Applied and Environmental Microbiology* 66, 4564-4567.

Lleó, M. M., Bonato, B., Tafi, M. C., Signoretto, M., Boaretti, M. and Canepari, P. (2001). Resuscitation rate in different enterococcal species in the viable but nonculturable state. *Journal of Applied Microbiology* 91, 1095-1102.

Long, R. A., Rowley, D. C., Zamora, E., Liu, J., Bartlett, D. H. and Azam, F. (2005). Antagonistic interactions among marine bacteria impede the proliferation of *Vibrio cholerae. Applied and Environmental Microbiology* 71, 8531–8536.

Lowder, M., Unge, A., Maraha, N., Jansson, J. K., Swiggett, J. and Oliver, J. D. (2000). Effect of starvation and the viable but nonculturable state on green fluorescent protein (GFP) fluorescence in GFP-tagged *Pseudomonas fluorescens* A506. *Applied and Environmental Microbiology* 66, 3160-3165.

Magariños, B., Romalde, J. L. and Toranzo, A. E. (1997). Viability of starved *Pasteurella piscicida* in seawater monitored by flow cytometry and the effect of antibiotics on its resuscitation. *Letters in Applied Microbiology* 24, 22-126.

Makinoshima, H., Nishimura, A. and Ishihama, A. (2002). Fractionation of *Escherichia coli* cell populations at different stages during growth transition to stationary phase. *Molecular Microbiology* 43, 269–279.

Makinoshima, H., Aizawa, S. I., Hayashi, H., Miki, T., Nishimura, A. and Ishihama, A. (2003). Growth phase-coupled alterations in cell structure and function of *Escherichia coli. Journal of Bacteriology* 185, 1338-1345.

Marco-Noales, E., Biosca, E. G. and Amaro, C. (1999). Effects of salinity and temperature on long-term survival of the eel pathogen *Vibrio vulnificus* biotype 2 (serovar E). *Applied and Environmental Microbiology* 65, 1117-1126.

Marco-Noales, E., Biosca, E. G., Rojo, C. and Amaro, C. (2004). Influence of aquatic microbiota on the survival in water of the human and eel pathogen *Vibrio vulnificus* serovar E. *Environmental Microbiology* 6, 364-376.

McCambridge, J. and McMeekin, T. A. (1980). Relative effects of bacterial and protozoan predators on survival of *Escherichia coli* in estuarine water samples. *Applied and Environmental Microbiology* 40, 907-911.

McCambridge, J. and McMeekin, T. A. (1981). Effect of solar radiation and predacious microorganisms on survival of fecal and other bacteria. *Applied and Environmental Microbiology* 41, 1083-1087.

McDougald, D. M., Rice, S. A. and Kjelleberg, S. (1999). New perspectives on the viable but nonculturable response. *Biologia Bratislava* 54, 617-623.

McFeters, G. A., Yu, F. P, Pyle, B. H. and Stewart, P. S. (1995). Psysiological assessment of bacteria using fluorochromes. *Journal of Microbiological Methods* 21, 1-13.

Menon, P., Billen, G. and Servais P. (2003). Mortality rates of autochthonous and fecal bacteria in natural aquatic ecosystems. *Water Research* 37, 4151-4158.

Morton, D. S. and Oliver, J. D. (1994). Induction of carbon starvation-induced proteins in *Vibrio vulnificus. Applied and Environmental Microbiology* 60, 3653-3659.

Muela, A., Santorum, P., Arana, I., García-Bringas, J. M. and Barcina, I. (1998). Discharge of disinfected wastewater in recipient aquatic systems: Fate of allochthonous bacterial and autochthonous protozoa populations. *Journal of Applied Microbiology* 85, 263-270.

Muela, A., Arana, I., Justo, J. I., Seco, C. and Barcina, I. (1999). Changes in DNA content and cellular death along a starvation-survival process of *Escherichia coli* in river water. *Microbial Ecology* 37, 62-69.

Muela, A., García-Bringas, J. M., Arana, I. and Barcina, I. (2000). The effect of simulated solar radiation on *Escherichia coli:* The relative roles of UV-B, UV-A, and photosynthetically active radiation. *Microbial Ecology* 39, 65-71.

Mukamolova, G. V., Kaprelyants, A. S., Young, D. I., Young, M. and Kell, D. B. (1998). A bacterial cytokine. Microbiology 95, 8916-8921.

Mukamolova, G. V. and Kormer, S. S. (1999). Stimulation of the multiplication of *Micrococcus luteus* by an autocrine growth factor. Archives of Microbiology 172, 9-14.

Mukamolova, G. V., Turapov, O. A., Kazarian, K., Telkov, M., Kaprelyants, A. S., Kell, D. B. and Young, M. (2002). The *rpf* gene of *Micrococcus luteus* encodes an essential secreted growth factor. *Molecular Microbiology* 46, 611-621.

Mukamolova, G. V., Murzin, A. G., Salina, E. G., Demina, G. R., Kell, D. B., Kaprelyants, A. S. and Young, M. (2006). Muralytic activity of *Micrococcus luteus* Rpf and its relationship to physiological activity in promoting bacterial growth and resuscitation. *Molecular Microbiology* 59, 84-98.

Munro, P. M., Flatau, G. N., Clement, R. L. and Gauthier, M. J. (1995). Influence of the RpoS (KatF) sigma factor on maintenance of viability and culturability of *Escherichia coli* and *Salmonella typhimurium* in seawater. *Applied and Environmental Microbiology* 61, 1853-1858.

Na, S. H., Miyanaga, K., Unno, H. and Tanji, Y. (2006). The survival response of *Escherichia coli* K12 in a natural environment. *Applied Microbiology and Biotechnology* 72, 386-392.

Nayak, B. B., Kamiya, E., Nishino, T., Wada, M., Nishimura, M. and Kogure, K. (2005). Separation of active and inactive fractions from starved culture of *Vibrio parahaemolyticus* by density dependent cell sorting. *FEMS Microbiology Ecology* 51, 179-186.

Nishino, T., Nayak, B. B. and Kogure, K. (2003). Density-dependent sorting of physiologically different cells of *Vibrio parahaemolyticus. Applied and Environmental Microbiology* 69, 3569-3572.

Noble, R. T., Lee, I. M. and Schiff, K. C. (2004). Inactivation of indicator micro-organisms from various sources of faecal contamination in seawater and freshwater. *Journal of Applied Microbiology* 96, 464-472.

Nyström, T. (2001). Not quite dead enough: on bacterial life, culturability, senescence, and death. *Archives of Microbiology* 176, 159-164.

Nyström, T. (2003). Nonculturable bacteria: programmed survival forms or cells at death's door? *BioEssays* 25, 204-211.

Ohtomo, R. and Saito, M. (2001). Increase in the culturable cell number of *Escherichia coli* during recovery from saline stress? Possible implication to resuscitation from VBNC state. *Microbial Ecology* 42, 208-214.

Oliver, J. D., Nilsson, L. and Kjelleberg, S. (1991). The formation of nonculturable cells of *Vibrio vulnificus* and its relationship to the starvation state. *Applied and Environmental Microbiology* 57, 2640-2644.

Oliver, J. D. (1993). Formation of viable but nonculturable cells. In S. Kjelleberg (Ed.), Starvation in bacteria (pp. 239-272). New York, USA: Kluwer Academic/Plenum Publishers.

Oliver, J. D. (2005a). The viable but nonculturable state in bacteria. *Journal of Microbiology* 43, 93-100.

Oliver, J. D., Dagher, M. and Linden, K. (2005b). Induction of *Escherichia coli* and *Salmonella typhimurium* into the viable but nonculturable state following chlorination in wastewater. *Journal of Water and Health* 3, 249-257.

Panutdaporn, N., Kawamoto, K., Asakura, H. and Makino, S. I. (2006). Resuscitation of the viable but non-culturable state of *Salmonella enterica* serovar Oranienburg by recombinant resuscitation-promoting factor derived from *Salmonella typhimurium* strain LT2. *International Journal of Food Microbiology* 106, 241-247.

Pertoft, H. (2000). Fractionation of cells *and subcellular particles with Percoll. Journal of Biochemical and Biophysical Methods* 44, 1-30.

Porter, K. and Feig, Y. (1980). The use of DAPI for identifying and counting aquatic microflora. *Limnology and Oceanography* 25, 943-948.

Ramaiah, N., Ravel, J., Straube, E. W. L., Hill, R. T. and Colwell, R. R. (2002). Entry of *Vibrio harveyi* and *Vibrio fischeri* into the viable but nonculturable state. *Journal of Applied Microbiology* 93, 108-116.

Richard, H. T. and Foster, J. W. (2003). Acid resistance in *Escherichia coli. Advances in Applied Microbiology* 52, 167-186.

Ritchie, J. M., Campbell, G. R., Shepherd, J., Beaton, Y., Jones, D., Killham, K. and Artz, R. R. (2003). A stable bioluminescent construct of *Escherichia coli* O157:H7 for hazard assessments of long-term survival in the environment. *Applied and Environmental Microbiology* 69, 3359-3367.

Rockne, K. J., Liang, W., Young, L. Y. and Taghon, G. L. (2003). Toxicity of density separation media to *Escherichia coli* and *Mycobacterium* strain PC01: implications for density-separation of soils and sediments. *FEMS Microbiology Ecology* 43, 185-189.

Rodríguez, G. G., Phipps, D., Ishiguro, K. and Ridgway, H. F. (1992). Use of fluorescent redox probe for direct visualization of actively respiring bacteria. *Applied and Environmental Microbiology* 58, 1801-1808.

Rozen, Y. and Belkin, S. (2001). Survival of enteric bacteria in seawater. FEMS Microbiology Reviews 725, 1-17.

Scheuerman, P. R., Schmidt, J. P. and Alexander, M. (1988). Factors affecting the survival and growth of bacteria introduced into lake water. *Archives of Microbiology* 150, 320-325.

Shapiro, H. (2000). Microbial analysis at the single-cell level: tasks and techniques. *Journal of Microbiological Methods* 42, 3-16.

Sherr, B. F., Sherr, E. B., Fallon, R. D. (1987). Use of monodispersed, fluorescently labelled bacteria to estimate in situ protozoan bacteriovory. *Applied and Environmental Microbiology* 53, 958-965.

Shleeva, M., Mukamolova, G. V., Young, M., Williams, H. D. and Kaprelyants, A. S. (2004). Formation of 'non-culturable' cells of *Mycobacterium smegmatis* in stationary phase in response to growth under suboptimal conditions and their Rpf-mediated resuscitation. *Microbiology* 150, 1687-1697.

Siegele, D. A., Almirón, M. and Kolter, R. (1993). Approaches to the study of survival and death in stationary-phase *Escherichia coli*. In S. Kjelleberg (Ed.), Starvation in bacteria (pp. 151-170). New York, USA: Kluwer Academic/Plenum Publishers.

Signoretto, C., Lleó, M. M., Tafi, M. C. and Canepari, P. (2000). Cell wall chemical composition of *Enteroccocus faecalis* in the viable but nonculturable state. *Applied and Environmental Microbiology* 66, 4564-4567.

Signoretto, C., Lleó, M. M. and Canepari, P. (2002). Modification of the peptidoglycan of *Escherichia coli* in the viable but nonculturable state. *Current Microbiology* 44, 125-133.

Sinclair, J. L. and Alexander M. (1984). Role of resistance to starvation in bacterial survival in sewage and lake water. *Applied and Environmental Microbiology* 48, 410-415.

Singh, A., Yu, F. P. and McFeters, G. A. (1990). Rapid detection of chlorine-induced bacterial injury by the direct viable count method using image analysis. *Applied and Environmental Microbiology* 56, 389-394.

Sinton, L. W., Hall, C. H., Lynch, P. A. and Davies-Colley, R. J. (2002). Sunlight inactivation of fecal indicator bacteria and bacteriophages from waste stabilization pond effluent in fresh and saline waters. *Applied and Environmental Microbiology* 68, 1122–1131.

Sjgren, R. E. and Gibson, M. J. (1981). Bacterial survival in dilute environment. *Applied and Environmental Microbiology* 41, 1331-1336.

Smith, B. and Oliver, J. D. (2006). *In situ* and *in vitro* gene expression by *Vibrio vulnificus* during entry into, persistence within, and resuscitation from the viable but nonculturable state. *Applied and Environmental Microbiology* 72, 1445-1451.

Srinivasan, S., Östling, J., Charlton, T., de Nys, R., Takayama, K. and Kjelleberg, S. (1998). Extracellular signal molecule(s) involved in the carbon starvation response of marine *Vibrio* sp. strain S14. *Journal of Bacteriology* 180, 201-209.

Steinert, M., Emody, L., Amann, R. and Hacker, J. (1997). Resuscitation of viable but nonculturable *Legionella pneumophila* Philadelphia JR32 by *Acanthamoeba castellanii*. *Applied and Environmental Microbiology* 63, 2047-2053.

Tisa, L. S., Koshikawa, T. and Gerhardt, P. (1982). Wet and dry bacterial spore densities determined by buoyant sedimentation. *Applied and Environmental Microbiology* 43, 1307-1310.

Tsien, R. Y. (1998). The green fluorescent protein. Annual Review of Biochemistry 67, 509-544.

Troussellier, M., Bonnefont, J. L., Courties, C., Derrien, A., Dupray, E., Gauthier, M., Gourmelon, M., Joux, F., Lebaron, P., Martin, Y. and Pommepuy, M. (1999). Responses of enteric bacteria to environmental stresses in seawater. *Oecanologica Acta* 21, 965-981.

Tryland, I., Pommepuy, M. and Fiksdal, L. (1998). Effect of chlorination on beta-D-galactosidase activity of sewage bacteria and *Escherichia coli*. *Journal of Applied Microbiology* 85, 51-60.

Turner, K., Porter, J., Pickup, R. and Edwards, C. (2000). Changes in viability and macromolecular content of long-term batch cultures of *Salmonella typhimurium* measured by flow cytometry. *Journal of Applied Mic*robiology 89, 90-99.

Vattakaven, T., Bond, P., Bradley, G. and Munn, C. B. (2006). Differential effects of temperature and starvation on induction of the viable-but-nonculturable state in the coral pathogens *Vibrio shiloi* and *Vibrio tasmaniensis*. *Applied and Environmental Microbiology* 72, 6508-6513.

Wai, S. N., Moriya, T., Kondo, K., Misumi, H. and Amako, K. (1996). Resuscitation of *Vibrio cholerae* O1 strain TSI-4 from a viable but nonculturable state by heat shock. FEMS Microbiology Letters 136, 187-191.

Wai, S. N., Mizunoe, Y., Takade, A. and Yoshida, S. (2000). A comparison of solid and liquid media for resuscitation of starvation- and low-temperature-induced nonculturable cells of *Aeromonas hydrophila*. *Archives of Microbiology* 173, 307-310.

Wang, G. and Doyle, M. P. (1998). Survival of enterohemorrhagic *Escherichia coli* O157:H7 in water. *Journal of Food Protection* 61, 662-667.

Wang, S., Zhu, R., Peng, B., Liu, M., Lou, Y., Ye, X., Xu, Z., Liu, D. and Peng, X. (2006). Identification of alkaline proteins that are differentially expressed in an overgrowth-mediated growth arrest and cell death of *Escherichia coli* by proteomic methodologies. *Proteomics* 6, 5212-5220.

Weichart, D. H. and Kell, D. B. (2001). Characterization of an autostimulatory substance produced by *Escherichia coli*. *Microbiology* 147, 1875-1885.

Weinbauer, M. G. and Höffe M. G. (1998). Significance of viral lysis and flagellate grazing as factors controlling bacterioplankton production in a eutrophic lake. *Applied and Environmental Microbiology* 64, 432-438.

Weiss, E., Dasch, G. A., Kang, Y. H. and Westfall, H. N. (1988). Substrate utilization by *Ehrlichia sennetsu* and Ehrlichia risticii separated from host constituents by Renografin gradient centrifugation. Journal of Bacteriology 170, 5012-5017.

Whiteley, A. S., Barer, M. R. and O'Donnell, A. G. (2000). Density gradient separation of active and non-active cells from natural environments. Antonie van Leeuwenhoek 77, 173-177.

Whitesides, M. D. and Oliver, D. B. (1997). Resuscitation of *Vibrio vulnificus* from the viable but nonculturable state. *Applied and Environmental Microbiology* 64, 3025-3028.

Whitman, R. L., Nevers, M. B., Korinek, G. C. and Byappanahalli, M. N. (2004). Solar and temporal effects on *Escherichia coli* concentration at a Lake Michigan swimming beach. *Applied and Environmental Microbiology* 70, 4276–4285.

Wong, H. C. and Wang, P. (2004). Induction of viable but nonculturable state in *Vibrio parahaemolyticus* and its susceptibility to environmental stresses. *Journal of Applied Microbiology* 96, 359-366.

Wong, H. C., Shen, C. T., Chang, C. N., Lee, Y. S. and Oliver, J. D. (2004a). Biochemical and virulence characterization of viable but nonculturable cells of *Vibrio parahaemolyticus*. *Journal of Food Protection* 67, 2430-2435.

Wong, H. C., Wang, P., Chen. S. Y. and Chiu, S. W. (2004b). Resuscitation of viable but non-culturable *Vibrio parahaemolyticus* in a minimum salt medium. *FEMS Microbiology Letters* 233, 269-275.

Xu, H. S., Roberts, N., Singleton, F. L., Attwell, R. W., Grimes, D. J. and Colwell, R. R. (1982). Survival and viability of nonculturable *Escherichia coli* and *Vibrio cholerae* in estuarine and marine environment. *Microbial Ecology* 8, 313-323

Yaron, S. and Matthews, K. (2002). A reverse transcriptase-polymerase chain reaction assay for detection of viable *Escherichia coli* O157:H7: investigation of specific genes. *Journal of Applied Microbiology* 92, 633-640.

Yukselen, M. A., Calli, B., Gokyay, O. and Saatci, A. (2003). Inactivation of coliform bacteria in Black Sea waters due to solar radiation. *Environment International* 29, 45-50.

Zaafrane, S., Maatouk, K., Gauthier, J. M. and Bakhrouf, A. (2004). Influence des conditions de culture préalables et de la présence du gène *rpoS* pour la survie de *Salmonella typhimurium* en eau de mer exposée à la lumière solaire. *Canadian Journal of Microbiology* 50, 341-350.

Zambrano, M. M., Siegele, D. A., Almiron, M., Tormo, A. and Kolter, R. (1993). Microbial competition: *Escherichia coli* mutants that take over stationary phase cultures. *Science* 259, 1757-1760.

Zambrano, M. M. and Kolter, R. (1996). GASPing for life in stationary phase. *Cell* 86, 181-184.

Zinser, E. R. and Kolter, R. (1999). Mutations enhancing amino acid catabolism confer a growth advantage in stationary phase. *Journal of Bacteriology* 181, 5800-5807.

Zinser, E. R. and Kolter, R. (2000). Prolonged stationary-phase incubation selects for *lrp* mutations in *Escherichia coli* K-12. *Journal of Bacteriology* 182, 4361-4365.

Zweifel, U. L. and Hagström, A. (1995). Total counts of marine bacteria include a large fraction of non-nucleoid-containing bacteria (ghosts). *Applied and Environmental Microbiology* 61, 2180-2185.

In: Microbial Ecology Research Trends
Editor: Thijs Van Dijk, pp. 139-154

ISBN: 978-1-60456-179-1

Chapter VI

Last Insights into the Role of Aquaporins in the Alleviation of Osmotic Stress by Arbuscular Mycorrhizal Symbiosis

Juan Manuel Ruiz-Lozano*[1] *and Ricardo Aroca
Departamento de Microbiología del Suelo y Sistemas Simbióticos, Estación Experimental del Zaidín (CSIC), Prof. Albareda, 1. 18008 Granada. Spain

Abstract

In this chapter we summarize the current information regarding alteration of aquaporin-encoding genes by the arbuscular mycorrhizal (AM) symbiosis, trying to correlate the modulation of aquaporin genes with the enhanced tolerance to water deficit conferred by the AM symbiosis. The results obtained so far show that the effects of the symbiosis on PIP gene expression depends on the own intrinsic properties of the osmotic stress. Under drought stress conditions, the AM symbiosis usually decreases or anticipates the decrease of PIP gene expression. Under salt stress, the trend is just the opposite since the AM symbiosis enhanced the expression of most of the PIP genes analyzed. The regulation of PIP gene expression under cold stress is less evident since one of the genes analyzed was down-regulated by the AM symbiosis, another was up-regulated and two genes were not affected by the symbiosis under such conditions. It seems also that the effects of the AM symbiosis on PIP gene expression depends on the endogenous levels of ABA in the host plant. In any case, the induction or inhibition of particular aquaporins by AM symbiosis should result in a better regulation of plant water status and contribute to the global plant resistance to the stressful conditions as evidenced by their better growth and water status under conditions of water deficit.

---

1 Corresponding autor: Dr. Juan Manuel Ruiz-Lozano.Departamento de Microbiología del Suelo y Sistemas Simbióticos, Estación Experimental del Zaidín (CSIC), Prof. Albareda, 1. 18008 Granada. Spain.Telph. +34 958 181600, Fax: + 34 958 129600 e-mail: juanmanuel.ruiz@eez.csic.es.

Keywords: *aquaporin, arbuscular mycorrhizal symbiosis, osmotic stress, tolerance.*

Introduction

Plants in nature are exposed to several stress conditions that retard their development and diminish their production, the commonest being drought, salinity and cold (Seki et al., 2003). All of those stresses share a common osmotic component since they cause a dehydration of plant tissues. Thus we refer to them as osmotic stresses. The dehydration caused by those stresses is a consequence of the imbalance between the water lost in the leaves and the water taken up by roots (Aroca et al., 2001). Indeed, drought caused by osmotic stresses is one of the most common environmental stress factors experienced by soil plants. It interferes with both normal development and growth and has a major adverse effect on plant survival and productivity (Kramer and Boyer, 1997; Bray, 2004). There is broad consensus that climate change continues to occur and that stresses from climatic extremes will continue, and possibly increase, and thus impose significant difficulties to plant and crop growth in many parts of the world. These difficulties will be particularly pronounced in currently semi-arid agricultural zones and/or under conditions of irrigation that often exacerbate soil salinization (Araus et al. 2003; Denby and Gehring, 2005).

The negative water potential in drying or saline soils obliges plants to face with the problem of acquiring sufficient amount of water (Ouziad et al. 2006), a process in which aquaporins participate (Luu and Maurel, 2005). Aquaporins are water channel proteins that facilitate and regulate the passive movement of water molecules down a water potential gradient (Kruse et al., 2006). These proteins belong to the large major intrinsic protein (MIP) family of transmembrane proteins and are represented in all kingdoms (Maurel, 1997). Two major classes of plant aquaporins, located in the plasma membrane (PIPs) or tonoplast (TIPs), respectively, have been identified so far. Other two classes of plant aquaporins are the homologues to the soybean Nodulin-26 aquaporin (NIPs) and the small basic intrinsic proteins (SIPs) (Johanson et al., 2001). The localization and function of SIPs are not clear at the moment (Luu and Maurel, 2005), although the membrane of endoplasmic reticulum seems to contain SIPs (Ishikawa et al., 2005).

The discovery of aquaporins in plants has caused a significant change in the understanding of plant water relations. In recent years, much effort has been concentrated on investigating the function and regulation of aquaporins. High levels of aquaporin expression were shown in tissues with high water fluxes across membranes, e.g. in fast growing regions, in shoots and leaves, but also in roots where water uptake occurs (Otto and Kaldenhoff, 2000). Thus aquaporins seem to play a specifically important role in controlling transcellular water transport in plant tissues (Javot and Maurel, 2002). However, the relationship that exists between aquaporins and plant responses to water deficit still remains elusive and with contradictory results (Aharon et al., 2003; Lian et al., 2004). In addition, although many aquaporins are highly selective for water, uptake experiments with *Xenopus laevis* oocytes clearly showed certain aquaporins to be permeable to small solutes such as glycerol, urea, amino acids, CO_2 and/or NH_3/NH_4 or even small peptides and ions (Uehlein et al., 2003, 2007), which opens many questions about the physiological roles of aquaporins.

Most terrestrial plants can establish a symbiotic association with a group of soil fungi called arbuscular mycorrhizal (AM) fungi. The AM symbiosis is present in all natural ecosystems, even in those affected by adverse environmental conditions (Smith and Read, 1997). Several eco-physiological studies investigating the role of AM symbiosis in protection against drought stress have demonstrated that the symbiosis often results in altered rates of water movement into, through and out of the host plants, with consequences on tissue hydration and plant physiology (for reviews see Augé, 2001, 2004; Ruiz-Lozano, 2003). Thus, it is accepted that AM symbiosis can protect host plants against the detrimental effects of water deficit and that the contribution of the AM symbiosis to plant drought tolerance results from a combination of physical, nutritional and cellular effects (Ruiz-Lozano, 2003). Studies carried out so far have suggested several mechanisms by which the AM symbiosis can alleviate drought stress in host plants. The most important are: direct uptake and transfer of water through the fungal hyphae to the host plant (Hardie, 1985; Ruiz-Lozano and Azcón, 1995; Marulanda et al., 2003), better osmotic adjustment of AM plants (Augé et al., 1992; Ruiz-Lozano et al., 1995; Kubikova et al., 2001), enhancement of plant gas exchange (Augé et al., 1992; Ruiz-Lozano et al., 1995, Goicoechea et al., 1997; Green et al., 1998), changes in soil water retention properties (Augé et al., 2001) and protection against the oxidative damage generated by drought (Ruiz-Lozano et al., 1996, 2001; Porcel et al., 2003; Porcel and Ruiz-Lozano, 2004).

The AM system is an excellent example for the extensive morphological alterations that plant root cells undergo in order to accommodate the presence of symbionts. Since most of the mycorrhiza-induced changes in plant root cells concern cytoplasmic or vacuolar membrane systems, a variation of expression patterns concerning genes that encode membrane-associated proteins such as aquaporins can be expected (Krajinski et al., 2000). In the following section we summarize the current information regarding alteration of aquaporin-encoding genes by the AM symbiosis trying to correlate this modulation of aquaporin genes with the enhanced tolerance to water deficit conferred by the AM symbiosis.

Regulation of Aquaporins by AM Symbiosis in Relation to Water-Deficit Stress Tolerance

The literature on aquaporins modulation by AM symbiosis has risen significantly in the last decade as reviewed recently by Uehlein et al. (2007). The first report on the modulation of aquaporin genes by AM symbiosis was provided by Roussel et al. (1997) followed by Krajinski et al. (2000), who found mycorrhiza-induced expression of TIP aquaporins in parsley and *Medicado truncatula*, respectively. Krajinski et al. (2000) related the changes in aquaporin gene expression to the changes in plant roots due to fungal colonization. In fact, during AM formation the plant plasma membrane extends to form a novel periarbuscular membrane, which closely surrounds the fungal hyphae resulting in an estimated 3- to 10-fold increase in the outer plant cell surface (Bonfante and Perotto, 1995; Gianinazzi-Pearson, 1996). It was proposed that the up-regulation of aquaporins by the AM symbiosis probably optimizes nutrient and water exchange between both symbiotic partners (Krajinski et al., 2000). However, the studies by Roussel et al. (1997) and Krajinski et al. (2000) were carried out under well-watered conditions and they did not test the expression of the aquaporin gene in AM plants under water deficit conditions.

Several aquaporin-encoding genes have been shown to be up-regulated in ectomycorrhizal poplar plants, and this was correlated with an increased water transport capacity of mycorrhizal poplar roots (Marjanovic et al., 2005). A PIP and a NIP aquaporin genes from *Medicago truncatula* were also shown to be induced by mycorrhization, while other four aquaporin genes analyzed did not change their expression patter as consequence of mycorrhization (Uehlein et al., 2007). Authors of this work related the mycorrhiza-induced change in expression of the two genes with physiological changes in the plant roots, i.e. the symbiotic exchange processes located at the pariarbuscular membrane (Uehlein et al., 2007). In contrast to the induction of aquaporin gene expression by mycorrhization, Ouziad et al. (2006) showed a decrease in the expression of PIP and TIP aquaporins by mycorrhizal colonization and salt stress in tomato plants.

The effects of reduced expression of the PIP aquaporin-encoding gene *NtAQP1* were investigated in mycorrhized *NtAQP1*-antisense tobacco plants under both, drought stress and well-watered conditions (Porcel et al., 2005). The objectives were to elucidate whether or not the impairment in *NtAQP1* gene expression affected the AM fungal colonization pattern and to find out if such impairment had any effect on the symbiotic efficiency of AM fungi. Reduction of *NtAQP1* expression had no effect on the colonization of the plant root by two AM fungi, suggesting that either *NtAQP1* function is irrelevant for the process of root colonization or that the impairment in *NtAQP1* gene expression has been compensated by changing the abundance or the activity of other aquaporins (Eckert et al., 1999; Johansson et al., 2000). In contrast, when Porcel et al. (2005) measured the symbiotic efficiency of two AM fungi (in terms of plant biomass production), they observed that under drought stress, mycorrhizal wildtype plants grew faster than mycorrhizal *NtAQP1* antisense plants. This indicates that the symbiotic efficiency of both AM fungi was greater with wildtype than with antisense plants and that the water transport meditated by NtAQP1 seems to be important for the efficiency of the symbiosis under drought stress conditions (Porcel et al., 2005). This may be related to the fact that *NtAQP1* allow CO_2 passage and is involved in plant growth promotion (Uehlein et al., 2003).

The results obtained in tobacco raised the question of what happened with aquaporin genes in AM plants when subjected to water deficit. In fact, mechanisms of osmotic adjustment and modulation of tissue hydraulic conductivity are required to maintain tissue water potential under water deficit conditions. Such mechanisms, which regulate water flux, are likely to be mediated, in part, by aquaporins (Maurel, 1997). Since aquaporins are regulated both at transcriptional and activity levels (Martre et al., 2002), we have studied whether the expression of aquaporin-encoding genes in roots is altered by the AM symbiosis as a mechanism to enhance host plant tolerance to water deficit. To achieve this, genes encoding plasma membrane aquaporins (PIPs) from soybean and lettuce were cloned and their expression pattern studied, in AM and nonAM plants cultivated under well-watered or drought stress conditions (Porcel et al., 2006). The starting hypothesis was that if AM fungi can transfer water to the root of the host plants, it is expected that the plant must increase its permeability for water and that aquaporin genes should be up-regulated in order to allow a higher rate of transcellular water flow (Javot and Maurel, 2002).

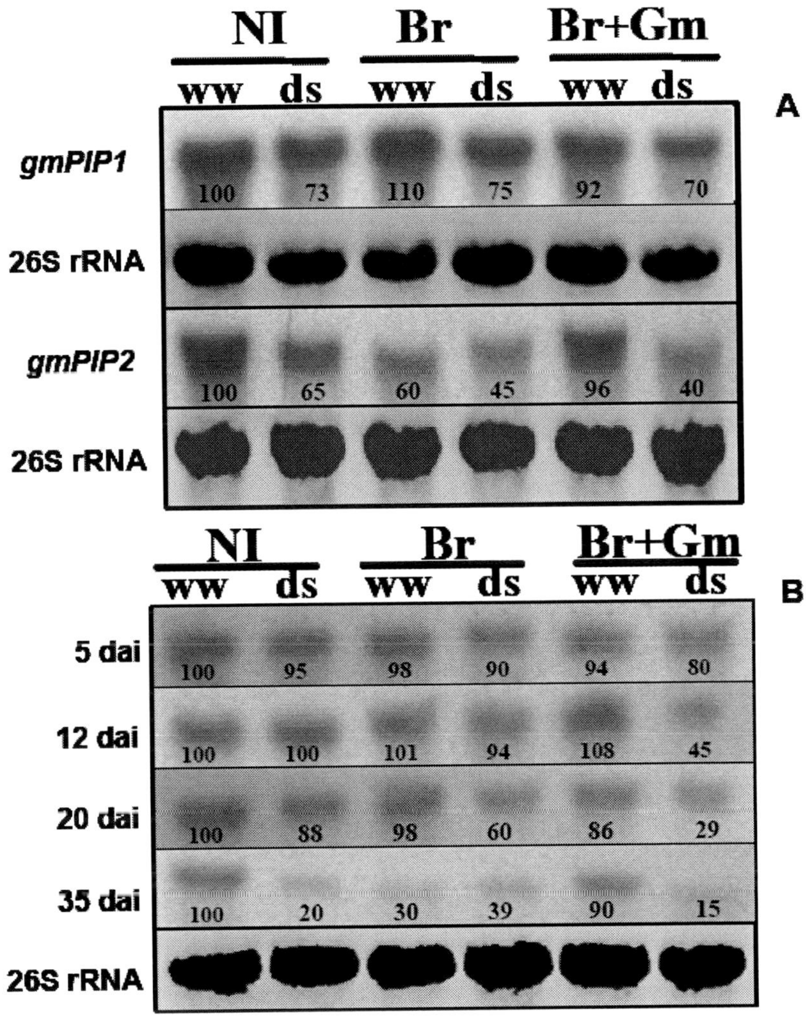

Figure 1. (A) Northern blot of total RNA (15 μg) from soybean roots using *gmPIP1* and *gmPIP2* gene probes. (B) Northern blot of total RNA (15 μg) from soybean roots harvested 5, 12, 20 or 35 days after inoculation (dai) using *gmPIP2* gene probe. Treatments are designed as NI, noninoculated controls; Br, *Bradyrhizobium japonicum*; Br+Gm, *B. japonicum* plus *G. mosseae*. Plants were either well-watered (ww) or drought stressed (ds) for 10 days. The percentage of gene expression is indicated by numbers close to each northern. The lower panel shows a representative example of the amount of 26S rRNA loaded for each treatment (methylene blue staining). Reproduced from Porcel et al. (2006), with kind permission of Springer Science and Business Media.

In contrast, to the above hypothesis, results obtained showed that the PIP genes studied were down-regulated both in soybean (Figure 1A) and lettuce (Figure 2) under drought stress and that such down-regulation was even more severe in plants colonized by *G. mosseae* than in nonAM plants (Porcel et al., 2006). A similar result was obtained by Ouziad et al. (2006) regarding the expression of PIP and TIP genes in roots of AM tomato plants subjected to salt stress. When the expression of *GmPIP2* gene from soybean was analyzed in a time-course (Figure 1B), it was clearly visible that AM plants already down-regulated that gene significantly at 5 days after inoculation (dai) and 12 dai, while both nonAM control plants still maintained *GmPIP2* gene expression almost unaltered. At 20 dai, the more intense down-regulation of that gene in AM plants than in both nonAM plants was still clearly visible. Finally, at 35 days all treatments had the same level of *GmPIP2* gene expression. This effect of the AM symbiosis anticipating the down-regulation of *GmPIP2* gene may have a physiological importance to help AM plants to cope with drought stress. In fact, according to Aharon et al. (2003), the over-expression of a PIP aquaporin in transgenic tobacco improves plant vigour under favourable growth conditions, but the over-expression of such PIP gene has no beneficial effect under salt stress, and has even negative effect during drought stress, causing fast wilting. A similar result has been obtained more recently by Jang et al. (2007) regarding two different PIP aquaporin genes in *Arabidopsis* and tobacco under dehydration conditions. Hence, the decreased expression of plasma membrane aquaporin genes during drought stress in AM plants can be a regulatory mechanism to limit the water lost from the cells (Barrieu et al., 1999). In support of this hypothesis data on leaf Ψ and relative water content (RWC) showed that AM plants (soybean and lettuce) had higher leaf Ψ and water content than nonAM plants (Porcel et al., 2006).

Data obtained with lettuce plants also colonized by *G. mosseae* point in the same direction (Figure 2), namely that under drought stress conditions there is a higher down-regulation of the PIP genes studied (and also at the protein level, as revealed by western blot) in AM plants than in nonAM plants. In contrast to *G. mosseae*, plants colonized by *G. intraradices* do not exhibit such down-regulation of PIP gene expression or protein accumulation. The expression of PIP genes under drought stress in these plants is similar to control nonAM plants.

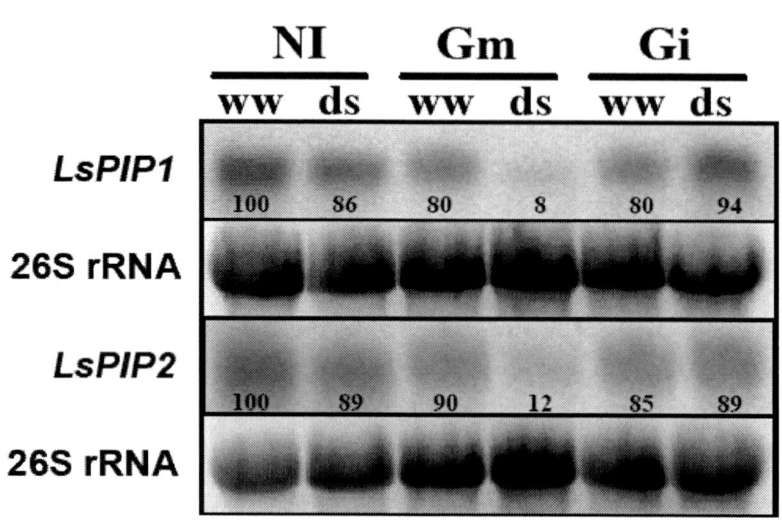

Figure 2. Northern blot of total RNA (15 μg) from lettuce roots, using *lsPIP1* and *lsPIP2* gene probes. Treatments are designed as NI, noninoculated controls; Gm, *Glomus mosseae* and Gi, *Glomus intraradices*. Plants were either well-watered (ww) or drought stressed (ds) for 10 days. The percentage of gene expression is indicated by numbers close to each northern. The lower panel shows the amount of 26S rRNA loaded for each treatment (methylene blue staining). Reproduced from Porcel et al. (2006), with kind permission of Springer Science and Business Media.

The exact reason for the different influence of *G. mosseae* and *G. intraradices* on lettuce *PIP* gene expression is not known. However, in a previous study, also with lettuce, we evaluated the ability of six AM fungal species, including *G. mosseae* and *G. intraradices*, to enhance the amount of soil water uptake by these plants (Marulanda et al., 2003). The study demonstrated that there were substantial differences among the six AM fungi used. One of the most efficient fungi stimulating water uptake by plants was *G. intraradices*, while *G. mosseae* showed a reduced ability to improve plant water uptake.

This may suggest that the strategy of both fungi to protect the host plant against water deficit is different. *G. intraradices* seems to have an important capacity to enhance the rate of water uptake by lettuce roots. This means that the water movement in these roots must be enhanced and thus, the root water permeability must also increase, maybe by maintaining high levels of PIP aquaporin gene expression as we observe in this study. Contrarily, *G. mosseae* seems to direct its strategy for plant protection against water deficit toward the conservation of the water existing in the plant and by that reason down-regulates the expression of PIP genes. Such down-regulation of PIP genes has been interpreted as a mechanism to decrease membrane water permeability and to allow cellular water conservation (Yamada et al., 1995; Smart et al., 2001). In any case, both strategies seems to protect the host plant in a similar way since lettuce plants had similar RWC and leaf Ψ regardless of the fungus colonizing their roots.

To further illustrate the complexity of the response of aquaporin genes to AM fungi we analyzed the responses of mycorrhizal lettuce plants (colonized by the same isolate of *G. intraradices*) to salt stress (Jahromi et al., 2007). Results showed that in the absence of salinity the expression of *LsPIP1* and *LsPIP2* genes was inhibited by mycorrhization (Figure 3), which agrees with the previous findings on these aquaporin genes (Porcel et al., 2006). Under saline conditions mycorrhizal plants maintained almost unaffected the expression of *LsPIP2* gene while up-regulated the expression of *LsPIP1* gene, mainly at 100 mM NaCl. This last result is just the opposite of that obtained for the same gene under drought stress conditions. Hence, these results clearly illustrates that the same aquaporin gene responds differently to each AM fungus analyzed and that the response depend also on the intrinsic characteristics of the osmotic stress applied. This highlights the complex regulation of aquaporin genes in response to the AM symbiosis (Jahromi et al., 2007).

Additional examples of such a complexity come from a study with mycorrhizal *Phaseolus vulgaris* plants (Aroca et al., 2007). In this study the expression of four PIP aquaporin genes from *P. vulgaris* (Aroca et al., 2006) was analyzed in mycorrhizal and nonmycorhizal plants subjected to three different osmotic stresses: drought, cold or salinity. Three of these PIP genes showed differential regulation by AM symbiosis under the specific conditions of each stress applied (Figure 4). In fact, *PvPIP1;1* was slightly inhibited by *G. intraradices* under drought stress conditions, while nonmycorrhizal plants did not change its expression pattern. Cold stress inhibited its expression similarly in AM and nonAM plants. Finally, salinity raised the gene expression in both groups of plants, but the enhancement was considerably higher in AM plants. The gene *PvPIP1;2* was inhibited by the three stresses in the same way in AM and nonAM plants. In contrast, *PvPIP1;3* showed important differences in AM and nonAM plants according to the stress imposed. Under drought stress it was clearly induced in nonAM plants and inhibited in AM plants. Under cold stress the behaviour was the opposite since it was inhibited in nonAM plants and induced in AM ones.

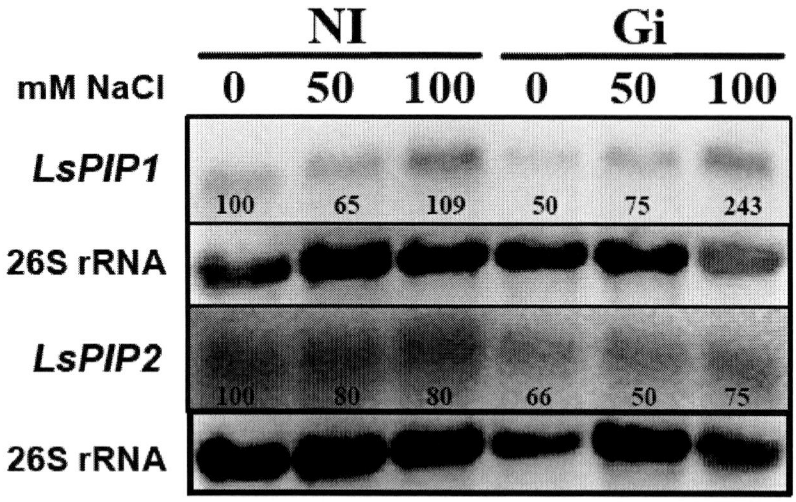

Figure 3. Northern blot of total RNA (15 μg) from lettuce roots using *LsPIP1* and *LsPIP2* gene probes. Treatments are designed as NI, noninoculated control or Gi, plants inoculated with *Glomus intraradices*. Plants were subjected to 0, 50 or 100 mM NaCl. The lower panels show the amount of 26S rRNA loaded for each treatment. Numbers close to each Northern represent the relative gene expression (after normalization to 26S rRNA) as a percentage of the value for control plants cultivated under non saline conditions. Reproduced from Jahromi et al. (2007), with kind permission of Springer Science and Business Media.

Finally, under salinity it was also induced in both groups of plants, especially in AM ones. The gene *PvPIP2;1* was induced in nonAM plants under drought stress but resulted inhibited in AM plants. The response of this gene to cold stress was not significant for any of the two plant groups and, again, the gene was considerably up-regulated under salinity, especially in AM plants.

The up- or down-regulation by drought stress of mRNAs encoding aquaporins homologues has been described in the roots of many plant species (Javot and Maurel, 2002). There are currently two opposite descriptions of the role of aquaporins in response to dehydration stress (Smart et al., 2001). The first is based on evidence that expression of some aquaporins is induced under dehydration stress (Barrieu et al., 1999; Jang et al., 2004), which is predicted to result in greater membrane water permeability and facilitated water transport. The second is based on the fact that aquaporin expression is down-regulated under dehydration stress, which should result in decreased membrane water permeability and may allow cellular water conservation (Yamada et al., 1995; Smart et al., 2001) during periods of dehydration stress.

The most interesting finding of results by Aroca et al. (2007) is that each *PIP* gene responded differently to each stress depending on the AM fungal presence. Valot et al. (2005) already found that several plasma membrane proteins were differently regulated by inoculation with *G. intraradices*, some of them were down-regulated and others were induced. Since *G. intraradices* has the capacity of altering root hydraulic properties (Marulanda et al., 2003; Khalvati et al., 2005), it is not strange that the fungus also changes *PIP* gene expression.

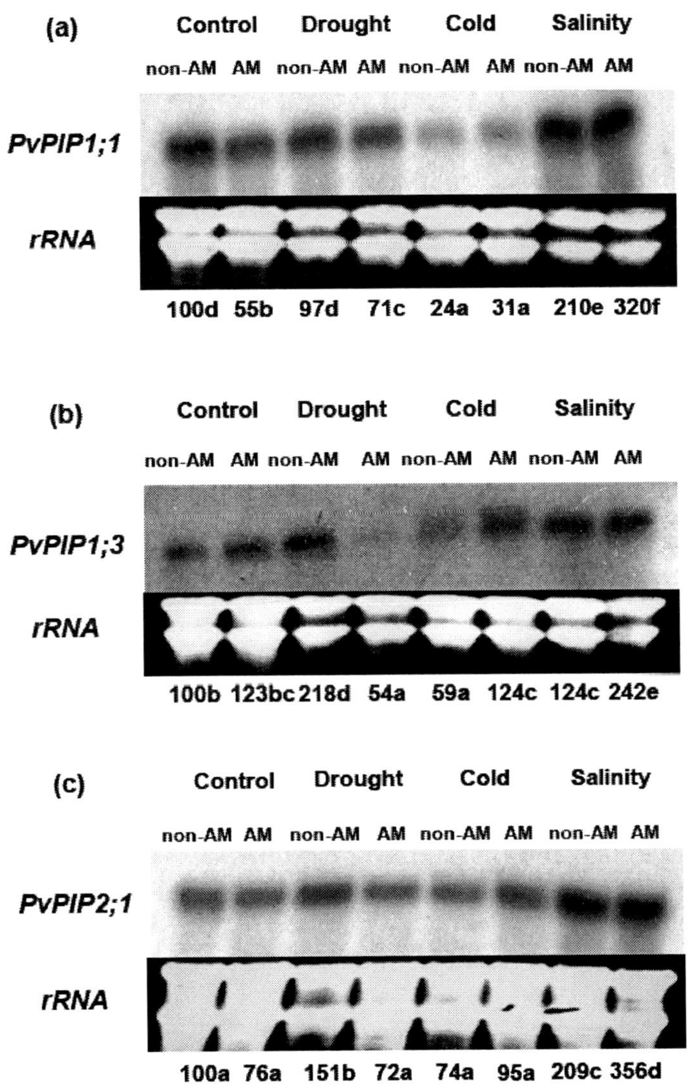

Figure 4. Northern blots analysis using 3' UTR as probes of *PvPIP1;1* (a), *PvPIP1;3* (b), and *PvPIP2;1* (c) in total RNA of *Phaseolus vulgaris* roots not inoculated (non-AM) or inoculated (AM) with AM fungi *Glomus intararadices*. Plants kept at 23°C and watered at full capacity with tap water were referred as Control. Plants kept at 23°C and subjected to no watering during 4 d were referred as Drought. Plants transferred to 4°C during 2 days and watered at full capacity with tap water were referred as Cold. Plants kept at 23°C and watered each 2 days during 6 days with 10 ml of 0.5 M NaCl solution were referred as Salinity. Quantification of the gene expression was performed by dividing the intensity value of each band by the intensity of corresponding rRNA stained with ethidium bromide. Control value of NI roots was referred as 100. Treatments with different letters are significant ($p<0.05$) different after ANOVA and Fisher LSD tests. n = 3. Reproduced from Aroca et al. (2007), with permission from New Phytologist.

It has been shown that the plant hormone ABA modulates the expression of some *PIP* genes in roots and leaves (Jang et al. 2004; Zhu et al. 2005; Aroca et al. 2006). Thus, we carried out a study with tomato and an ABA-deficient mutant (sitiens) to analyze the expression of four *PIP* aquaporin genes depending on mycorrhizal presence, exogenous ABA application and the plant ABA phenotype used (Aroca et al. 2008). In this study, we observed differential expression of some of the genes in AM and nonAM plants after ABA or drought treatments depending on the plant ABA phenotype. As examples of that, the application of exogenous ABA under well-watered conditions enhanced the expression of *SlPIP1-4* gene in root of wild-type plants (both in AM, line 4 vs line 2 and in nonAM plants, line 3 vs line 1) (Figure 5). In contrast, in *sitiens*, the application of ABA decreased the expression in nonAM roots (line 11 vs line 9) and did not affect the expression of AM roots (line 12 vs line 10). Also, drought duplicated the expression of *SlPIP1-4* in roots of wild-type plants (both AM, line 6 vs line 2 and nonAM plants, line 5 vs line 1), but in *sitiens* plants, drought decreased the expression of the gene in nonAM plants (line 13 vs line 9) and did not change the expression in AM plants (line 14 vs line 10). Similarly, the *SlPIP1-5* gene was induced by drought in roots of wild-type nonAM plants (line 5 vs line 1) and remained unchanged in wild-type AM ones (line 6 vs line 2) (Figure 5). In contrast, in *sitiens* plants, drought decreased the expression of this gene in roots of both AM (line 14 vs line 10) and nonAM plants (line 13 vs line 9).

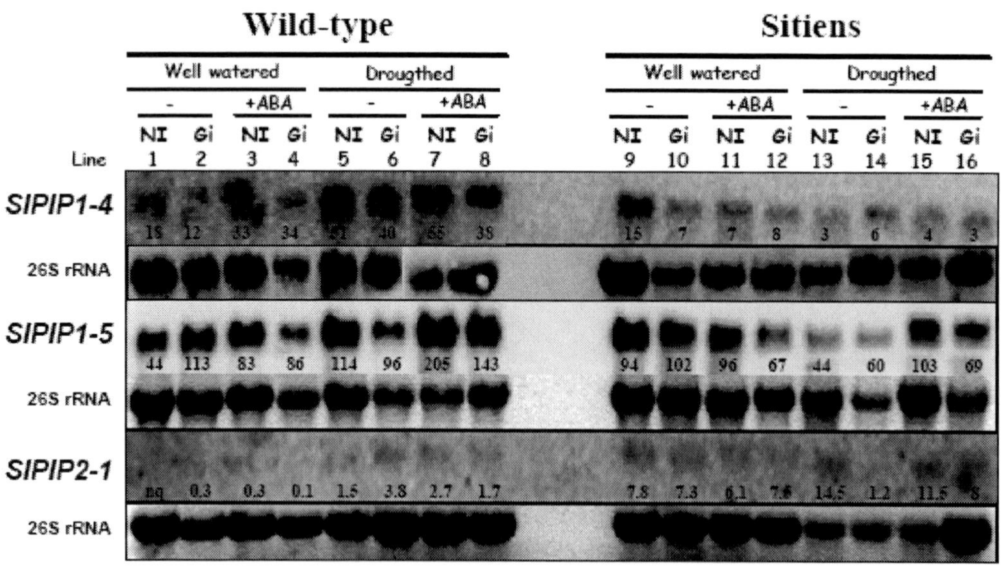

Figure 5. Northern blot of total RNA (15 µg) from tomato roots (wild-type and sitiens) using *SlPIP1-4* (Accession AF218774), *SlPIP1-5* (Accession X73848) and *SlPIP2-1* (Accession BI929127) as gene probes. Treatments are designed as NI, noninoculated controls or Gi, plants inoculated with *Glomus intraradices*. Plants were cultivated under well-watered conditions or subjected to drought stress with or without addition of exogenous ABA. The lower panels show the amount of 26S rRNA loaded for each treatment. Numbers close to each northern represent the relative gene expression after normalization to rRNA. Reproduced from Aroca et al. (2008), with kind permission of Springer Science and Business Media.

This suggests that this gene is regulated by drought, but also by ABA and needs high levels of ABA to induce its expression. If the levels of ABA are low (as in *sitiens* plants) its expression is down-regulated by drought. Finally, drought induced the expression of *SlPIP2-1* gene in roots of wild-type plants AM (line 6 vs line 2) and nonAM (line 5 vs line 1) (Figure 5). In contrast, in *sitiens* plants, drought enhanced the expression only in nonAM roots (line 13 vs line 9) and decreased the expression of this gene in AM roots (line 14 vs line 10).

The reasons for such effects are currently unknown. It may be possible that AM fungal presence can directly regulate gene expression independently of ABA, as has been evidenced for a variety of genes (for reviews see Gianinazzi-Pearson and Brechenmacher 2004; Balestrini and Lanfranco 2006). However, differences in compartmentation of ABA within the cell or tissues or differences in the rate of ABA metabolism (Wilkinson and Davies 2002; Hartung et al., 2005; Zhang et al., 2006) between AM and nonAM plants can also account for such a differential gene expression. In any case, these results showed that mycorrhization regulated differently the expression of the PIP aquaporin genes analyzed during drought stress and after exogenous ABA application and this effect was dependent on the plant phenotype studied. This agrees with results by Lian et al. (2006) who found that PIP genes in rice responded in a different way to water stress and ABA, indicating that during water deficit the regulation of PIP genes involves both ABA-dependent and ABA-independent signalling pathways. These results suggest that the AM symbiosis exerts a differential control on expression of aquaporin genes, inducing or inhibiting particular genes, and this depends on the endogenous ABA content in the host plant.

Conclusion

The results obtained so far on regulation of PIP aquaporin gene expression by the AM symbiosis show that the effects of the symbiosis on PIP gene expression depends on the own intrinsic properties of the osmotic stress. Under drought stress conditions, the AM symbiosis usually decreases or anticipates the decrease of PIP gene expression. Under salt stress, the trend is just the opposite since the AM symbiosis enhanced the expression of most of the PIP genes analyzed. The regulation of PIP gene expression under cold stress is less evident since one of the genes analyzed was down-regulated by the AM symbiosis, another was up-regulated and two genes were not affected by the symbiosis under such conditions. It seems also that the effects of the AM symbiosis on PIP gene expression depends on the endogenous levels of ABA in the host plant. In any case, the induction or inhibition of particular aquaporins by AM symbiosis should result in a better regulation of plant water status and contribute to the global plant resistance to the stressful conditions (Yamada et al., 1997; Barrieu et al., 1999; Jang et al., 2004) as evidenced by their better growth and water status under conditions of water deficit.

Acknowledgments

This work was carried out in the frame of a CICYT-FEDER Project (AGL2005-01237).

References

Aharon R, Shahak Y, Wininger S, Bendov R, Kapulnik Y and Galili G. (2003). Overexpression of a plasma membrane aquaporins in transgenic tobacco improves plant vigour under favourable growth conditions but not under drought or salt stress. *Plant Cell* 15, 439-447.

Araus JL, Bort J, Steduto P, Villegas D and Royo C. (2003). Breeding cereals for Mediterranean conditions: ecophysiology clues for biotechnology application. *Annals of Applied Biology* 142,129–141.

Aroca R, Alguacil MM, Vernieri P and Ruiz-Lozano JM. (2008). Plant responses to drought stress and exogenous ABA application are differently modulated by mycorrhization in tomato and an ABA-deficient mutant (*sitiens*). *Microbial Ecology* (In press).

Aroca R, Amodeo G, Fernádez-Illescas S, Herman EH, Chaumont F and Chrispeels MJ. (2005). The role of aquaporins and membrane damage in chilling and hydrogen peroxide induced changes in the hydraulic conductance of maize roots. *Plant Physiology* 137, 341-353.

Aroca R, Ferrante A, Vernieri P and Chrispeels MJ. (2006). Drought, abscisic acid, and transpiration rate effects on the regulation of PIP aquaporin gene expression and abundance in *Phaseolus vulgaris* plants. *Annals of Botany* 98, 1301-1310.

Aroca R, Porcel R and Ruiz-Lozano JM. (2007). How does arbuscular mycorrhizal symbiosis regulate root hydraulic properties and plasma membrane aquaporin in *Phaseolus vulgaris* under drought, cold or salinity stresses? New Phytologist 173: 808-816.

Aroca R, Tognoni F, Irigoyen JJ, Sánchez-Díaz M and Pardossi A. (2001). Different root low temperature response of two maize genotypes differing in chilling sensitivity. *Plant Physiology and Biochemistry* 39, 1067-1073.

Augé RM. (2001). Water relations, drought and vesicular-arbuscular mycorrhizal symbiosis. *Mycorrhiza* 11, 3-42.

Augé RM. (2004). Arbuscular mycorrhizae and soil/plant water relations. *Canadian Journal Soil Sciences* 84, 373-381.

Augé RM, Foster JG, Loescher WH and Stodola AW. (1992). Symplastic sugar and free amino acid molality of Rosa roots with regard to mycorrhizal colonization and drought. *Symbiosis* 12, 1-17.

Augé RM, Stodola AW, Tims JE and Saxton AM. (2001). Moisture retention properties of a mycorrhizal soil. *Plant and Soil* 230, 87-97.

Balestrini R and Lanfranco L. (2006). Fungal and plant gene expression in arbuscular mycorrhizal symbiosis. *Mycorrhiza* 16, 509-524

Barrieu F, Marty-Mazars D, Thomas D, Chaumont F, Charbonnier M and Marty F. (1999). Desiccation and osmotic stress increase the abundance of mRNA of the tonoplast aquaporin *BobTIP26-1* in cauliflower cells. *Planta* 209, 77-86.

Bonfante P and Perotto S. (1995). Strategies of arbuscular mycorrhizal fungi when infecting host plants. *New Phytologist* 130, 3-21.

Bray EA. (2004). Genes commonly regulated by water-deficit stress in Arabidopsis thaliana. *Journal of Experimental Botany* 55, 2331-2341.

Denby K and Gehring C. (2005). Engineering drought and salinity tolerance in plants: lessons from genome-wide expression profiling in *Arabidopsis*. *Trends in Biotechnology* 23, 547-552.

Eckert M, Biela A, Siefritz F and Kaldenhoff R. (1999). New aspects of plant aquaporin regulation and specificity. *Journal of Experimental Botany* 50, 1541-1545.

Gianinazzi-Pearson V. (1996). Plant cell response to arbuscular mycorrhizal fungi: getting to the roots of the symbiosis. *Plant Cell* 8, 1871-1883.

Gianinazzi-Pearson V and Brechenmacher L. (2004). Functional genomics of arbuscular mycorrhiza: Decoding the symbiotic cell programme. *Canadian Journal of Botany* 82, 1228-1234.

Goicoechea N, Antolin MC and Sánchez-Díaz M. 1997. Gas exchange is related to the hormone balance in mycorrhizal or nitrogen-fixing alfalfa subjected to drought. *Physiologia Plantarum* 100, 989-997.

Green CD, Stodola A and Augé RM. (1998). Transpiration of detached leaves from mycorrhizal and nonmycorrhizal cowpea and rose plants given varying abscisic acid, pH, calcium and phosphorus. *Mycorrhiza* 8, 93-99.

Hardie K. (1985). The effect of removal of extraradical hyphae on water uptake by vesicular-arbuscular mycorrhizal plants. *New Phytologist* 101, 677-684.

Hartung W, Schraut D and Jiang F. (2005). Physiology of abscisic acid (ABA) in roots under stress - a review of the relationship between root ABA and radial water adn ABA flows. *Australian Journal of Agricultural Research* 56, 1253-1259.

Ishikawa F, Suga S, Uemura T, Sato MH and Maeshima M. (2005). Novel type aquaporin SIPs are mainly localized to the ER membrane and show cell-specific expression in *Arabidopsis thaliana*. *FEBS Letters* 579, 5814-5820.

Jahromi F, Aroca R, Porcel R and Ruiz-Lozano JM. (2007). Influence of salinity on the *in vitro* development of *Glomus intraradices* and on the *in vivo* physiological and molecular responses of mycorrhizal lettuce plants. *Microbial Ecology* (Published online: DOI: 10.1007/s00248-007-9249-7).

Jang JY, Kim DG, Kim YO, Kim JS andKang H. (2004). An expression analysis of a gene family encoding plasma membrane aquaporins in response to abiotic stresses in *Arabidopsis thaliana*. *Plant Molecular Biology* 54, 713-725.

Jang JY, Lee SH, Rhee JY, Chung GC, Ahn SJ and Kang H. (2007). Transgenic *Arabidopsis* and tobacco plants overexpressing an aquaporin respond differently to various abiotic stresses. *Plant Molecular Biology* 64, 621-632.

Javot H and Maurel C. (2002). The role of aquaporins in root water uptake. *Annals of Botany* 90, 301-313.

Johanson U, Karlsson M, Johanson I, Gustavsson S, Sjövall S, Fraysse L, Weigh AR and Kjellbom P. (2001). The complete set of genes encoding major intrinsic proteins in *Arabidopsis* provides a framework for a new nomenclature for major intrinsic proteins in plants. *Plant Physiology* 126, 1358-1369.

Johansson I, Karlsson M, Johansson U, Larsson C and Kjellbom P. (2000). The role of aquaporins in cellular and whole plant water balance. *Biochimica and Biophysica Acta* 1465, 324-342

Khalvati MA, Hu Y, Mozafar A and Schmidhalter U. (2005). Quantification of water uptake by arbuscular mycorrhizal hyphae and its significance for leaf growth, water relations, and gas exchange of barley subjected to drought stress. Plant Biology 7: 706-712.

Krajinski F, Biela A, Schubert D, Gianinazzi-Pearson V, Kaldenhoff R and Franken P. (2000). Arbuscular mycorrhiza development regulates the mRNA abundance of *Mtaqp1* encoding a mercury-insensitive aquaporin of *Medicago truncatula*. *Planta* 211, 85-90.

Kramer PJ and Boyer JS. (1997). Water relations of plants and soils. Academic Press. San Diego.

Kruse E, Uehlein N and Kaldenhoff R. (2006). The aquaporins. *Genome Biology* 7, 206.

Kubikova E, Moore JL, Ownlew BH, Mullen MD and Augé RM. (2001). Mycorrhizal impact on osmotic adjustment in *Ocimum basilicum* during a lethal drying episode. *Journal of Plant Physiology* 158, 1227-1230.

Lian HL, Yu X, Lane D, Sun WN, Tang ZC and Su WA. (2006). Upland rice and lowland rice exhibit different PIP expression under water deficit and ABA treatment. *Cell Research* 16, 651-660

Lian HL, Yu X, Ye Q, Ding XS, Kitagawa Y, Swak SS, Su WA and Tang ZC. (2004). The role of aquaporin RWC3 in drought avoidance in rice. *Plant and Cell Physiology* 15, 481. 489.

Luu DT and Maurel C. (2005). Aquaporins in a challenging environment: molecular gears for adjusting plant water status. *Plant, Cell and Environment* 28, 85-96.

Marjanovic Z, Uehlein N, Kaldenhoff R, Zwiazek JJ, Weiβ M, Hampp R and Nehls U. (2005). Aquaporins in poplar: what a difference a symbiont makes! *Planta* 222, 258-268.

Martre P, Morillon R, Barrieu F, North GB, Nobel PS and Chrispeels MJ. (2002). Plasma membrane aquaporins play a significant role during recovery from water deficit. *Plant Physiology* 130, 2101-2110.

Marulanda A, Azcón R and Ruiz-Lozano JM. (2003). Contribution of six arbuscular mycorrhizal fungal isolates to water uptake by *Lactuca sativa* L. plants under drought stress. *Physiologia Plantarum* 119, 526-533.

Maurel C. (1997). Aquaporins and water permeability of plant membranes. *Annual Review of Plant Physiology and Plant Molecular Biology* 48, 399-429.

Otto B and Kaldenhoff R. (2000). Cell-specific expression of the mercury-insensitive plasma-membrane aquaporin NtAQP1 from *Nicotiana tabacum*. *Planta* 211, 167-172.

Ouziad F, Wilde P, Schmelzer E, Hildebrandt U and Bothe H. (2006). Analysis of expression of aquaporins and Na+/H+ transporters in tomato colonized by arbuscular mycorrhizal fungi and affected by salt stress. *Environmental and Experimental Botany* 57, 177-186.

Porcel R, Aroca R, Azcón R and Ruiz-Lozano JM. (2006). PIP aquaporin gene expression in arbuscular mycorrhizal *Glycine max* and *Lactuca sativa* plants in relation to drought stress tolerance. *Plant Molecular Biology* 60, 389-404.

Porcel R, Barea JM and Ruiz-Lozano JM. (2003). Antioxidant activities in mycorrhizal soybean plants under drought stress and their possible relationship to the process of nodule senescence. *New Phytologist* 157, 135-143.

Porcel R, Gómez M, Kaldenhoff R and Ruiz-Lozano JM. (2005). Impairment of *NtAQP1* gene expression in tobacco plants does not affect root colonization pattern by arbuscular mycorrhizal fungi but decreases their symbiotic efficiency under drought. *Mycorrhiza* 15, 417-423.

Porcel R and Ruiz-Lozano JM. (2004). Arbuscular mycorrhizal influence on leaf water potential, solute accumulation and oxidative stress in soybean plants subjected to drought stress. *Journal of Experimental Botany* 55, 1743-1750.

Roussel H, Bruns S, Gianinazzi-Pearson V, Hahlbrock K and Franken P. (1997). Induction of a membrane intrinsic protein-encoding mRNA in arbuscular mycorrhiza and elicitor-stimulated cell suspension cultures of parsley. *Plant Science* 126, 203-210.

Ruiz-Lozano JM. (2003). Arbuscular mycorrhizal symbiosis and alleviation of osmotic stress. New perspectives for molecular studies. *Mycorrhiza* 13, 309-317.

Ruiz-Lozano JM and Azcón R. (1995). Hyphal contribution to water uptake in mycorrhizal plants as affected by the fungal species and water status. *Physiologia Plantarum* 95, 472-478.

Ruiz-Lozano JM, Azcón R and Gómez M. (1995). Effects of arbuscular mycorrhizal *Glomus* species on drought tolerance: physiological and nutritional plant responses. *Applied and Environmental Microbiology* 61, 456-460.

Ruiz-Lozano J., Azcón R and Palma JM. (1996). Superoxide dismutase activity in arbuscular-mycorrhizal *Lactuca sativa* L. plants subjected to drought stress. *New Phytologist* 134, 327-333.

Ruiz-Lozano JM, Collados C, Barea JM and Azcón R. (2001). Arbuscular mycorrhizal symbiosis can alleviate drought-induced nodule senescence in soybean plants. *New Phytologist* 151, 493-502.

Seki M, Kamei A, Yamaguchi-Shinozaki K and Shinozaki K. (2003). Molecular responses to drought, salinity and frost: common and different paths for plant protection. *Current Opinion in Biotechnology* 14, 194-199.

Smart LB, Moskal WA, Cameron KD and Bennett AB. (2001). Mip genes are down-regulated under drought stress in *Nicotiana glauca*. *Plant and Cell Physiology* 42, 686-693.

Smith SE and Read DJ. (1997). Mycorrhizal symbiosis. San Diego, CA; Academic Press.

Uehlein N, Fileschi K, Eckert M, Bienert GP, Bertl A and Kaldenhoff R. (2007). Arbuscular mycorrhizal symbiosis and plant aquaporin expression. *Phytochemistry* 68, 122-129.

Uehlein N, Lovisolo C, Siefritz F and Kaldenhoff R. (2003). The tobacco aquaporin NtAQP1 is a membrane CO_2 pore with physiological functions. *Nature* 425, 734-737.

Valot B, Dieu M, Recorbet G, Raes M, Gianinazzi S and Dumas-Gaudot E. (2005). Identification of membrane-associated proteins regulated by the arbuscular mycorrhizal symbiosis. *Plant Molecular Biology* 59, 565-580.

Wilkinson S and Davies WJ. (2002). ABA-based chemical signalling: the co-ordination of responses to stress in plants. *Plant, Cell and Environment* 25, 195-210.

Yamada S, Katsuhara M, Kelly WB, Michalowski CB and Bonhert HJ. (1995). A family of transcripts encoding water channel proteins: tissue-specific expression in common ice plant. *Plant Cell* 7, 1129-1142.

Zhang J, Jia W, Yang J and Ismail AM. (2006). Role of ABA in integrating plant responses to drought and salt stresses. *Field Crop Research* 97, 111-119.

Zhu C, Schraut D, Hartung W and Schäffner AR. 2005. Differential responses of maize MIP genes to salt stress and ABA. *Journal of Experimental Botany* 56, 2971-2981.

In: Microbial Ecology Research Trends
Editor: Thijs Van Dijk, pp. 155-168

ISBN: 978-1-60456-179-1

Chapter VII

External and Internal Spines in Plants Insert Pathogenic Microorganisms into Herbivore's Tissues for Defense

Simcha Lev-Yadun[1] *and Malka Halpern*[1]
Department of Biology, Faculty of Science and Science Education,
Haifa University - Oranim, Tivon 36006, Israel

Abstract

Recently, it has been proposed that plants use pathogenic aerobic and anaerobic bacteria, as well as pathogenic fungi that harbor spines, thorns and prickles, for defense from herbivores, especially vertebrates. These sharp defensive organs introduce various pathogenic agents into the body of the herbivores through the outer defensive layer of the skin in a type of biological warfare. This paper provides a review of our data regarding the various bacterial taxa found on spines, as well as the medical literature regarding infections by bacteria and fungi related to spine injuries. In addition, we propose that many plant species which have internal microscopic defensive sharp structures, such as raphides (calcium oxalate needles) and sharp silica bodies, may also introduce pathogenic microorganisms into the tissues of herbivores through the microscopic wounds that these defensive structures induce, especially in the mouth and digestive system. The internal microscopic spines can wound large herbivores and also insects and other very small herbivores that usually attack plants in between spines and thorns. The large spines and the microscopic sharp structures may introduce not only the microorganisms that harbor them, but also those that pre-exist on the skin or inside the digestive system of the herbivores and on the surface of non-spiny consumed plant parts. A majority of the spiny plants visually advertise their spiny nature, a characteristic known as aposematism (warning coloration). Since the pathogenic microorganisms may sometimes be much more dangerous than the physical wounds inflicted by the spines, we propose that the cooperation with the pathogenic microorganisms contributes to the evolution of aposematism in spiny plants (including those with raphides and silica needles) and possibly also in spiny animals.

---

1 Corresponding authors: E-mail: levyadun@research.haifa.ac.il , E-mail: mhalpern@research.haifa.ac.il.

Introduction

Plants suffer from ceaseless pressure of herbivory by various types of herbivores (Crawley, 1983), thereby creating a permanent evolutionary arms race between plants and their herbivores. On an evolutionary time scale, plants acquire better defenses and herbivores either partly or fully overcome them (Cornell and Hawkins, 2003). Thorns, spines and prickles, here collectively termed spines, are a common anti-herbivory defense in thousands of plant species (Grubb, 1992). English botanical usage distinguishes between a prickle (a sharp emergence from the cortex), a spine (a sharp part of a leaf), and a thorn (a sharp branch). They are known to provide physical protection against herbivory (e.g., Janzen and Martin, 1982; Cooper and Owen-Smith, 1986; Janzen, 1986; Myers and Bazely, 1991; Grubb, 1992; Gowda, 1996; Rebollo et al., 2002) because they can wound mouths, digestive systems (Janzen and Martin, 1982; Janzen, 1986), and other body parts of herbivores.

Microorganisms seem to occupy virtually every living and non-living niche on earth. In the past few decades, plant scientists have begun to realize that all the aerial plant surfaces, including leaves, stems, flowers, and fruits, are inhabited by diverse assemblages of microorganisms, including filamentous fungi, yeasts, bacteria, and bacteriophages (Lindow and Leveau, 2002; Lindow and Brandl, 2003). This paper addresses the potential defensive nature of this phenomenon, specifically the defense by external sharp plant organs.

Pathogenic Bacteria and Thorns

In two recent publications, we showed that spines from *Phoenix dactylifera* (date palm) and thorns from *Crataegus aronia* (common hawthorn), *Sarcopoterium spinosum* (thorny burnet), and *Alhagi graecorum* (manna tree) harbor an array of pathogenic bacteria (Halpern et al., 2007a, 2007b). Spines from several date palm (*Phoenix dactylifera*) trees were sampled in the Jordan Valley in northern Israel. In addition, thorns from common hawthorn (*Crataegus aronia*) trees and two thorny shrub species, thorny burnet (*Sarcopoterium spinosum*) and manna tree (*Alhagi graecorum*), were sampled in Mount Carmel, Israel. Every typical mature individual of these trees and shrubs carries hundreds and even thousands of conspicuous aposematic spines or thorns.

Although we sampled only a small fraction of the spines and thorns found on each tree or shrub, we were able to isolate and identify various pathogenic bacterial species from these thorns, which proved it to be a common phenomenon. Bacterial isolates were selected in both aerobic and anaerobic growth conditions and identified by means of 16S rRNA gene analysis. Eighty-five bacterial isolates from the thorny trees and shrubs belonged to 25 different species, of which 15 are known to be pathogenic to animals or humans: *Clostridium perfringens*, *C. sordellii*, *C. sardiniens*, *Bacillus anthracis*, *B. cereus*, *B. circulans*, *B. thuringiensis*, *B. licheniformis*, *B. megaterium*, *B. pumilus*, *Enterococcus faecalis*, *E. faecium*, *Rahnella aquatilis*, *Shigella boydii*, and *Pantoea agglomerans* (Halpern et al., 2007a, 2007b).

Members of the *Clostridium* species, isolated and identified from spines of the date palm and thorns of the common hawthorn (Halpern et al., 2007a), are Gram positive, endospore-forming obligate anaerobic bacteria that cause infections associated with wounds. *Clostridium perfringens* is known to be a flesh-eater in that it can produce a necrotizing infection of the skeletal muscle called gas gangrene (Shimizu et al., 2002). *Clostridium sordellii* and *C. sardiniense* are considered *C. perfringens*-like strains and have been isolated from infected tissues in cases of gas gangrene (Masaki et al., 1988; el Sanousi and Musa, 1989).

The *Bacillus cereus sensu lato* group comprises three Gram-positive, spore-forming species: *B. anthracis*, *B. thuringiensis*, and *B. cereus sensu stricto* (Jensen et al., 2003). Recent molecular data suggest that all three are members of a single species, *B. cereus sensu lato* (Daffonchio et al., 2000). Spines and thorns from three of the sampled plant species (date palm, common hawthorn, and manna tree) were found to be the habitat for *B. anthracis*, *B. cereus*, and *B. thuringiensis*. *Bacillus cereus* is an opportunistic pathogen causing food poisoning manifested by diarrhea or emetic symptoms (Ivanova et al., 2003), and *B. thuringiensis* is a well-known insect pathogen (Aronson and Shai, 2001). *Bacillus anthracis* is the etiological agent of anthrax, a notorious acute fatal disease in animals (domesticated and wild, particularly herbivorous) and humans (Jensen et al., 2003). The cutaneous form of the disease is usually acquired through injured skin or mucous membranes, a typical thorn injury. Leendertz et al. (2004) reported sudden deaths of wild chimpanzees caused by *B. anthracis* in the tropical rainforest of Tay National Park, Ivory Coast. The source of the infection was not clarified and remains a mystery, leaving open the possibility that plant thorns were the cause of that incident.

Enterococcus faecalis and *E. faecium* were isolated from common hawthorn thorns (Halpern et al., 2007a). *Enterococci* are Gram-positive, aerotolerant anaerobes that can cause various complicated infections (abdominal, skin, urinary tract, and blood) (Ruoff et al., 1990). *Pantoea agglomerans*, *Rahnella aquatilis*, and *Shigella boydii* are all members of Enterobacteriaceae and were isolated from common hawthorn thorns (Halpern et al., 2007a). *Pantoea agglomerans*, a common colonizer of plant surfaces (Lindow and Leveau, 2002), has been reported in the medical literature as the cause of septic arthritis after palm spine injury (Kratz et al., 2003), as well as the cause of osteomyelitis after rose prickle injury (Vincent and Szabo, 1988). *Rahnella aquatilis* is an uncommon human pathogen, and the sources for most *Rahnella*-related human illnesses remain unknown (Oh and Tay, 1995; Janda, 2002). Most of its human infections have involved episodes of bacteraemia, urinary tract infections, post-surgical wound infections, and endocarditis. *Shigella* is a common cause of diarrheal illness that can cause food-borne outbreaks. Recently, *S. boydii* was implicated in an outbreak of food-borne illness in the USA (Chan and Blaschek, 2005) and identified as a cause of necrotizing enterocolitis (a serious gastrointestinal disease in neonates) (Sawardekar, 2005).

The published medical literature reviewed in Halpern et al. (2007a) includes medical case reports indicating that injuries from plant thorns can result in septic inflammation (Sugarman et al., 1977; Cahill and King, 1984; Vincent and Szabo, 1988; Hodes and Teferedegne, 1990; Freiberg et al., 1993; De Champs et al., 2000; Ergonul et al., 2003; Kratz et al., 2003; Pascual et al., 2003). None of the published data discussed ecological or evolutionary issues or aposematism, but rather were published only in the interests of medical practice. However, those data showed that plant thorns may regularly harbor various toxic or pathogenic bacteria. *Clostridium tetani*, for instance, is the etiological agent of tetanus, a serious disease in humans and animals that can be fatal when left untreated. Thorn injuries have been known to cause tetanus in the USA, Ethiopia, and Turkey (Hodes and Teferedegne, 1990; Ergonul et al., 2003; Pascual et al., 2003). Other inflammatory states of the musculoskeletal system have similarly been associated with specific pathogens introduced into animal tissues by various thorn injuries, such as *Pantoea agglomerans*, *Staphylococcus aureus*, *S. albus*, *Streptococcus β hemolyticus*, and *Nocardia pyarthrosis* (Sugarman et al., 1977; Cahill and King, 1984; Vincent and Szabo, 1988; Freiberg et al., 1993; De Champs et al., 2000; Kratz et al., 2003).

Pathogenic Fungi and Thorns

In their review of the medical literature, Halpern et al. (2007b) found that septic inflammation caused by plant thorn injury can result not only from bacteria, but also from thorns, spines or prickles, which can introduce pathogenic fungi into animals or humans. Dermatophytes that cause subcutaneous mycoses are unable to penetrate the skin and must be introduced into the subcutaneous tissue by a puncture wound (Willey et al., 2008).

One type of subcutaneous mycosis is chromoblastomycosis, caused by pigmented or dematiaceous saprophytic molds. The most common etiologic agents are *Fonsecaea pedrosoi* and *Cladophialophora carrionii*, both of which can be isolated from plants. Infection is acquired by inoculation of the etiologic agent into the subcutaneous tissues of the subject by penetrating thorns or spines of diverse plants (Martínez and Tovar, 2007). Salgado et al. (2004) reported the isolation of *Fonsecaea pedrosoi* from thorns of the plant *Mimosa pudica* at the place of infection in one of their patients. Another subcutaneous mycosis is sporotrichosis, caused by the fungus *Sporothrix schenckii*. This disease occurs throughout the world and is the most common subcutaneous mycotic disease in the USA. The disease is an occupational hazard for florists, gardeners, and forestry workers; it is also known as the rose-gardener's disease, as it is commonly transmitted by a prick from rose prickles (Ware et al., 1999; Engle et al., 2007; Haldar et al., 2007; Willey et al., 2008). Mycetoma is a chronic, specific, granulomatous, progressive subcutaneous inflammatory disease with a worldwide geographical distribution. The disease is caused by true fungi or by filamentous bacteria and hence is classified into Eumycetoma and Actinomycetoma, respectively (Fahal, 2004). These fungi or bacteria gain access to the tissues via wooden splinters or thorns (Fahal, 2004; Sheikh and Amr, 2007), and abscesses in the skin can then spread to the bones and muscles.

Do Spiny Plants Harbor Microbial Pathogens on their Spines while non-Spiny Plants do not?

Given that microorganisms are generally ubiquitous, there is no reason to assume that only specific plants or specific plant organs will be rich with microorganisms. Despite this ubiquitous occurrence, however, certain plants or plant organs may have specific chemical components or structures on their surfaces that either reduce or increase the possibility of survival for microorganism taxa. Microorganisms can grow on plant surfaces in biofilms (assemblages of bacterial cells attached to a surface and enclosed in adhesive polysaccharides excreted by the cells). Within the biofilm matrix, several different microenvironments can exist, including anoxic conditions that facilitate the existence of anaerobic bacteria. Considering the findings of Halpern et al. (2007a) in regard to spines and thorns, it is clear that anaerobic bacteria can survive on these defensive structures. Although it is assumed that an array of biofilm types is formed on plant surfaces, this issue should be systematically studied in relation to defense from herbivory in order to gain a better understanding of the anti-herbivory role of microorganisms.

Silica Needles and Raphids Made of Calcium Oxalate

We propose that thousands of plant species without thorns, spines, or prickles have an alternative of two types of usually internal, but sometimes external, sharp defensive microscopic structures: silica needles and raphids, which are needles made of calcium oxalate.

Silica Bodies (Phytoliths)

Silica bodies in plants are formed by an ordered biological deposition of silicon that enters the plant via the roots (Richmond and Sussman, 2003; Liang et al., 2005). Recent studies have shown that the action of specific genes is involved in silicon biology in plants (Ma et al., 2004, 2006, 2007). Silica bodies have several known functions: structural, serving as co-factors in the detoxification of heavy metals, and defense from herbivory (e.g., Neumann et al., 1997; Richmond and Sussman, 2003; Wang et al., 2004; Ma et al., 2006). We will focus our discussion on a specific potential defensive function: enabling the penetration of microorganisms into the body of herbivores.

Phytoliths are chemically and physically similar to sandpaper and act like it. Silica bodies in plants are found in a rich repertoire of shapes (Rapp and Mulholland, 1992; Meunier and Colin, 2001; Prychild et al., 2004), some of which are needle-shaped or otherwise sharp. The best known defensive function of silica bodies in plants is their ability to cause teeth wear in mammals and mouth parts wear in insects (Baker et al., 1959; Walker et al., 1978; McNaughton and Tarrants, 1983; McNaughton et al., 1985; Massey and Hartley, 2006; Massey et al., 2007). However, the potential defensive function of silica bodies in plants is not limited to teeth wear, as demonstrated in several studies (e.g., Wadham and Parry, 1981; Gali-Muhtasib et al., 1992). There is clear evidence for the considerable ability of silica needles to wound animal and human tissues and eventually even cause cancer (e.g., O'Neill et al., 1980; Bhatt et al., 1984; Newman, 1986; Hodson et al., 1994).

Raphids

Thousands of plant species belonging to many families produce raphids made of calcium-oxalate (Franceschi and Horner, 1980). Usually, raphids are formed in specific parenchymatic cells that differ from their neighboring cells and are called idioblasts (Fahn, 1990). The raphids are formed in idioblasts in large numbers and are packed compactly, aligned parallel to each other, but spread when the tissue is wounded. Raphids are always elongated, needle-shaped, and have two sharp pointed ends. This, however, is not the whole story. Studies conducted with a scanning electron microscope have revealed that in many cases, the raphids may be barbed or may have deep grooves along them. The grooves serve as channels through which plant toxins are introduced into the tissues of the herbivores (Sakai et al., 1972; Franceschi and Horner, 1980).

There are many shapes of calcium oxalate crystals in plants (Franceschi and Horner, 1980; Franceschi and Nakata, 2005), and raphids in the shape of needles are only one general type. Like silica in plants, calcium oxalate has several functions in plants, including tissue calcium regulation, defense from herbivory, metal detoxification, and structural functions (Franceschi and Horner, 1980; Ruiz et al., 2002; Nakata, 2003; Franceschi and Nakata, 2005). Raphids made of calcium oxalate provide a well-known defensive mechanism. However, calcium oxalate crystals in plants may serve as defense from herbivory through other means besides being sharp when they are in the shape of raphids. They may also serve as mechanical barriers, such as in the bark of conifers (Hudgins et al., 2003). Plants that have raphids are known to cause dermatitis, to be acrid, and to act as an irritant (e.g., Pohl, 1965; Sakai and Hanson, 1974; Arditti and Rodriguez, 1982; Evans, 1987; Gardner, 1994; Bradbury and Nixon, 1998). Focused studies on field behavior of herbivores, such as dorcas gazelles (*Gazelle dorcas*), have shown that they avoid plant tissues with raphides (Ward et al., 1997; Salts and Ward, 2000).

Raphides and Silica Needles may Introduce Pathogenic Microorganisms into Herbivores

In addition to the ability of both types of internal microscopic spines (raphides and silica needles) to introduce plant toxins into the wounded tissues of the herbivore by causing a mechanical irritation, we propose that they are also able to introduce pathogenic microorganisms. Because of their small size, raphides and silica needles can internally wound the mouth and digestive system of insects and other small herbivores that manage to escape from thorns by passing between them. Through the wounds inflicted by the silica needles and raphides, microorganisms found on the plant surfaces themselves, and in the mouth and digestive tract of the herbivore may cause infection. Like spines, thorns, and prickles, the raphides and silica needles actually inject the pathogenic microorganisms into the sensitive mouth and digestive tract of the herbivore.

The use of such microorganisms to harm animals by wounding them is already known from zoology. For instance, the predaceous huge lizard *Varanus komodoensis* (the Komodo dragon) seems to use the pathogenic bacteria found in its saliva as an additional advantage in hunting, like snake's venom. Animals wounded by a bite from the Komodo dragon commonly suffer from bacteremia and thus will be caught later after being incapacitated if the primary attack did not kill them (Montgomery et al., 2002).

Aposematic Thorns, Spines, and Prickles

The physical defense by spines in thousands of plant species involves an additional component of visual signaling. Many thorns, spines, and prickles are aposematic, that is, marked by various types of warning coloration (Lev-Yadun, 2001, 2003a, 2003b, 2006; Midgley et al., 2001; Lev-Yadun and Ne'eman, 2004, 2006; Midgley, 2004; Rubino and McCarthy, 2004; Ruxton et al., 2004; Speed and Ruxton, 2005; Halpern et al., 2007a, 2007b). This coloration was recently proposed to deter large herbivores (Lev-Yadun, 2001, 2006). The special character of the possible aposematic function of colorful spines is that, unlike poisonous organisms, the spines directly advertise their defensive quality (Ruxton et al., 2004; Speed and Ruxton, 2005). Aposematic coloration in animals and its mimicry is described and discussed in thousands of publications (e.g., Cott, 1940; Wickler, 1968; Edmunds, 1974; Komárek, 1998; Ruxton et al., 2004). Interestingly, animal thorns are also commonly colorful and aposematic (Ruxton et al., 2004; Inbar and Lev-Yadun, 2005; Speed and Ruxton, 2005).

Spiny plants, such as cacti, *Agave*, *Euphorbia*, and *Aloa,* have colorful spines with red, orange, yellow, black, or brown colors that may deter herbivores (Lev-Yadun, 2001). These colorful thorns, like other aposematic types, are proposed to confer a selective advantage because predators learn to associate conspicuous coloration with unpleasant qualities (e.g., Cott, 1940; Gittleman and Harvey, 1980; Harvey and Paxton, 1981; Wiklund and Järvi, 1982; Lev-Yadun, 2001, 2006; Ruxton et al., 2004; Speed and Ruxton, 2005). An important question is whether the relevant herbivorous animals can see such color signals. This issue was in debate until recent reviews provided enough data to conclude that these animals are able to see colors (Jacobs, 1993; Kelber et al., 2003). It is very likely that large mammalian herbivores do not have the ability to see colors in the same way that trichromatic humans see them. Still, even with herbivores' dichromatic vision, colorful spines or other plant parts may look different from regular green tissues because of their hue, saturation, or brightness (see Kelber et al., 2003; Motoyoshi et al., 2007). Moreover, in many plant species, spines and thorns are associated with white coloration, which increases their conspicuousness even for color-blind animals or under low illumination (Lev-Yadun, 2001, 2003a, 2006).

Plant Biological Warfare: Thorns Inject Pathogenic Bacteria into Herbivores

The physical defense provided by thorns, spines, prickles, silica needles, and raphides against herbivores might be only the tip of the iceberg in a much more complicated story. All of these sharp plant structures may inject bacteria into herbivores by wounding them and in so doing may cause them severe infections that are much more dangerous and painful than the mechanical wounding itself. We have previously presented evidence that thorns harbor an array of pathogenic bacteria which are much more dangerous to herbivores than the painful mechanical wounding itself (Halpern et al., 2007a, 2007b). We have also presented evidence that thorns, spines, or prickles can introduce pathogenic fungi into animals or humans (Halpern et al., 2007b).

In this review, we propose that microscopic spines, such as raphides and silica needles, may also provide defense via the introduction of pathogenic microorganisms. This novel hypothesis is relevant to thousands of plant species in many types of habitats that operate as a biological warfare facility. These thorn-inhabiting or thorn-injected microorganisms have a considerable potential role in anti-herbivory and may have uniquely contributed to the common evolution of aposematism (warning coloration) in thorny plants or on surfaces of plants that have internal microscopic spines.

Another theoretical aspect is the delay between the thorn's contact and wounding and the microorganism's action. While the pain from contacting thorns is immediate, the microorganism's action is delayed. Yet, the same is true for the delayed action of poisons in aposematic poisonous organisms. Nevertheless, there is a general agreement that poisonous and colorful organisms are aposematic (e.g., Cott, 1940; Edmunds, 1974; Gittleman and Harvey, 1980; Harvey and Paxton, 1981; Ruxton et al., 2004). Therefore, there is no reason to view microorganism's contamination and its delayed action any differently.

Conclusion

We propose that, at least on a theoretical basis, thorns, spines and prickles, or surfaces of plants that have internal microscopic spines may have specific chemical/structural characteristics which facilitate the existence of both aerobic and anaerobic bacteria on them. Moreover, there is a possibility that the microorganisms inhabiting the plant thorns secrete toxins while multiplying on the plant and that those toxins can also potentially harm herbivores. The various external sharp defensive plant structures act by wounding and inserting pathogenic microorganisms (bacteria and fungi) into the body of the herbivores as a sort of natural injection. The injury enables the microorganisms to pass the animal's first line of defense (the skin) and to cause a disease. In Israel, for instance, the severity and frequency of infections among orchard workers following date-palm thorn wounding has necessitated the costly practice of removal of all the millions of spines from many of the orchards by mechanical saws. Indeed, the phenomenon described by Halpern et al. (2007a, 2007b) is a natural biological warfare by thorny plants against their herbivores.

Many plants that do not have regular thorns have one of the two types of microscopic sharp structures (silica needles and raphides made of calcium oxalate) which wound the mouth and digestive system of the herbivore. It is well established that such microscopic spines enable plant toxins to penetrate the herbivore's tissues, resulting in irritation, inflammation, and even cancer. We propose that in parallel, pathogenic microorganisms may be introduced into the herbivore's body by these microscopic wounds. To the best of our knowledge, this issue has never been discussed. The microbe-thorn combination seems to be an important factor in the common evolution of the aposematic coloration of thorny plants

(Halpern et al., 2007a, 2007b). Hence, bacteria that harbor thorns or spines in plants and animals alike seem to have enhanced the common, convergent evolution of aposematism in these spiny organisms. While it is now clear that thorny plants are aposematic, the issue of potential aposematism in plants with microscopic internal spines in the shape of raphides and silica needles has not yet been systematically addressed.

References

Arditti, J. and Rodriguez, E. (1982). *Dieffenbachia*: uses, abuses and toxic constituents: A review. *J. Ethnopharmacology* 5, 293-302.

Aronson, A. I., and Shai, Y. (2001). Why *Bacillus thuringiensis* insecticidal toxins are so effective: unique features of their mode of action. *FEMS Microbiol. Lett.* 195, 1-8.

Baker, G., Jones, L. H. P. and Wardrop, I. D. (1959). Cause of wear in sheeps' teeth. *Nature* 184, 1583-1584.

Bhatt, T., Coombs, M. and O'Neill, C. (1984). Biogenic silica fiber promotes carcinogenesis in mouse skin. *Int. J. Cancer* 34, 519-528.

Bradbury, J. H. and Nixon, R. W. (1998). The acridity of raphides from the edible aroids. *J. Sci. Food Agric.* 76, 608-616.

Cahill, N. and King, J. D. (1984). Palm thorn synovitis. *J. Pediatr. Orthop.* 4, 175-179.

Chan, Y. C. and Blaschek, H. P. (2005). Comparative analysis of *Shigella boydii* 18 foodborne outbreak isolate and related enteric bacteria: role of rpoS and adiA in acid stress response. *J. Food Prot.* 68, 521-527.

Cooper, S. M. and Owen-Smith, N. (1986). Effects of plant spinescence on large mammalian herbivores. *Oecologia* 68, 446-455.

Cornell, H. V. and Hawkins, B. A. (2003). Herbivore responses to plant secondary compounds: A test of phytochemical coevolution theory. *Amer. Nat.* 161, 507-522.

Cott, H. B. (1940). *Adaptive coloration in animals*. London, Methuen and Co Ltd.

Crawley, M. J. (1983). *Herbivory: The dynamics of animal-plant interactions*. Oxford, Blackwell Scientific Publications.

Daffonchio, D., Cherif, A. and Borin, S. (2000). Homoduplex and heteroduplex polymorphisms of the amplified ribosomal 16S-23S internal transcribed spacers describe genetic relationships in the '*Bacillus cereus* group'. *Appl. Environ. Microbiol.* 66, 5460-5468.

De Champs, C., Le Seaux, S., Dubost, J. J., Boisgard, S., Sauvezie, B. and Sirot, J. (2000). Isolation of *Pantoea agglomerans* in two cases of septic monoarthritis after plant thorn and wood sliver injuries. *J. Clin. Microbiol.* 38, 460-461.

Edmunds, M. (1974). *Defence in animals: A survey of anti-predator defences*. Harlow, Longman Group Ltd.

el Sanousi, S. M. and Musa, M. T. (1989). Note on an association of *Clostridium novyi* type A and *Clostridium sordellii* with a case of gas-gangrene in a Zebu cow. *Rev. Elev. Med. Vet. Pays. Trop.* 42, 391-392.

Engle, J., Desir, J. and Bernstein, J. M. (2007). A rose by any other name. *Skinmed* 6, 139-141.

Ergonul, O., Erbay, A., Eren, S. and Dokuzoguz, B. (2003). Analysis of the case fatality rate of tetanus among adults in a tertiary hospital in Turkey. *Eur. J. Clin. Microbiol. Infect. Dis*. 22, 188-190.

Evans, C. R. H. (1987). Oral ulceration after contact with the houseplant *Dieffenbachia*. *British Dental J.* 162, 467-468.

Fahal, A. H. (2004). Mycetoma: A thorn in the flesh. *Trans. R. Soc. Trop Med. Hyg.* 98, 3-11.

Fahn, A. (1990). *Plant Anatomy* (4th ed.). Oxford, Pergamon Press.

Franceschi, V. R. and Horner, H. T. Jr. (1980). Calcium oxalate crystals in plants. *Bot. Rev.* 46, 361-427.

Franceschi, V. R. and Nakata, P. A. (2005). Calcium oxalate in plants: formation and function. *Annu. Rev. Plant Biol.* 56, 41-71.

Freiberg, A. A., Herzenberg, J. E. and Sangeorzan, J. A. (1993). Thorn synovitis of the knee joint with *Nocardia pyarthrosis*. *Clin. Orthop.* 287, 233-236.

Gali-Muhtasib, H. U., Smith, C. C. and Higgins, J. J. (1992). The effect of silica in grasses on the feeding behavior of the prairie vole, *Microtus ochrogaster. Ecology* 73, 1724-1729.

Gardner, D. G. (1994). Injury to the oral mucous membranes caused by the common houseplant, diffenbachia. A review. *Oral. Surg. Oral Med. Oral Pathol.* 78, 631-633.

Gittleman, J. L. and Harvey, P. H. (1980). Why are distasteful prey not cryptic? *Nature* 286, 149-150.

Gowda, J. H. (1996). Spines of *Acacia tortilis*: what do they defend and how? *Oikos* 77, 279-284.

Grubb, P. J. (1992). A positive distrust in simplicity - lessons from plant defences and from competition among plants and among animals. *J. Ecol.* 80, 585-610.

Haldar, N., Sharma, M. K. and Gugnani, H. C. (2007). Sporotrichosis in north-east India. *Mycoses* 50, 201-204.

Halpern, M., Raats, D. and Lev-Yadun, S. (2007a). Plant biological warfare: Infecting pathogenic bacteria into herbivores by thorns. *Environ. Microbiol.* 9, 584-592.

Halpern, M., Raats, D. and Lev-Yadun, S. (2007b). The potential anti-herbivory role of microorganisms on plant thorns. *Plant Signaling and Behavior.* 2, 503-504.

Harvey, P. H. and Paxton, R. J. (1981). The evolution of aposematic coloration. *Oikos* 37, 391-396.

Hodes, R. M. and Teferedegne, B. (1990). Tetanus in Ethiopia: Analysis of 55 cases from Addis Ababa. *E. Afr. Med. J.* 67, 887-893.

Hodson, M. J., Smith, R. J., van Blaaderen, A., Crafton, T. and O'Neill, C. H. (1994). Detecting plant silica fibres in animal tissue by confocal fluorescence microscopy. *Ann. Occup. Hyg.* 38, 149-160.

Hudgins, J. W., Krekling, T. and Franceschi, V. R. (2003). Distribution of calcium oxalate crystals in the secondary phloem of conifers: a constitutive defense mechanism? *New Phytol.* 159, 677-690.

Inbar, M. and Lev-Yadun, S. (2005). Conspicuous and aposematic spines in the animal kingdom. *Naturwiss* 92, 170-172.

Ivanova, N., Sorokin, A., Anderson, I., Galleron, N., Candelon, B., Kapatral, V., Bhattacharyya, A., Reznik, G., Mikhailova, N., Lapidus, A., Chu, L., Mazur, M., Goltsman, E., Larsen, N., D'Souza, M., Walunas, T., Grechkin. Y., Pusch, G., Haselkorn,

R., Fonstein, M., Ehrlich, S. D., Overbeek, R. and Kyrpides, N. (2003). Genome sequence of *Bacillus cereus* and comparative analysis with *Bacillus anthracis*. *Nature* 423, 87-91.

Jacobs, G. H. (1993). The distribution and nature of colour vision among the mammals. *Biol. Rev.* 68, 413-471.

Janda, M. J. (2002). New Members of the Family Enterobacteriaceae. In *The Prokaryotes*, Dworkin, M., Falkow, S., Rosenberg, E., Schleifer, K. H. and Stackebrandt, E. (eds.), Release 3.9. Available: http://141.150.157.117:8080/prokPUB/index.htm.

Janzen, D. H. and Martin, P. S. (1982). Neotropical anachronisms: The fruits the gomphotheres ate. *Science* 215, 19-27.

Janzen, D. H. (1986). Chihuahuan Desert nopaleras: Defaunated big mammal vegetation. *Annu. Rev. Ecol. Syst.* 17, 595-636.

Jensen, G. B., Hansen, B. M., Eilenberg, J. and Mahillon, J. (2003). The hidden lifestyles of *Bacillus cereus* and relatives. *Environ. Microbiol.* 5, 631-640.

Kelber, A., Vorobyev, M. and Osorio, D. (2003). Animal colour vision - behavioural tests and physiological concepts. *Biol. Rev.* 78, 81-118.

Komárek, S. (1998). Mimicry, aposematism and related phenomena in animals and plants: bibliography 1800 - 1990. Prague, Vesmir.

Kratz, A., Greenberg, D., Barki, Y., Cohen, E. and Lifshitz, M. (2003). *Pantoea agglomerans* as a cause of septic arthritis after palm tree thorn injury; case report and literature review. *Arch. Dis. Child* 88, 542-544.

Leendertz, F. H., Ellerbrok, H., Boesch, C., Couacy-Hymann, E., Matz-Rensing, K., Hakenbeck, R., Bergmann, C., Abaza, P., Junglen, S., Moebius, Y., Vigilant, L., Formenty, P. and Pauli, G. (2004). Anthrax kills wild chimpanzees in a tropical rainforest. *Nature* 430, 451-452.

Lev-Yadun, S. (2001). Aposematic (warning) coloration associated with thorns in higher plants. *J. Theor. Biol.* 210, 385-388.

Lev-Yadun, S. (2003a). Weapon (thorn) automimicry and mimicry of aposematic colorful thorns in plants. *J. Theor. Biol.* 224, 183-188.

Lev-Yadun, S. (2003b). Why do some thorny plants resemble green zebras? *J. Theor. Biol.* 244, 483-489.

Lev-Yadun, S. (2006). Defensive coloration in plants: a review of current ideas about anti-herbivore coloration strategies. In: *Floriculture, ornamental and plant biotechnology: advances and topical issues. Vol. IV*, Teixeira da Silva, J. A. (ed.), pp. 292-299. London, Global Science Books.

Lev-Yadun, S. and Ne`eman, G. (2004). When may green plants be aposematic? *Biol. J. Linn. Soc.* 81, 413-416.

Lev-Yadun, S. and Ne`eman, G. (2006). Color changes in old aposematic thorns, spines, and prickles. *Isr. J. Plant Sci.* 54, 327-333.

Liang, Y., Si, J. and Römheld, V. (2005). Silicon uptake and transport is an active process in *Cucumis sativus*. *New Phytol.* 167, 797-804.

Lindow, S. E. and Leveau, J. H. (2002). Phyllosphere microbiology. *Curr. Opin. Biotechnol.* 13, 238-243.

Lindow, S. E. and Brandl, M. T. (2003). Microbiology of the Phyllosphere. *Appl. Environ. Microbiol.* 69, 1875-1883.

Martínez, R. L. and Tovar, L. J. M. (2007). Chromoblastomycosis. *Clint. Dermatol.* 25, 188-94.

Ma, J. F., Mitani, N., Nagao, S., Konishi, S., Tamai, K., Iwashita, T. and Yano, M. (2004). Characterization of the silicon uptake system and molecular mapping of the silicon transporter gene in rice. *Plant Physiol.* 136, 3284-3289.

Ma, J. F., Tamai, K., Yamaji, N., Mitani, N., Konishi, S., Katsuhara, M., Ishiguro, M., Murata, Y. and Yano, M. (2006). A silicon transporter in rice. *Nature* 440, 688-691.

Ma, J. F., Yamaji, N., Mitani, N., Tamai, K., Konishi, S., Fujiwara, T., Katsuhara, M. and Yano, M. (2007). An efflux transporter of silicon in rice. *Nature* 448, 209-212.

Masaki, T., Umehashi, H., Miyazaki, H., Takano, M., Yamakawa, K. and Nakamura, S. (1988). *Clostridium absonum* from gas gangrene. *Jap. J. Med. Sci. Biol.* 41, 27-30.

Massey, F. P. and Hartley, S. E. (2006). Experimental demonstration of the antiherbivore effects of silica in grasses: impacts on foliage digestibility and vole growth rates. *Proc. R. Soc. Lond. B.* 273, 2299-2304.

Massey, F. P., Ennos, A. R. and Hartley, S. E. (2007). Grasses and the resource availability hypothesis: the importance of silica-based defenses. *J. Ecol.* 95, 414-424.

McNaughton, S. J. and Tarrants, J. L. (1983). Grass leaf silicification: Natural selection for an inducible defense against herbivores. *Proc. Natl. Acad. Sci. USA* 80, 790-791.

McNaughton, S. J., Tarrants, J. L., McNaughton, M. M. and Davis, R. H. (1985). Silica as a defense against herbivory and a growth promotor in African grasses. *Ecology* 66, 528-535.

Meunier, J. D. and Colin, F. (eds.). (2001). *Phytoliths: application in earth sciences and human history.* Lisse, A. A. Balkema Publishers.

Midgley, J. J. (2004). Why are spines of African *Acacia* species white? *Afr. J. Range and For Sci.* 21, 211-212.

Midgley, J. J., Botha, M. A. and Balfour, D. (2001). Patterns of thorn length, density, type and colour in African Acacias. *Afr. J. Range and For. Sci.* 18, 59-61.

Montgomery, J. M., Gillespie, D., Sastrawan, P., Fredeking, T. M. and Stewart, G. L. (2002). Aerobic salivary bacteria in wild and captive Komodo dragons. *J. Wildl. Dis.* 38, 545–551.

Motoyoshi, I., Nishida, S., Sharan, L. and Adelson, E. H. (2007). Image statistics and the perception of surface qualities. *Nature* 447, 206-209.

Myers, J. H. and Bazely, D. (1991). Thorns, spines, prickles, and hairs: are they stimulated by herbivory and do they deter herbivores? In: *Phytochemical induction by herbivores.* Tallamy, D. W. and Raupp, M. J. (eds.), pp. 325-344. New York, John Wiley and Sons, Inc.

Nakata, P. A. (2003). Advances in our understanding of calcium oxalate crystal formation and function in plants. *Plant Sci.* 164, 901-909.

Neumann, D., Lichtenberger, O., Schwieger, W. and zur Nieden, U. (1997). Silicon storage in selected dicotyledons. *Botanica Acta* 110, 282-290.

Newman, R. H. (1986). Fine biogenic silica fibres in sugar cane: a possible hazard. *Ann. Occup. Hyg.* 30, 365-370.

Oh, H. M. L. and Tay, L. (1995). Bacteraemia caused by *Rahnella aquatilis*: Report of two cases and review. *Scand. J. Infect. Dis.* 27, 79-80.

O'Neill, C. H., Hodges, G. M., Riddle, P. N., Jordan, P. W., Newman, R. H., Flood, R. J. and Toulson, E. C. (1980). A fine fibrous silica contaminant of flour in the high oesophageal cancer area of north-east Iran. *Int. J. Cancer* 26, 617-628.

Pascual, F. B., McGinley, E. L., Zanardi, L. R., Cortese, M. M. and Murphy, T. V. (2003). *Tetanus surveillance, United States, 1998-2000. 52, SS03 1-8* Available from: URL: http://www.cdc.gov/mmwr/preview/mmwrhtml/ss5203a1.htm.

Pohl, R. W. (1965). Contact dermatitis from the juice of *Ornithogalum caudatum*. *Toxicon* 3, 167-168.

Prychild, C. J., Rudall, P. J. and Gregory, M. (2004). Systematics and biology of silica bodies in Monocotyledons. *Bot. Rev.* 69, 377-440.

Rapp, G. Jr. and Mulholland, S. C. (1992). *Phytolith systematics. Emerging issues.* New York, Plenum Press.

Rebollo, S., Milchunas, D. G., Noy-Meir, I. and Chapman, P. L. (2002). The role of spiny plant refuge in structuring grazed shortgrass steppe plant communities. *Oikos* 98, 53-64.

Richmond, K. E. and Sussman, M. (2003). Got silicon? The non-essential beneficial plant nutrient. *Curr. Opin. Plant Biol.* 6, 268-272.

Rubino, D. L. and McCarthy, B. C. (2004). Presence of aposematic (warning) coloration in vascular plants of southeastern Ohio. *J. Torrey Bot. Soc.* 131, 252-256.

Ruiz, N., Ward, D. and Saltz, D. (2002). Calcium oxalate crystals in leaves of *Pancratium sickenbergeri*: constitutive or induced defence? *Funct. Ecol.* 16, 99-105.

Ruoff, K. L., de la Maza, L., Murtagh, M. J., Spargo, J. D. and Ferraro, M. J. (1990). Species identities of enterococci isolated from clinical specimens. *J. Clin.. Microbiol.* .28, 435-437.

Ruxton, G. D., Sherratt, T. N. and Speed, M. P. (2004). *Avoiding attack: The evolutionary ecology of crypsis, warning signals and mimicry.* Oxford, Oxford University Press.

Sakai, W. S. and Hanson, M. (1974). Mature raphid and raphid idioblast structure in plants of the edible aroid genera *Colocasia*, *Alocasia*, and *Xanthosoma*. *Ann. Bot.* 38, 739-748.

Sakai, W. S., Hanson, M. and Jones, R. C. (1972). Raphids with barbs and grooves in *Xanthosoma sagittifolium* (Araceae). *Science* 178, 314-315.

Salgado, C. G., da Silva, J. P., Diniz, J. A., da Silva, M. B., da Costa, P. F., Teixeira, C. and Salgado, U. I. (2004). Isolation of *Fonsecaea pedrosoi* from thorns of *Mimosa pudica*, a probable natural source of chromoblastomycosis. *Rev. Inst. Med. Trop. Sao Paulo* 46, 33-36.

Salts, D. and Ward, D. (2000). Responding to a three-pronged attack: desert lilies subject to herbivory by dorcas gazelles. *Plant Ecol.* 148, 127-138.

Sawardekar, K. P. (2005). Shigellosis caused by *Shigella boydii* in a preterm neonate, masquerading as necrotizing enterocolitis. *J. Pediatr. Infect. Dis.* 24, 184-185.

Sheikh, S. S. and Amr, S. S. (2007). Mycotic cysts: Report of 21 cases including eight pheomycotic cysts from Saudi Arabia. *Int. J. Dermatol.* 46, 388-392.

Shimizu, T., Ohtani, K., Hirakawa, H., Ohshima, K., Yamashita, A., Shiba, T., Ogasawara, N., Hattori, M., Kuhara, S. and Hayashi, H. (2002). Complete genome sequence of

Clostridium perfringens, an anaerobic flesh-eater. *Proc. Natl. Acad. Sci. USA* 99, 996-1001.

Speed, M. P. and Ruxton, G. D. (2005). Warning displays in spiny animals: one (more) evolutionary route to aposematism. *Evolution* 59, 2499-2508.

Sugarman, M., Stobie, D. G., Quismorio, F. P., Terry, R. and Hanson, V. (1977). Plant thorn synovitis. *Arthritis Rheum.* 20, 1125-1128.

Vincent, K. and Szabo, R. M. (1988). *Enterobacter agglomerans* osteomyelitis of the hand from a rose thorn. A case report. *Orthopedics* 11, 465-457.

Wadham, M. D. and Parry, D. W. (1981). The silicon content of *Oryza sativa* L. and its effect on the grazing behaviour of *Agriolimax reticulatus* Müller. *Ann. Bot.* 48, 399-402.

Walker, A., Hoeck, H. N. and Perez, L. (1978). Microwear of mammalian teeth as an indicator of diet. *Science* 201, 908-910.

Wang, Y., Stass, A. and Horst, W. J. (2004). Apoplastic binding of aluminum is involved in silicon-induced amelioration of aluminum toxicity in maize. *Plant. Physiol.* 136, 3762-3770.

Ward, D., Spiegel, M. and Saltz, D. (1997). Gazelle herbivory and interpopulation differences in calcium oxalate content of leaves of a desert lily. *J. Chem. Ecol.* 23, 333-346.

Ware, A. J., Cockerell, C. J. and Skiest, D. J. (1999). Kussman HM Disseminated sporotrichosis with extensive cutaneous involvement in a patient with AIDS. *J. Am. Acad. Dermatol.* 40, 350-355.

Wickler, W. (1968). *Mimicry in plants and animals*. London, Weidenfeld and Nicolson.

Wiklund, C. and Järvi, T. (1982). Survival of distasteful insects after being attacked by naive birds: A reappraisal of the theory of aposematic coloration evolving through individual selection. *Evolution* 36, 998-1002.

Willey, J. M., Sherwood, L. M. and Woolverton, C. J. (2008). *Prescot, Harley, and Klein's Microbiology* (7th ed.). New York, McGraw Hill.

In: Microbial Ecology Research Trends
Editor: Thijs Van Dijk, pp. 169-205
ISBN: 978-1-60456-179-1

Chapter VIII

The Structure and Stability of Microbial Communities in Metallophytic Grassland Soils that Have Been Subjected to Long Term Contamination by Heavy Metals

Stephen P. Cummings * ***and Lisa Bamborough***
School of Applied Sciences, University of Northumbria,
Newcastle-upon-Tyne, UK

Abstract

Heavy metals are a significant source of pollution in soils that have been demonstrated to exert significant toxic effects on soil microbial communities. Here, we investigate the impact of the heavy metals lead and zinc on the bacterial community of an upland metallophytic grassland the UK, contaminated by historical lead mining activity. We focus on the total microbial diversity and that of the Actinobacteria, a significant component of the aerobic soil community that demonstrate a broad metabolic activity and represent useful indicators of soil health.

Two experimental approaches were adopted, the first employed microcosms containing soil from the site of an abandoned lead mine spoil heap, bacterial diversity was investigated using denaturing gradient gel electrophoresis (DGGE) of PCR amplified 16S rRNA DNA with universal primers to amplify the total community and more specific Actinobacteria primers. The Actinobacteria specific PCR primers showed that this taxon represented a diverse and metabolically active element of the microbial community of lead contaminated upland soil. Subsequently, we investigated how the

* Correspondents address: Stephen Cummings, School of Applied Sciences, Ellison Building, University of Northumbria, Newcastle-upon-Tyne. NE1 8ST; Tel/ fax. 0191 227 3176. Email; Stephen.cummings @unn.ac.uk

stability of the bacterial communities responded to an increase in stress by adding additional Pb and Zn to the soil. There is significant interest in determining whether communities under stress are more or less robust in the face of additional environmental stress. Our data was equivocal, the total community diversity responded negatively to additional heavy metal input into the soil, whereas, the Actinobacteria community did not show any substantial change in its diversity.

A second experiment characterised the bacterial soil community along a transect of a metallophytic grassland site that had significant differences in the Pb, Zn, organic C and N content across it. In this study the metabolically active bacterial population was studied using reverse transcription of mRNA from16S rRNA gene. Universal primers showed a low diversity of metabolically active bacteria, and both organic C and heavy metal concentrations in the soil seemed to impact on the community diversity. In contrast Actinobacteria showed higher diversities and seemed to be able to maintain this diversity even in soils with low organic C and high heavy metal concentrations suggesting that they are competitive in this stressed environment with other soil flora. The data presented here may prove useful in determining the effect of heavy metal concentrations on soil microbial diversity and community stability, particularly in managing rare metallophytic grassland sites and more widely in remediating contaminated soils.

Introduction

Heavy metals are a poorly defined group of elements that lack a coherent scientific definition, however, the term is commonly applied to a group of toxic metallic elements including As, Cd, Cr, Cu, Pb, Hg, Ni, Se and Zn that exert biological toxicity (Duffus 2002). These metals find their way into the environment through a range of anthropogenic activities and cause in significant problems due to their toxicity and persistence. In soil, high concentrations of heavy metals often arises due to anthropogenic activities such as mining (Bamborough and Cummings, 2006), the smelting of ore (Almås et al. 2004), from vehicle exhaust emissions and agricultural practices such as sewage sludge or fertiliser application to arable land (Kozdrój and van Elsas, 2001). The impact of heavy metals on soil fertility can be significant, affecting crop plants, livestock, and the indigenous microbial community (Giller et al. 1998). Such effects are of concern because in the case of crops and livestock, there is a potential for human exposure through the food chain. While the microbial community is responsible for key nutrient cycling activities in the soil essential for its continuing fertility. As a consequence there is an extensive literature on the impact of heavy metals on soil microbial communities and the processes that they drive.

Here we present a recent study that examines the effect of long term heavy metal contamination on soil, focussing particularly on Actinobacteria. These are a group of soil prokaryotes that have been reported to be the most important sub-population of bacteria in soils with high heavy metal content (Gremion et al. 2003). Our aim was to determine whether these bacteria are indeed significant contributors soil microbial communities exposed to long term heavy metal contamination and how the population responds in such soil to additional heavy metal stress. We contextualise this study within the recent literature describing the interactions between microbial communities and heavy metals in soil. This discussion highlights a number of problems that remain to be solved before we can accurately describe the impact of heavy metal contamination on soil processes. Amongst the most important of these are (i) how we relate the complexity of the microbial community at both a phylogenetic, and functional level in soil to the effects of heavy metal disturbance. (ii) How the toxicity of a particular heavy metal in soil measured?

Many studies on the impact of heavy metals on soils have attempted to measure how the microbial community structure changes in response to their addition (Sandaa et al. 2001; Hery et al 2003; Nakatsu et al. 2005). However, caution must be exercised in describing both the significance of such changes and whether they have consequences for the ecological processes driven by the community under scrutiny. Recent work on mathematically modelling soil bacterial populations to predict their diversity from small sample sizes, typical of those derived using clone libraries, indicated a high probability of diversities of between 10^5 to10^6 taxa in pristine soils (Lunn et al. 2004; Gans et al. 2005). In addition, related work has illustrated gaps in our understanding, in particular, the microbial community structure *in situ* that is being sampled and the appropriate scale to represent and characterise such communities (Sloan et al. 2007). Moreover, inherent problems in the methodology used, particularly PCR biases (Narang and Dunbar, 2004) and the acceptance that 16S rRNA genes are conservative markers for bacterial diversity (Sloan et al. 2007) inject further uncertainty into estimates of microbial diversity and community structure.

As a consequence, determining the result of heavy metal contamination on the structure of a microbial soil population is not straightforward. Using molecular approaches to identify the impact of heavy metal addition on microbial diversity and function may be limited in that they lack the sensitivity to identify small populations whose activity may be of high functional significance. Recent estimations of the impact of metal pollution on the bacterial populations of soil suggests that they reduce diversity by 99.9% (Gans et al. 2005) affecting rarer taxa most especially. Functional studies that seek to identify how heavy metals impact on community function also have limitations. For example, measures of changes in total biomass, respiratory rates, enzyme activities and substrate utilisation profiles can demonstrate significant responses to heavy metal addition (Kelly et al. 1999; Lazzaro et al. 2006a), and can act as robust biological indicators of the ecological health of the soil (Dawson et al, 2007). However, they do not necessarily correspond to changes in specific microbial processes that are important in maintaining soil fertility. Despite these methodological problems this remains an area of significant research activity, as improved tools become available the impact of heavy metals on soils may be more clearly elucidated, however, the heterogeneity and complexity of soil and soil communities make this a formidable challenge.

The Impact of Environmental Stress on Microbial Communities in Soil

The soil is a complex and heterogeneous environment and as yet we have no clear idea as to its phylogenetic and functional diversity (Curtis et al 2006) . In order to understand the impact of anthropogenic pollution on soil health a greater understanding of the relationship between the soil community, the pollutant and the physical and chemical parameters of the soil are essential. The microbial community is a key driver in the cycling of nutrients and energy through the system and is therefore a vital component in determining the resilience of the soil. Soil resilience can be defined as the ability of the soil to recover after a disturbance, this term has been subdivided into resistance (to disturbance) and resilience (to disturbance) that in combination define soil ecosystem stability (Griffiths et al. 2001).

Recent studies on the resilience of soil microbial communities have been directed at testing how environmental stresses affect the functional stability of the ecosystem. Two theories have been postulated. Firstly, that systems under stress are more stable than those which are not, as they have acquired physiological adaptations, that enable them to maintain function the face of additional stresses. The second states that non-stressed systems have a greater stability as they have more resources at their disposal that can be deployed to maintain ecosystem function in the face of stress (Tobor-Kaplan et al. 2005). These hypotheses were tested by Griffiths et al. (2001) using grassland, agricultural and industrial soil and Kuan et al. (2006) with grassland soils subject to various disturbances (biocide, sewage sludge application). Both studies challenged the soils with heat and Cu stress and measured the effects on soil functional stability. In each study, less stressed soils had greater stability supporting the hypothesis that non stressed soils have more resistance to stresses that those previously exposed to environmental challenge.

The work described above illustrates a growing awareness amongst microbial ecologists that the study of complex microbial systems in soil may benefit from the use of applied classical ecological theory to enable them to frame testable hypotheses on the role of phylogenetic and functional diversity on soil ecosystem functioning (Prosser et al. 2007). There has been significant interest in applying ecological theory to help describe the relationship between ecosystem sustainability and biodiversity, a number of studies have highlighted a positive correlation between ecosystem functioning and biodiversity or the insurance hypothesis of biodiversity. This theory states that diverse systems are more stable (Wardle and Zackrisson, 2005). However, there are few studies that have addressed this relationship within soil microbial communities. Wertz et al. (2006) showed that the dilution of soil diversity by four orders of magnitude had no appreciable impact on carbon mineralisation, denitrification and nitrification. This data suggests that the huge biodiversity of soil makes the functioning of this ecosystem largely insensitive to the erosion of biodiversity, even for specific functions such as nitrification, so long as all functional groups are not eliminated and their abundance can be recovered. The corollary of this observation is that anthropogenic pollution that affects soil microbial diversity should not affect ecosystem functioning. However, studies in the impact of disturbance on soil ecosystems have reported a significant loss of diversity and effects on the functional response of the ecosystem (Lazzaro et al. 2006b; Griffiths et al 2001; Kuan et al. 2006) this suggests that the qualifications relating to the persistence of all functional groups and the recovery of abundance are frequently not met in stressed soil. Superficially these observations that reduced biodiversity negatively affect ecosystem functioning would appear to lend support to the insurance hypothesis of biodiversity. However, other factors could also be significant, such as the energy costs required to cope with stress, which may be accommodated more readily by unstressed communities were maintenance energy is lower (Tobor-Kaplan et al. 2005).

Heavy Metals in Soil

The broad responses to environmental stress observed in soil communities highlights some of the important issues that need to be addressed in resolving how and why soil microbial ecosystems respond in the ways we observe to disturbance. From here on we shall draw more specifically on the literature that deals with the particular impacts of heavy metals on the soil ecosystem. Almost ten years ago the impact of heavy metals on microorganisms and microbially driven processes in agricultural soils was extensively reviewed (Giller et al 1998). In this study, the authors identified that there was an enormous variability, up to two orders of magnitude, in literature on the concentrations of heavy metals that exerted toxicity on various soil microbial communities. Subsequent to the publication of that review, there has been the widespread application of molecular techniques to study the impact and effects of heavy metal contamination on soil communities. However, many of these studies have simply confirmed the earlier work identifying the variability of the response of soil communities to heavy metal application (Feris et al. 2004; Griffiths et al 2005). It is clear that to understand heavy metal pollution of soil, a multi-disciplinary approach is essential, embracing the interactions between the biological, chemical and physical parameters within the soil. Perhaps the most significant factors to understand are those that modify the toxicity of heavy metals by affecting their bioavailability and the differential sensitivity of the microbial community exposed to them.

The bioavailability of heavy metals in soils determines what impact they will have on soil communities and contributes to the high degree of variability between soils, of heavy metal concentrations that exert toxicity. Differences in bioavailability may arise for a number of reasons and may reflect a range of diverse and heterogeneous interactions between heavy metals the soil matrix and the microbial biomass. The bioavailability of heavy metals is dose dependent (Almås et al. 2004), however, it has been shown that the total concentration of a heavy metal in a soil is not a reliable indicator of its bioavailability (Prokop et al. 2003). The toxicity of a metal is dependent on how it speciates in the soil (Knight and McGrath, 1995) with the free metal form being the most biologically active. Determining what fraction of the heavy metal in a soil is bioavailable has been measured using a range of techniques, in the case of Cd, it has correlated well with the fraction that could be extracted using chelating agents such as DPTA (Prokop et al. 2003). However, another study suggested that soil pore water concentration may give the best prediction of metal toxicity to soil communities (Almås et al. 2004).

The composition of the soil matrix can also influence the toxicity of heavy metals. For example, in soils with clay contents less than 50%, heavy metals exhibited higher inhibiton of microbial activity than those with higher proportions of clay (Lazzaro et al. 2006b). This may reflect the observation that clay minerals such as sodium and calcium bentonite and zeolite have been shown immobilise and decrease the water extractable concentration of heavy metals including Cd, Zn, Cu and Ni (Usman et al. 2005). Similarly, the organic carbon content of the soil can also exert an significant effect on the availability of heavy metals, Cd has been shown to adsorb strongly to organic carbon (Prokop et al 2003), therefore, one would expect the proportion of these two components of the soil to exert significant effects on the bioavailability of heavy metals in a soil. Recent work has demonstrated the significance of the heterogeneity of the soil environment across small (a few centimetre) distances. Most studies have monitored bioavailability using homogenised bulk soil, however, it has been shown that the spatial heterogeneity of soil matrices can result in a 30–fold differences in metal concentrations at sites less than 1 cm apart (Becker et al. 2006). Given that soil microbial communities demonstrate high phylogenetic and functional heterogeneity (Fierer et al. 2005; Rondon et al. 2000), large differences across relatively short distances may be explained at least in part by these steep gradients in heavy metal concentrations. The bioavailability of heavy metals also changes with time reflecting changes in the soil environment often driven by microbial activity, the metabolism of organic substrates by the microbial flora of the soil has been shown to increase the concentration of dissolved organic carbon content in soil and reduce the pH which enhanced both metal mobility and bioavailability. Concurrently, the adsorption to biomass and the sedimentation with biologically generated phosphate or other anions will immobilise heavy metals (Wu et al 2006).

The differential response to heavy metals by microbial communities in the presence of equivalent bioavailable concentrations, reflect the different sensitivities of these communities to heavy metal exposure (Giller et al. 1998). In many studies the presence of high concentrations of heavy metals has been associated with a decrease in the biomass (Wang et al. 2007; Liao et al, 2007; Vaasquez-Murrieta et al. 2006; Li et al. 2006; Frey et al. 2006), indeed some data suggests that heavy metals even in relatively low concentrations can depress the microbial biomass even during long term exposure and this is accompanied by changes in the structure of the microbial community (Witter et al 2000). Decreases in biomass may relate to the reduction in fungal, particularly mycorhizzal biomass as plant roots are frequently absent in highly contaminated soils, particularly forest soil are undisturbed (Frostegård et al. 1993). However, such effects are not always observed and in a number of field based studies no significant effects of heavy metals on biomass have been noted (Griffiths et al. 2005; Ranjard et al. 2006).

In addition, heavy metals have had measurable impacts on a range of other parameters used to measure microbial activity, such as the production of carbon dioxide, which correlates with the basal respiration rate of the soil microbes, and enzymatic activities that are involved in key metabolic cycles or substrate utilisation profiles. The effect of high concentrations of heavy metals on basal respiration rates and enzymic activities has also produced conflicting data in the literature. Enzyme activities have been shown to be significantly decreased (Wang et al, 2007; Kandeler et al. 2000), whereas, other studies report no significant changes in soil contaminated with Zn and Cu compared to controls (Niklínska et al. 2006).

A number of studies have, in addition, to identifying changes in biomass and functional activities of the soil community (Abaye et al. 2005) also reported that these responses were frequently accompanied by changes in the microbial community structure (Giller et al. 1998). It is not clear whether these changes are related to changes in the functional activity of the community or whether they represent adaptation of the community to the presence of heavy metals (Griffiths et al. 2001). The overarching issue is whether we can apply a more profound understanding of the impact of heavy metals to the microbial communities in soil to more effectively manage the toxicity and environmental degradation that they cause.

Most work has focussed on how the diversity of microbial community structure changes in response to heavy metals. Ecological theory tells us that environmental stress reduces biodiversity and intuitively, one would expect that the presence of heavy metals in soils will have a negative impact on the diversity of the microbial community such responses have been observed (Li et al. 2006; Joynt et al. 2006) but not in all cases (Feris et al 2003). Moreover, it is logical to assume that the presence of a toxic stress such as heavy metals will also lead over time to the adaptation of the community to tolerate that stress, some studies have seen such an increase in heavy metal resistance among microbial populations (Schmidt et al. 2005; Frey et al. 2006), however, others have reported no such development of tolerant communities (Feris et al. 2003). It has been pointed out that the duration of the exposure that is being studied will affect the response on thc microbial community, laboratory studies tend to be short term, whereas field studies tend to have significantly longer durations (Bååth et al. 2005). Short term addition of heavy metals soils have been reported to have quite different consequences to long term exposure. Renella et al. (2002) studied the impact of the addition of individual and combination additions of Cu, Zn and Cd salts to arable and grassland soils. They found that reductions in microbial biomass and respiration rate were variable small (5-12 %) compared to the effects of chronic exposure which gave reductions in these parameters of up to 50 % (Brookes and McGrath 1984). In summary, the community structure of heavy metal contaminated soils, particularly as a result of long term exposure, appears to change significantly, although this does not necessarily mean the diversity will decrease (Lorenz et al 2006) or that the community will be more tolerant of heavy metals (Feris et al. 2003; Giller et al 1998; Lorenz et al 2006). Often these changes have been related to specific taxa, such as Actinobacteria and fungi becoming more prevalent (Kelly et al 2003; Hery et al. 2003), however, the variability and conflicting data in the literature suggest that such changes are affected by a range of inter-related variables including the nature of the soil matrix, its chemistry and the composition of the soil flora. Therefore, each soil should be considered on a case by case basis.

The Response of Actinomycetes to Heavy Metals in Soil

Actinobacteria are a broad taxonomic group within the domain *Eubacteria.* This division has been divided into six orders and fifteen suborders (Stackebrandt et al. 1997) that embraces a large number of genera. The group describes a diverse phylogenetic and functional assemblage of bacteria that share a high proportion of guanine and cytosine nucleotides in their DNA, greater than 50%, and sequence similarities of their 16S ribosomal ribonucleic acid (rRNA). These bacteria are also described in the literature as high G+C Gram positives and Actinomycetes, names that have no standing in the nomenclature of bacterial classification (Eusby 2007) but remain widely used. They are of significant biotechnological interest as some taxa within the group, notably the within the genus *Streptomyces*, produce secondary metabolites such as the antibiotics erythromycin and tetracycline that are of great economic and pharmaceutical significance (Paradkar et al. 2003). In addition, they are of interest from the evolutionary perspective and recently, this group of bacteria have been proposed to belong to a novel paraphyletic division of the

Eubacterial domain called the Posibacteria from which the Eukaryota and Archaebacteria emerged (Cavalier-Smith, 2002).

Within the Actinobacteria there is huge phlyogenetic and functional diversity, and individual taxa will perform a variety of, probably quite diverse, functional roles within soil. However, their ubiquity in soil, their metabolic diversity and their significance in ecology, biotechnology, and human and veterinary medicine make them a useful candidate taxa to study to determine how bacterial populations responds to the perturbation of the soil environment by the presence of heavy metals.

Actinobacteria Resistant to Multiple Heavy Metals

Bacteria are thought to tolerate metals by means of several systems. These include precipitation at the cell surface, biosorption to extracellular polymeric substances (EPS), including polysaccharides, and proteins, or to the cell surface (Valls *et al.*, 2000; Kefala *et al.*, 1999). There are also a number of active heavy metal efflux systems which are widely distributed amongst bacteria, such as the mercury efflux (Mer) proteins and P-type ATPases that transport heavy metals across the cellular membrane (Neis, 2003). Actinobacteria have been shown to employ both avoidance of heavy metals by biosorption (Rho and Kim, 2002) and resistance through their active efflux (Schmidt et al. 2005). For example, *Frankia* strains were shown to exhibit resistance to lead and copper through binding to an extracellular matrix composed of polysaccharides and proteins that prevented metal ions from entering the cell (Richards *et al.*, 2002). In contrast, Ravel *et al.*, (1998) isolated *Streptomyces* spp. from polluted sediment samples that demonstrated high levels of mercury resistance. This was mediated through the expression of a plasmid borne mercury efflux (Mer) system. Similarly, evidence for high affinity nickel transporters has been demonstrated in this genus (Amoroso et al. 2000)

Actinobacteria in Heavy Metal Contaminated Soil

The significance of Actinobacteria within the microbial communities of soils contaminated with heavy metals remains unclear, Gremion et al (2003) studying the bacterial community of a soil contaminated with Cd, Cu and Zn showed that Actinobacteria dominated clone libraries derived from both the bulk soil and the rhizosphere of *Thlaspi caerulescens* a heavy metal hyperaccumulator. Moreover, 70% of the Actinobacterial clones belonged to the genus *Rubrobacter*, a widespread taxa in soils, that has as yet few cultured representatives (Holmes et al. 2000). Similarly, a study on an agricultural soil amended with sewage sludge containing elevated levels of Zn produced a clone library in which the most common OTUs were high G+C Gram positive clones belonging to the *Rubrobacter* genus (Moffett et al. 2003). High abundance of Actinobacteria have also been found in soils that contain naturally high concentrations of heavy metals such as Ni, Cr and Co, culture based studies identified 80% of the bacteria recovered as belonging to this group (Amir and Pineau 1998) and actinobacterial sequences in clone libraries from nickel mine spoils comprised 96 % of the community (Hery et al, 2005). Studies using PFLA techniques also showed that although total biomass decreased on these soils compared to non-serpentine soils the proportion of actinobacterial marker fatty acids to be significantly increased (DeGrood et al. 2005).

However, the abundance of Actinobacteria is probably determined by a range of factors in the soil. For example, studies on the bacterial communities of rhizosphere soil from grain legumes indicated Gram positive organisms made up almost 45% of the community derived from the analyses of a clone library. Furthermore, Actinobacteria contributed between 70-95% of the clones from this fraction. The authors speculated that the predominance of Actinobacteria could be attributed to the timing of sampling which occurred after fruiting of the legumes. At this point the root exudation would diminish resulting in organisms that adopted a K-strategy predominating (Sharma et al. 2005). They have also been observed to increase as a result of remediation of contaminated soils, such changes have been linked to to increases in the total N, pH and porosity of the soil as it is re-vegetated (Zhang et al. 2006). Actinobacteria have also been shown to be dominant within particular microenvironments within the soil. A study of the bacterial communities found inside and outside of soil micro-aggregates in disturbed and undisturbed soils showed them to be distinctly different, while *Proteobacteria* clones were prevalent in libraries derived from whole microaggregates. The community inside micro-aggregates, in undisturbed soil, had a high abundance of Actinobacteria that consisted entirely of taxa from the order *Actinobacteridae*. Whereas, in micro-aggregates from disturbed soil, Actinobacteria were dominant but 53% of the clones were attributed to the order *Rubrobacteridae* (Mummey and Stahl, 2004). This work demonstrated that there is a stratification of bacterial taxa in relation to the communities inside and outside soil micro-aggregates, furthermore, the inner micro-aggregate community was also significantly different between different soil types, indicating the actinobacterial community structure is affected by soil disturbance even in these sequestered environments.

In contrast to the assertion by Gremion et al (2003) a number of studies on the impact of heavy metal contamination on bacterial community structure have reported that there was a significant decline in the contribution of Actinobacteria to the bacterial community. In a forest soil contaminated with cadmium, copper zinc and lead there was a significant decrease in the Actinobacterial marker using a PFLA protocol compared to control soils. The microbial community also showed more general changes in response to heavy metal amendment, one that was more tolerant of heavy metals but with a compromised functional activity (Frey et al. 2006). Soils subjected to amendment with zinc in a microcosm study, and those from a soil in the vicinity of a Zn smelter also showed significant changes in the microbial community that included a decrease in the Actinobacteria (Kelly et al. 1999; Kelly et al. 2003). However, in the former case there was no significant change in the culturable bacterial population or the microbial biomass of the soil compared to the control from the vicinity

In this study we have examined the bacterial communities present in two soils subjected to long-term heavy metal contamination, principally lead and zinc, as a result of historical lead mining activity. The first site was a metallophytic grassland, a habitat that contains a high diversity of flora capable of tolerating heavy metals in the soil (Balabane et al. 1999). The site had been previously surveyed for vegetation coverage (Leonard, 2001). Here we characterised the extractable lead and zinc from the soil in eight distinct areas defined by the vegetation. Heavy metal concentrations differed significantly across these areas and we characterised the bacterial soil community in each to understand how the community structure differed under the different metal concentrations. The second site was a spoil heap from an abandoned lead mine. Here we were interested in seeing how a bacterial community that had developed under high concentrations of heavy metals responded to additional metal stress. In particular whether such communities exhibited higher or lower stability to additional environmental disturbance.

Materials and Methods

Field Sites and Sampling Strategy

Two sites were used in this study both were metallophytic grassland sites (Balabane et al, 1999) in the Pennine uplands of northern England. The Garrigill site, located at grid reference 714 435 (OS Landranger 87) has thin, poorly sorted soil, with sparse but species rich vegetation, comprising of grasses, predominantly sheep fescue (*Festuca ovina*) and species typical of metallophytic grasslands (table 1). The site is roughly triangular with distinct boundaries of the South Tyne to the south, a dry stone wall to the north and a tributary to the river defining the eastern boundary (figure 1). The site has been extensively grazed by rabbits. Heavy metals on the site has arrived through alluvial deposition via the South Tyne river as a consequence of intensive lead and zinc mining in the nineteenth and early twentieth centuries in the North Pennines (Macklin *et al* 1997). The site represents a river shingle in the floodplain of the South Tyne that has, due to decreased flooding since the turn of the nineteenth century, become vegetated. The community has developed over the last 150 years and has areas of grassland, scrub and some mixed deciduous and coniferous woodland. The vegetation of the site was surveyed in an earlier study (Leonard, 2001) that identified eight areas (figure 1).

Table 1. The differentiation of the Garrigill field site into eight distinct areas based on the most abundant plants, lichens and Bryophytes in each after Leonard (2001)

Area	Vegetation
1	*Festuca ovina* (abundant); *Viola lutea**
2	*Gunneracea*; Scrub (*Crataegus monogyna)*
3	*Festuca ovina* (abundant), *Plantago lanceolata, Hyloconium splendens*
4	*Festuca ovina* (sparse); *Plantago lanceolata, Hyloconium splendens*
5	*Festuca ovina* (sparse); *Crepis aurea*
6	*Minuartia verna*, Thalspi caerulescens*, Peltigera leucophlebia**
7	*Armeria maritime*, Minuartia verna*, Thalspi caerulescens**
8	Mixed woodland (Pine; Beech); *Ulex europaeus*

* Species diagnostic of a metallophytic grassland community.

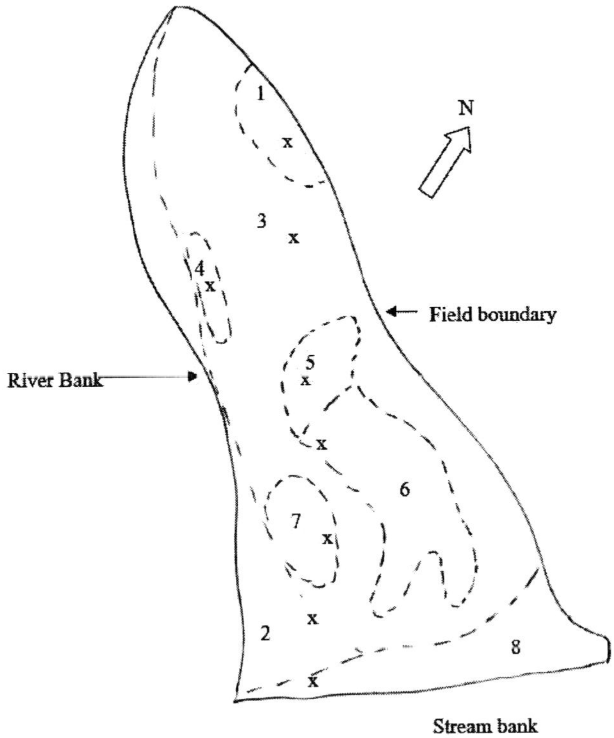

Figure 1. A plan of the Garrigill site showing the eight areas identified by vegetation survey (Leonard, 2001). The position of the 1 m^2 quadrats used for sampling are marked by 'x' in each area. All except that for area 4 were taken along a transect 3 m wide.

The site was divided into 8 areas on the basis of the vegetation cover (Leonard, 2001). To sample each area a transect 3 m wide was plotted across the site that covered all areas except area 4. A 1 m^2 quadrat divided into 10 squares was then placed along the transect within each area. Each of the 8 areas was then randomly sampled from 3 squares within the quadrat by collecting soil cores from the top 10 cm of topsoil in May 2005. Area 4 was sampled identically.

The second site at Bardon Mill, grid reference 779 644 (OS Landranger 87) was derived from an abandoned mine spoil heap. The soil had very sparse vegetation, including grasses, however, most of the site had no vegetation cover. Mining occurred from the mid sixteenth century to the early twentieth century, therefore the site is no more than a few hundred years old. Bardon Mill soils were collected from non-vegetated areas by collecting top soil from the top 10 cm of soil. The area was much more limited and homogeneous than the Garrigill site. Therefore, triplicate samples of topsoil were excavated from the non-vegetated area of the site.

Immediately after collection soil was transported to the laboratory and stored at 4° C until required. Prior to being used in microcosms, soil was sieved (2mm) and air dried overnight. When required, soils were sterilised by autoclaving the soil (121° C, 15 mins) on three consecutive days with incubations at 37° C in between and a final incubation at 100° C for 30 min to dry (Baxter et al. 2006).

Analyses of Soil Samples

Total metal content was determined with duplicate soil samples (0.5 g). Samples were air dried and digested in 10 ml of boiling *aqua regia* (HNO_3: HCl, 1:3, v/v).Once the sample solutions were cool, they were filtered through Whatmans No. 1 filter papers to remov any insoluble sediment or organic material. The solution was then made up to 100 ml with distilled water. Sample were tested for total metal content by Flame Atomic Absorption Spectoscopy (Flame Atomic Absorption Spectroscope) (Perkin Elmer / AAnalyst 100), with "Calib" software. Samples were calibrated against Pb and Zn solutions.

Carbon and Nitrogen analyses were commercially done with duplicate samples of soil (1g) using a Carlo Erba 1108 Elemental Analyser controlled with CE Eager 200 software, run in accordance with the manufacturer's instructions and weighed using a Mettler MT 5 Microbalance by the Advanced Chemical and Material Analysis, University of Newcastle-upon-Tyne, Newcastle, UK.

The contamination index (CI) was calculated for Pb and Zn in the Garrigill site soils according to the method of Feris et al. (2003). The index measures the contamination in the soil relative to the most uncontaminated area in the site (area 8) according to the formula

$$CI = \Sigma\{\log [HM_n / HM_{n \text{ in control}}]\} / \text{Number of metals in index}$$

n = Pb and Zn

(Feris et al. 2004)

Soil Microcosms

For each treatment, namely, controls containing unmodified soil and test soil with additional heavy metal added, three microcosms, consisting of 500 g of soil in each replicate, were established in 600ml glass beakers. The soil was rewetted to a moisture content of 10% (w/w, dry weight basis). To increase the heavy metal stress of the soils to five times that of the background levels $PbNO_3$ and $ZnSO_4$ were added to the soil in aqueous solution. A carrier system was used to ensure consistent metal application. Appropriate concentrations of $PbNO_3$ and $ZnSO_4$ dissolved in dH_2O were applied to 1 g aliquots of sterile soil subsequently these were incubated overnight to allow evaporation of the water. The treated aliquot was cut into the rewetted soil by thorough mixing, giving soil microcosms with final concentrations of $PbNO_3$ and $ZnSO_4$.

Samples (1g) of soil were taken from each microcosm for DGGE analysis. Soil water content was measured and found to be constant during the experiment. Initial samples (day 0) were taken immediately after the microcosms were set up. Between samplings microcosms were incubated at 18° C in a water-saturated atmosphere (Rasmussen and Sorensen 2001).

DNA Extraction and PCR Amplification of 16S rRNA Genes from Soil Microcosms

DNA was extracted from 1g of soil with an UltraClean Soil DNA kit (MoBio) and stored at –20°C. PCR amplification of V3 region of the bacterial 16S rRNA gene was performed using Primer 1 (5'-CCTACGGGAGGCAGCAG-3') and primer 2 (5'-CGCCCGCCGCGC GCGGCGGGCGGGGCGGGGGCACGGGGGGCCTACGGGAGGCAGCAG-3') (Muyzer et al., 1993). Amplification of DNA was performed with 0.5 µM each primer, 2 x *Pfx* amplification buffer, 1 µl template DNA, 0.3mM each dNTP, 1mM $MgSO_4$, 1 x PCR enhancement solution, 500 µg/ml BSA and 1.25 U Platinum *Pfx* DNA polymerase (Invitrogen) in a 50 µl PCR reaction mixture. The thermocycling conditions used were as follows: 3 min. at 94°C (1 cycle), 94°C for 1 min. (10 cycles), 60°C for 1 min. (with reduction of 1°C every cycle), 68°C for 1 min. (10 cycles), then 94°C for 3 min., 50°C for 1 min. and 68°C for 1 min. (25 cycles) with a final extension step of 68°C for 10 min (Baxter and Cummings 2006). All reactions were carried out using the Eppendorf Mastercycler Gradient PCR machine. PCR products were analysed by agarose (1.5% w/v) gel electrophoresis and ethidium bromide staining.

Reverse Transcription PCR (RT-PCR)

Bacterial RNA was extracted from soil using with an Ultraclean™ Microbial RNA Kit (MoBio) according to the manufacturer's instructions, and stored at -20° C.

RNA was amplified using the Omniscript RT kit (Qiagen). RT PCR was performed in a final volume of 20µl, containing 5µl RNA, 10U RNAseOUT (Invitrogen), 10µM Random Hexamers (Qiagen), 0.5µM each dNTP, 10U Omniscript reverse transcriptase, and 1x Buffer RT. The mixture was incubated at 37° C for 60min then stored at -20° C.

Selective amplification of Actinomycetes was performed using Primer F243 (5'-GGAT GAGCCCGCGGCCTA-3') and primer R513 (5'-CGGCCGCGGCTGCTGGCACGTA-3') (Heuer *et al.*, 1997). Amplification of soil Actinomycete DNA was performed as previously described by Heuer *et al.* 0.5µM each primer, 1µl of template DNA, 2 x *Pfx* amplification buffer, 1mM $MgSO_4$, 1 x PCR enhancement solution, 25µg BSA, and 1.25U Platinum *Pfx* DNA polymerase (Invitrogen). PCR conditions were 5 min at 94° C; 35 cycles 1 min at 94 ° C, 1 min at 63 ° C, and 2 min at 72° C; and a final extension for 10 min at 72°C.

PCR products were analysed by agarose (1.5% w/v) gel electrophoresis and ethidium bromide staining.

Denaturing Gradient Gel Electrophoresis

Denaturing Gradient Gel Electrophoresis (DGGE) analysis was performed using the D-Code DGGE system (Bio-Rad) as described by Baxter and Cummings (2006). For analysis of bacterial populations, PCR products were loaded on to polyacrylamide gels (12%) with a denaturant gradient ranging from 35-55% denaturant (with 100% denaturant corresponding to

7M urea plus 40% v/v formamide). Gels were run at 60° C for 10.5 hours at 80 V. Gels were stained in SYBR Green I (1:10000 dilution in 1x TAE buffer; Molecular Probes) for 30 min. and viewed with UV transillumination and photographed using the Gel Doc 2000 gel documentation system (Bio-Rad).

Statistical Analysis of DGGE Profiles

For each treatment, PCR products, derived from DNA extracted from 3 independently derived communities, were analysed by DGGE. The patterns obtained from these replicates were highly reproducible. To facilitate sample processing, PCR products were pooled for each treatment. The pooled PCR products were then subjected to triplicate DGGE analyses to confirm that the band patterns were reproducible with those derived from the independently derived community replicates (Sigler and Turco, 2002; Brodie *et al.* 2003).

DGGE profiles were analysed using Quantity One 4.1.1 software (Bio-Rad). The lane background was subtracted by using a rolling disc algorithm and the lanes were 'normalised' to ensure all lanes contained the same amount of total signal. Lane profiles were corrected for differences in migration rate by manually assigning R_f lines to marker lane bands (Griffiths et al. 2003). To facilitate the efficient handling of samples and data, each sample from each replicate were screened with PCR-DGGE to assess replicate variability in DGGE profiles. When the reproducibility of PCR profiles within a treatment had been ensured, all PCR products derived from a common treatment were then pooled for the final analysis in order to accommodate the electrophoresis of samples on a single gel.

Band patterns were analysed using three methods. Correspondence analysis (CA) was performed to visualise the relationships among band patterns using Statistica 6.0 software (StatSoft Inc.). Cluster analyses of band patterns were also performed with the unweighted pair group method using averages (UPGMA) based on the Morisita index of Similarity. Band pattern data was analysed using Raup and Crick's index of similarity The Raup and Crick index of similarity was used to test if similarities observed between soil cultures profile patterns were significantly higher or lower than would be observed by chance (Rowan et al. 2003). Raup and Crick and cluster analyses were calculated using the PAST statistical analysis software.

Results

The Effect of Additional Heavy Metal Stress on Bacterial Communities in Soils from Lead Mining Spoil

To study how bacterial community structure responded to the input of additional heavy metals in contaminated soils, microcosm experiments were performed using top soil from the Bardon Mill site. The concentrations of extractable Pb and Zn in this soil were determined to be ~3 mg g^{-1} and 1 mg g^{-1} respectively. In experimental microcosms these levels were supplemented with $PbNO_3$ and Zn $ZnSO_4$ to give an increase in the concentration of these two metals to 15mg g^{-1} $PbNO_3$ and 5 mg g^{-1} $ZnSO_4$. Subsequently, the impact of these additions were monitored over a 51 day period and compared with control microcosms that had received no additional heavy metal inputs, using DGGE analyses of the bacterial community.

Samples of soil (1 g) were collected immediately after setting up the microcosms (day 0) and then on days 2, 7, 10, 16, 23, 30, 37, 44 and 51. Total DNA was extracted from each and amplified with universal eubacterial (Eub)primers and primers to amplify actinobacteria 16S rRNA genes. Amplified products from each microcosm were run on DGGE gels (figure 2). Visual comparison of the gels showed that there were qualitative and quantitative differences between the control microcosms and those that had been supplemented with additional lead and zinc. The samples of community DNA amplified with universal Eub primers showed that across thc timc coursc of thc cxpcriment (51 days) the diversity calculated using the Shannon Weiner index, within each treatment did not change significantly. However, when the control and heavy metal amended microcosms were compared there was a significant ($P > 0.05$) decrease in the diversity of the soil that had received additional heavy metal input compared to the control microcosm (table 2).

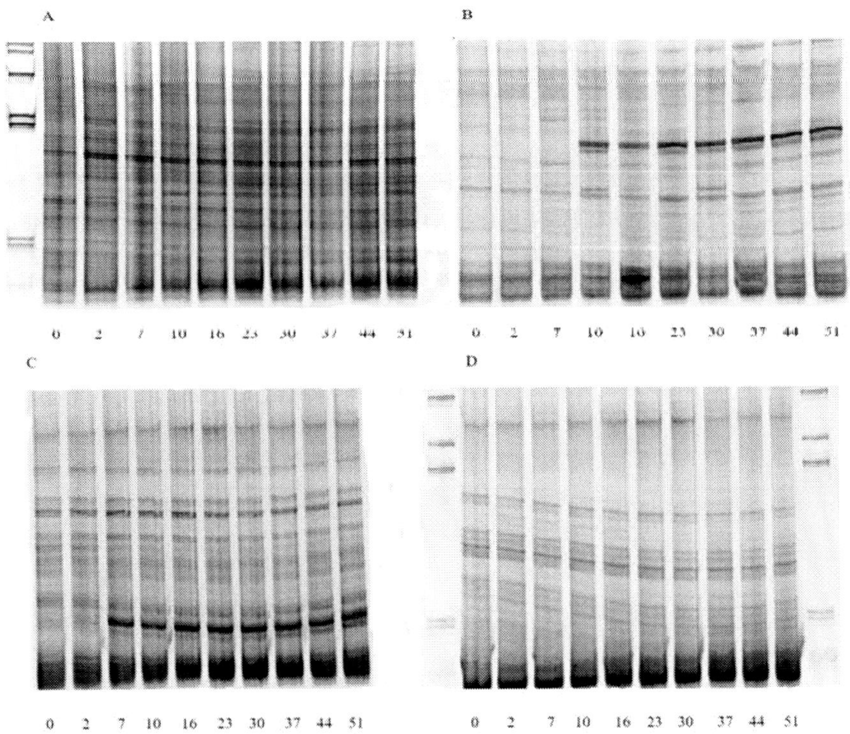

Figure 2. 16S rDNA DGGE analysis of bacterial communities in soil microcosms derived from the Bardon Mill field site. A. Eubacterial profile from non-amended soil. B. Eubacterial profile from soil spiked with a fivefold increase in Pb and Zn. C. Actinobacterial profile from non-amended soil. D. Actinobacterial profile from soil spiked with a fivefold increase in Pb and Zn. Each gel was produced from pooled DNA derived from triplicate microcosms after the reproducibility of the banding patterns was confirmed between replicates, and replicates and pooled DNA. Lane 1: day 0 profile; Lane 2: day 2; Lane 3: day 7; Lane 4: day 10; Lane 5: day 16; Lane 6: day 23; Lane 7: day 30; Lane 8: day 37; Lane 9: day 44; Lane 10: day 51.

Table 2. A comparison of the Shannon-Weiner diversity indices (H') of the Eubacterial (Eub) and Actinobacterial (Act) communities in microcosms derived from the Bardon Mill field site. Microcosm soil was either left untreated or amended with a fivefold increase in Pb and Zn (x5) concentrations

		H'		
	Eub	Eub x5	Act	Act (x5)
day 0	3.81	2.68	3.28	2.92
day 2	3.71	2.57	3.01	3.01
day 7	3.88	2.81	2.97	2.99
day 10	3.83	3.03	3.01	2.96
day 16	3.88	2.75	2.91	3.03
day 23	3.90	2.89	2.70	2.97
day 30	3.84	2.96	2.86	3.04
day 37	3.74	2.96	2.78	3.09
day 44	3.79	2.90	2.85	3.00
day 51	3.78	3.00	2.94	2.95
Mean	3.82 ± 0.06	2.86 ± 0.15	2.93 ± 0.16	3.00 ± 0.05

In relation to the actinobacterial community structure, control microcosms showed a decrease in the diversity after the initial sample, but subsequently the H' values remained similar throughout the course of the experiment (table 2). In the amended microcosm, there was no significant change over the course of the experiment and the mean diversities measured between control and amended soils was not significantly different (table 2). In contrast to the reduced diversity in amended soils determined using V3 primers sets.

The DGGE band patterns generated at each time point were subjected to statistical analyses to identify any measurable changes in the community structure across the time course of the experiment and in response to heavy metal addition. Cluster analyses using the UPGMA algorithm of lane band patterns produced using the eubacterial primers (figure 3 and figure 4) were performed. The eubacterial community in non-supplemented soil gave two clusters one consisting of the earlier samples (days 0 to16), and the second comprised of the later samples. The similarity between the two clusters was about 80%. A similar partition of the profiles was also seen in the microcosm augmented with fivefold more Pb and Zn (figure 3). Although these clustering patterns were observed in both microcosm treatments, the similarity between the clusters was less pronounced (73 %) in the microcosm soil supplemented with additional Pb and Zn. The use of the Raup and Crick (S_{RC}) index to quantify the relationships between the community profiles indicated that the eubacterial communities changed over time, irrespective of the presence of additional heavy metal and became significantly similar to each other (table 3). However, additional heavy metal appeared to inhibit this drift toward significant similarity, so for example, in microcosms without added Pb and Zn, the profiles from days 23 to day 51 showed a similarity to each other suggesting greater than could be explained by chance matching of bands. This is indicative of deterministic selection was acting on the community (table 3a). In contrast, this trend towards similar population profiles was much less pronounced, occurring from day 37 onwards, in microcosms amended with fivefold more Pb and Zn added.

Table 3. A comparison of similarity values (S_{RC}) for the community profiles from Bardon Mill soil microcosms over a 51 day incubation. A. Eubacterial communities B. Eubacterial communities amended with a fivefold increase in Pb and Zn concentrations. C. Actinobacterial communities D. Actinobacterial communities mended with a fivefold increase in Pb and Zn concentrations. 0.95> S_{RC}> 0.05 similarity no greater than expected by chance (null hypothesis upheld), S_{RC}< 0.05 significant dissimilarity, S_{RC} > 0.95 significant similarity (null hypothesis fails)

A	**day 0**	**day 2**	**day 7**	**day 10**	**day 16**	**day 23**	**day 30**	**day 37**	**day 44**	**day 51**
day 0	1	1	0.98	0.74	0.56	0.84	0.77	0.94	0.94	0.43
day 2		1	1	0.99	0.95	0.96	0.89	0.95	0.81	0.79
day 7			1	0.79	0.92	0.97	0.92	0.53	0.56	0.33
day 10				1	0.97	0.67	0.49	0.88	0.74	0.34
day 16					1	0.98	0.93	0.90	0.57	0.76
day 23						1	1	1	0.99	0.99
day 30							1	1	1	1
day 37								1	1	1
day 44									1	1
day 51										1
B										
day 0	1	0.97	0.76	0.34	0.84	0.89	0.92	0.90	0.84	0.84
day 2		1	0.99	0.32	0.99	0.98	0.63	0.98	0.99	0.8
day 7			1	0.50	0.98	0.82	0.80	0.80	0.92	0.34
day 10				1	0.82	0.28	0.28	0.002	0.24	0.33
day 16					1	0.98	0.86	0.64	0.84	0.14
day 23						1	0.99	0.99	0.88	0.51
day 30							1	0.93	0.89	0.54
day 37								1	0.99	0.97
day 44									1	0.99
day 51										1

Table 3. (Continued)

C	day 0	day 2	day 7	day 10	day 16	day 23	day 30	day 37	day 44	day 51
day 0	1	0.59	0.06	0.27	0.05	0.28	0.16	0.34	0.02	0.22
day 2		1	0.79	0.63	0.46	0.89	0.87	0.94	0.65	0.95
day 7			1	0.99	0.8	0.77	0.80	0.97	0.99	0.96
day 10				1	0.98	0.88	0.88	0.99	0.99	0.99
day 16					1	0.99	0.9	1	1	0.97
day 23						1	1	1	0.98	0.99
day 30							1	0.99	0.95	0.98
day 37								1	1	1
day 44									1	0.99
day 51										1
D										
day 0	1	0.92	0.69	0.88	0.56	0.61	0.50	0.28	0.40	0.60
day 2		1	0.96	0.99	0.66	0.97	0.64	0.45	0.83	0.93
say 7			1	0.99	0.99	0.99	0.91	0.70	0.96	0.99
day 10				1	0.98	0.99	0.99	0.91	0.95	0.99
day 16					1	0.97	0.89	0.84	0.88	0.97
day 23						1	0.99	0.98	1	1
day 30							1	1	0.97	0.99
day 37								1	0.95	0.91
day 44									1	1
day 51										1

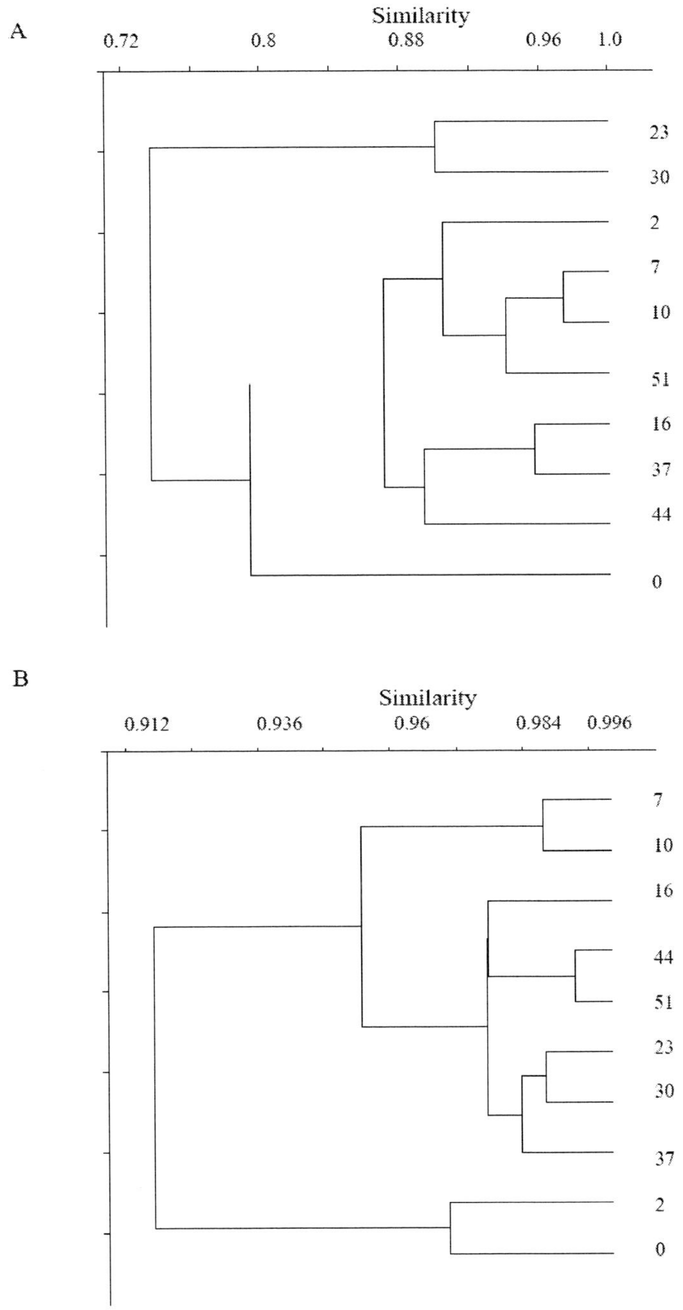

Figure 3. Cluster analyses using UPGMA algorithm of DGGE profiles from microcosm soils. A. Eubacterial profiles from non-amended soil B. Eubacterial profiles from soil spiked with a fivefold increase in Pb and Zn. Numbers refer to the day that the soil was sampled.

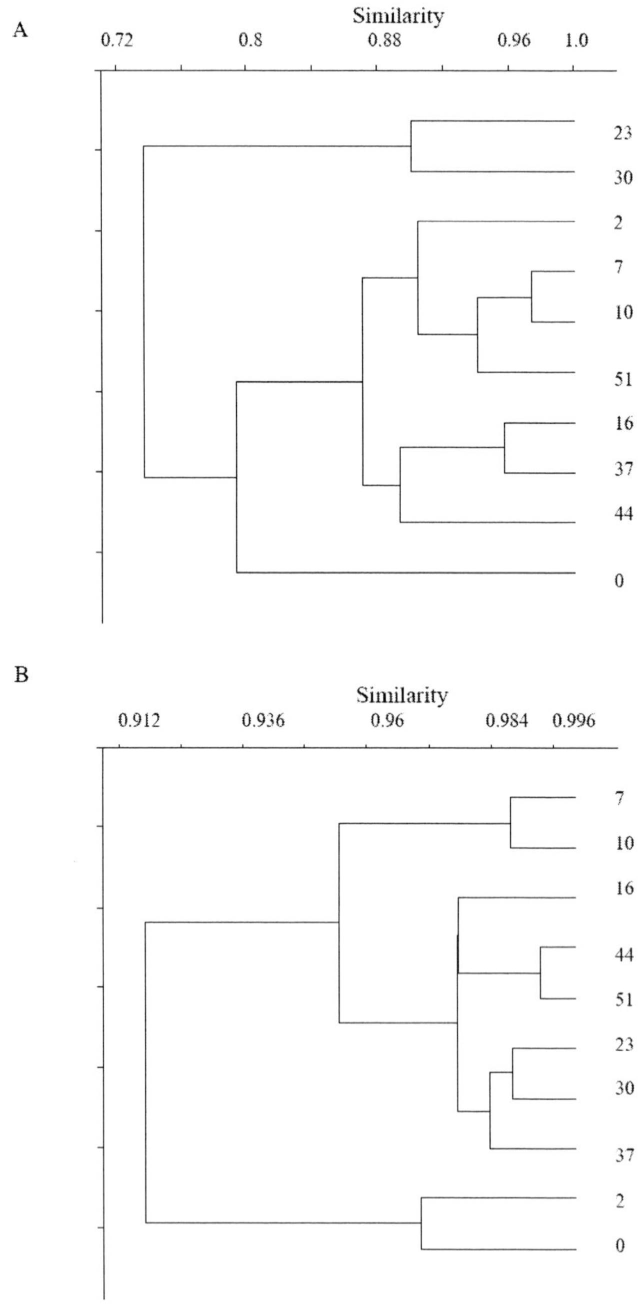

Figure 4. Cluster analyses using UPGMA algorithm of DGGE profiles from microcosm soils. A. Actinobacterial profiles from non-amended soil. B. Actinobacterial profiles from soil spiked with a fivefold increase in Pb and Zn. Numbers refer to the day that the soil was sampled.

The actinobacterial profiles gave less clear time related clusters. In non supplemented microcosm soil two distinct clusters were seen, however, one consisted of the profiles derived from days 23 and 30, the other cluster contained the remaining profiles that appeared interdigitated, with the exception of the day 0 profile, the similarity between the two was

73%, in addition. The day 0 profile formed an outlier in this second cluster, being 79% similar to the other profiles within it (figure 4a). In supplemented soils the first two samples points (days 0 and 2) gave a distinct cluster, the remaining profiles formed a second but the similarity between the two was 91.6 %. In contrast to the eubacterial profiles, it appeared that adding additional heavy metal to the soil increased the similarity between profiles. The Raup and Crick analyses showed actinobacterial communities under both treatments showed a drift towards similar community profiles, such that the comparison of profiles from day 7 onwards under both treatments shows high, frequently significant, similarity between the community profiles (tables 3c,d).

Bacterial Community Analyses from Metallophytic Grasslands of the South Tyne Valley

The Garrigill field site was located on the bank of the upland river the South Tyne in the Pennine uplands of northern England. The site had been subjected to a vegetation survey (Leonard, 2001) that had divided the site into eight areas on the basis of predominant vegetation and its coverage (figure 1). The majority of the site was considered to be a metallophytic grassland that included heavy metal tolerant indicator species, however, areas 2 and 8 on the SE border of the site were predominantly woodland and scrub (table 1). The analyses of the bacterial community was undertaken by sampling top soil from each of the 8 areas, soil was characterised in terms of its composition (sand 84.2%, silt 10.5 % and clay 5.3 %) total extractable lead and zinc content and total organic carbon and nitrogen (table 4).

The amounts of Pb and Zn in the soil was found to vary significantly across the site. Area 1 had the highest concentration of heavy metals and the wooded area 8 the lowest (table 4). The contamination index (CI) was determined with respect to Pb and Zn for each area (Feris et al., 2003).

We extracted RNA from soil samples to determine the diversity of the metabolically active eubacterial and actinobacterial communities. The metabolically active eubacterial community amplified (figure 5a) produced a limited number of bands. The diversity calculated with the Shannon Weiner (H') index was correspondingly low (table 4). In the most contaminated area 1 only a single band was visible on the gel and the H' value was significantly ($P>0.05$) lower than that of the other areas. The limited diversity was not consistently linked to the contamination index, the most diversity was found in area 5 (CI = 0.53) and this was significantly higher than that of the other areas in the field site. The soil in this area had the highest organic C and N content (8 %) and these elements were also shown to be relatively high in areas 6 and 7 and to a lesser extent area 2 which exhibited the next highest eubacterial diversities (table 4). Conversely, diversity was lowest in 1, 4 and 8 which had the lowest amounts of measured organic C and N. While area 8 had the lowest CI, that area of 4 was amongst the highest.

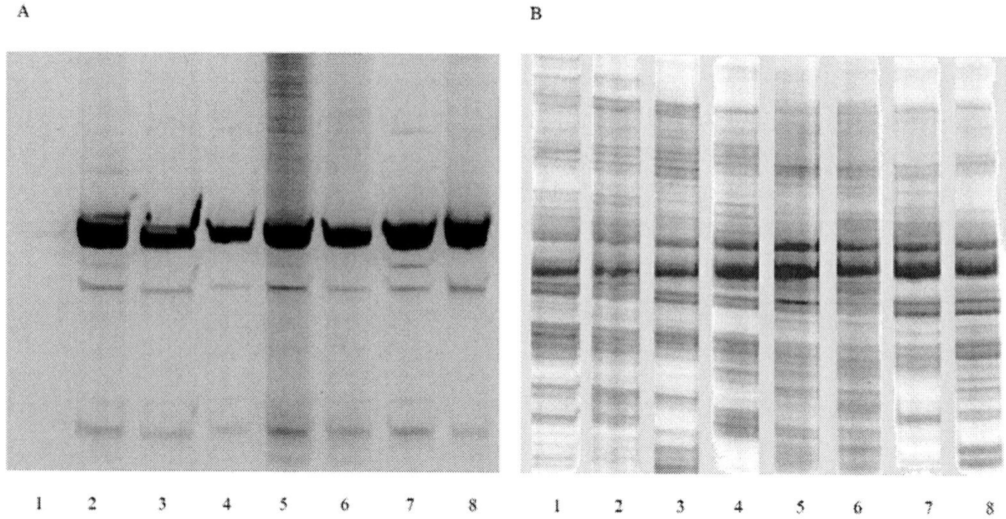

Figure 5. 16S rRNA DGGE analysis of bacterial communities in soil from the Garrigill site A. Eubacterial profiles. B. Actinobacterial profiles. Each gel was produced from pooled DNA derived from triplicate 1 g soil samples after the reproducibility of the banding patterns was confirmed between replicates, and replicates and pooled DNA. Lane 1: area 1 profile; Lane 2: area 2; Lane 3: area3; Lane 4: area 4; Lane 5: area 5; Lane 6: area 6; Lane 7: area 7; Lane 8: area 8. Areas correspond to those described in figure 1.

Using more specific primers to amplify the metabolically active Actinobacteria community produced a large number of bands and a high diversity index in each area (figure 5b). In contrast to the community identified using the eubacterial primers area 1, with the highest CI, also had the highest diversity of any area on the site. However, as with the eubacteria there was no clear cut relationship between the CI and the bacterial diversity. Similarly, the amount of increased organic C and N in the soil did not give consistent responses in actinobacterial diversity (table 4).

Correspondence analyses of the DGGE band patterns of eubacterial and actinobacterial communities for each area enabled visualization of the relationship between the bacterial communities from the different areas (figure 6). Correspondence analyses is a multivariate technique that enables the transformation of a complex data set by constructing a new set of variables that are linear combinations of the original variables, i.e. the 16S rDNA gene bands and the soil from the 8 areas of the site. For the total bacterial analyses using the V3 primers the samples with the lowest CI (areas 2, 3 and 8) clustered very closely on the first axis that explained 48.5 % of the inertia but were separated across the second axis that explained almost 23% of the inertia figure 6a). Areas 3 and 8 also formed a tight cluster with areas 4, 7 and 6 across both axes, areas 6 and 7 had amongst the highest organic C and total N content, however, area 4 in contrast had low values but all sites shared a similar CI (table 4). Area 5 that had the significantly highest diversity was separated from all other profiles across the first axis, and area 1 that had the highest CI lay between area 5 and the remaining profiles along the first axis.

The profiles of the metabolically active Actinobacteria indicated that the profile from area 1 with the highest CI was distinct from the other profiles across the first axis that accounted for 29.6% of the inertia (figure 6b). In addition areas 2 and 3 formed a tight group, on both axes, these soil samples exhibited, along with the woodland area 8, the lowest CI and moderate to low levels of organic C and total N. The latter area was separated form these profiles along the first axis but clustered closely with them along the second. Indeed all the low CI soil profiles (2, 3, 5 and 8) formed a discrete group along axis two bit were separated along axis 1 (figure 6b). The profiles 4, 5, 6, 7 and 8 were very tightly clustered across the first dimension but were separated across the second axis that described over 17% of the inertia in the sample (figure 6b).

A

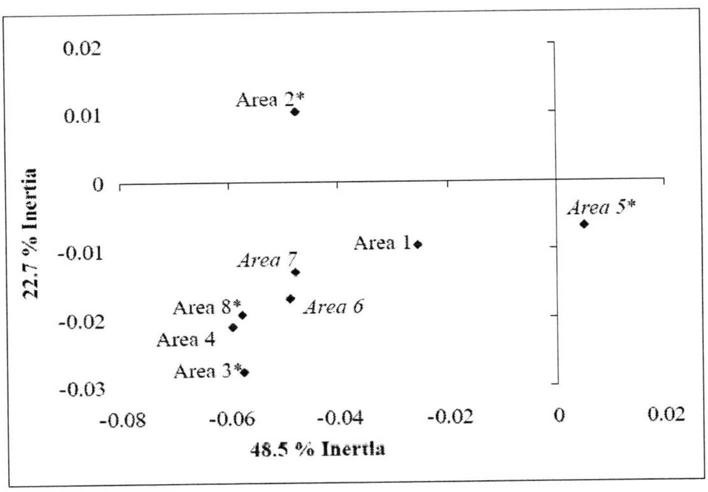

B

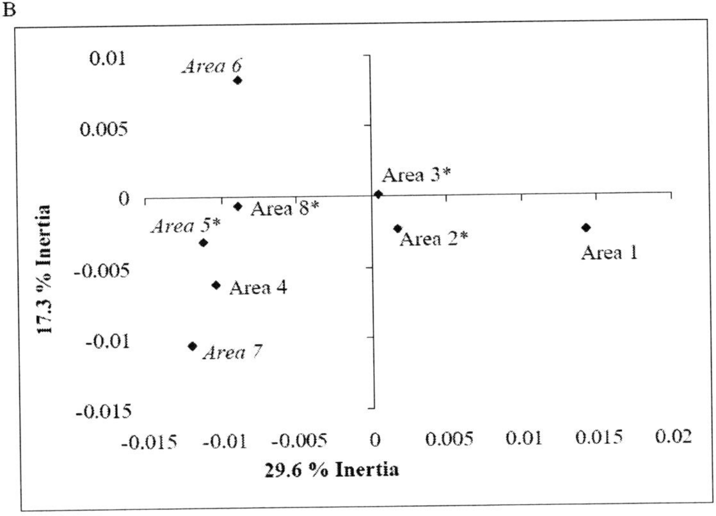

Figure 6. Correspondence analysis of the 16S rRNA DGGE community profiles from the 8 areas within the Garrigill field site. A. profiles of Eubacterial communities. B. Profiles of Actinobacterial communities. Those areas italicised have soil with high organic C and total N content. * Indicates a low contamination index (see table 4).

Raup and Crick analyses was undertaken to determine whether any of the profiles across the site shared a statistically meaningful relationship (table 5). the analyses of the profiles determined from the metabolically active eubacteria indicate that two sites areas 3 and 5 have bacterial community profiles that are significantly different to each other (table 5a), the CI for these two areas are both on the lower side of the range for the site but have a significant difference in their organic C and total N content (table 4). The vegetation at both sites was predominantly grass (*F. ovina*), but in area 5, the coverage of grass was sparse and typical heavy metal tolerant plant species of a metallophytic grassland were evident (table 1). The community profile of area 4 was significantly similar to that of areas 6, 7 and 8 suggesting a deterministic selection for these populations (table 5a), vegetation between areas 4, 6 and 7 was similar, with comparable CI values for each soil, but area 4 had significantly lower amounts of organic C and N present in the soil. In contrast, area 8 was woodland and had the lowest CI index on the site (table 4). Areas 2 and 7 showed significant similarity in population profiles (table 5). However, the organic C and N content of the soil and the vegetation was distinct between these two sites, area 2 being largely scrub, whereas area 7 had a large coverage within the quadrat of typical metallophytic flora, however the two sites were physically linked (figure 1).

Table 5. A comparison of similarity values (S_{RC}) for the metabolically active community profiles from soil derived from the 8 areas (Fig. 1) of the Garrigill field site. A. Eubacterial communities B. Actinobacterial communities. 0.95> S_{RC}> 0.05 similarity no greater than expected by chance (null hypothesis upheld), S_{RC}< 0.05 significant dissimilarity, S_{RC} > 0.95 significant similarity (null hypothesis fails)

A	Area 1	Area 2	Area 3	Area 4	Area 5	Area 6	Area 7	Area 8
Area 1	1	0.7725	0.88	0.865	0.5875	0.755	0.855	0.8725
Area 2		1	0.2675	0.92	0.1425	0.375	0.995	0.865
Area 3			1	0.605	0.005	0.2525	0.46	0.55
Area 4				1	0.8225	0.985	0.99	0.9875
Area 5					1	0.2025	0.6125	0.4725
Area 6						1	0.925	0.8
Area 7							1	0.92
Area 8								1
B	**Area 1**	**Area 2**	**Area 3**	**Area 4**	**Area 5**	**Area 6**	**Area 7**	**Area 8**
Area 1	1	0.9775	1	0.3825	0.1325	0.41	0.565	0.81
Area 2		1	0.975	0.305	0.405	0.375	0.345	0.8925
Area 3			1	0.6025	0.315	0.7975	0.64	0.6275
Area 4				1	0.6175	0.7475	0.985	0.3575
Area 5					1	0.705	0.995	0.4375
Area 6						1	0.6425	0.8675
Area 7							1	0.835
Area 8								1

The profiles derived for the metabolically active Actinobacteria showed that areas 1, 2 and 3 had statistically similar community profiles (table 5b), while areas 1 and 3 shared common features in terms of vegetation coverage of predominantly grass and heavy metal tolerant flora, area 2 was in contrast scrub (figure 1). Organic carbon and N content of the soils were similar, and while areas 2 and 3 shared similar CI values, amongst the lowest on the site, area 1 was the most contaminated soil. The profile from area 7 was significantly similar to that of areas 4 and 5 (table 5b). Each of these areas shared a moderate to high CI value, however the vegetation cover and organic carbon content of the soil was different in each area.

Conclusion

There have been many studies investigating the impact of long term heavy metal addition on the community structure on soil bacterial populations. However, this is the first report to our knowledge of the bacterial community of the metallophytic grassland of the Pennines. These habitats were formed as a result of a short period mining activity between the mid-eighteenth and mid twentieth centuries. They comprise of spoil heaps from abandoned mine workings and also alluvial deposits, on the flood plains of the local watercourses that have been contaminated by water from abandoned mine workings. They represent a unique habitat, containing a highly diverse flora, many nationally scarce within the UK, that are described as metallophytes (high metal tolerance). These rare habitats are increasingly threatened by natural events, such as succession by birch and gorse, and a range of anthropogenic activities, including, farming, spoil reclamation, gravel extraction and leisure uses. Our aim was to investigate the stability of the bacterial soil communities associated with these sites and to characterise the bacterial community structure under such a metallophytic grassland subjected to different heavy metal concentrations and vegetation cover.

The first soil studied was obtained from a historical lead mine spoil heap at the Bardon Mill site, The hypothesis we wished to test was whether a bacterial community previously exposed to high heavy metal concentrations would tolerate additional heavy metal, due to adaptation and functional changes within the community predisposing it to cope with the additional stress (Odum, 1981). We predicted that additional heavy metal addition to the soil would not lead to a significant decrease in on the community diversity and structure, compared to the control microcosm if the hypothesis was correct.

Visual inspection of the DGGE gels derived from the microcosm soils suggested that using universal eubacterial (Eub) primers the impact of a fivefold increase in Pb and Zn was to reduce the diversity of the bacterial soil community. This was quantified by determining the Shannon–Weiner index which demonstrated a significant decrease in the diversity of the soil when Pb and Zn concentrations were increased fivefold (table 2). In contrast, the actinobacterial community in microcosms supplemented with additional Pb and Zn , showed no significant change in diversity. These observations are of interest, because the decrease in diversity determined with the Eub primer set suggests that the community is not able to tolerate additional heavy metal stress. Caution must be exercised in the interpretation of DGGE data as the primers used may have a significant effect on the profiles obtained. Often

PCR primers can show a bias towards particular sequences resulting in differential amplification (Joynt et al 2006). This can lead to underestimations of diversity because the more abundant sequences are preferentially amplified masking sequences of lower copy number. Nevertheless, DGGE with Eub primers can demonstrate whether environmental disturbance does cause gross changes in community structure.

The results of our experiment examining the eubacterial populations support the observation by Tobor-Kaplan et al. (2005) that ecosystem stability to heavy metals is not enhanced by long term adaptation to this stress, because diversity significantly declined in the soils challenged with additional Pb and Zn (table 2). Several other studies have found similar reductions in bacterial diversity as a result of heavy metal addition (Sandaa et al. 1999, Joynt et al. 2006; Wu et al. 2006). However, this response is not universal as for example, Lazzaro et al. (2006b) found that diversity was not significantly reduced in forest soils and perhaps more significantly, in a study of a hyporheic (sediment) communities from heavy metal contaminated watercourses no correlation between microbial diversity and metal concentration was apparent (Feris et al. 2003).

In contrast to the eubacterial populations, our Actinobacteria community data was more in line with these latter studies. The Shannon-Weiner diversity indices did not change significantly between the two soil treatments. This suggests that the Actinobacteria taxa are not negatively affected by the additional heavy metal stress. There are conflicting reports of the response of Actinobacteria to heavy metal input into soil. A number of studies have reported that its abundance and diversity decreases in the presence of heavy metal contamination in soils (Frey et al. 2006; Lorenz et al. 2006; Kelly et al. 1999). Whereas, our data supports the observations that they are a significant and diverse element of the bacterial community in soils that have high natural concentrations of heavy metals or has been subjected to long term contamination (DeGrood et al. 2005; Gremion et al. 2003).

The impact of increasing the heavy metal concentration on the bacterial community structure was explored using cluster analyses of the DGGE profiles derived from the microcosm experiments (Li et al. 2006; Kandeler et al. 2000). This analyses indicated that the similarity of the eubacterial profiles showed a time dependency with two clusters, one consisting of profiles from early sample times, and the second comprised of the later time points of the experiment. A similar pattern was observed in microcosms irrespective of the addition of extra Pb and Zn but the difference between early and later clusters was higher than in controls (figure 3). The S_{RC} values showed these changes were significant and suggested a deterministic selection (Rowan et al 2003) was acting on the soil communities in the microcosms (table 3a and 3b). Further heavy metal addition seemed to reduce this deterministic selection. In comparison, the actinobacterial community profiles in the control microcosms showed less time related changes in community profiles. The S_{RC} profiles indicated the addition of heavy metals increased the frequency of significant similarities occurring between profiles. The implication of these data are that the Actinobacteria are subject to deterministic selection in the microcosm and that adding additional heavy metals increases this selection pressure (figure 4a).

Our data suggests that the short term response to additional heavy metal exposure did affect the eubacterial and actinobacterial community structure in the microcosms, but not dramatically. Renella et al. (2002) showed related modest changes to community parameters in short term exposure experiments. It has been widely reported that heavy metals can change the community structure of bacterial populations in soil. Even prior to the widespread application of molecular techniques changes in the genetic structure of communities exposed to long term heavy metal exposure had been noted (Giller et al. 1998). Frey et al. (2006) analysed the impact on community structure of heavy metal contamination and noted a shift in the community structure measured using both PLFA and T-RFLP and that the resulting community had a reduced functional diversity compared to control soils. They speculated that the changes caused by the heavy metal application were to shift to more metal tolerant species. However, the work of Feris et al. (2003) does not support this they found no correlation between heavy metal concentration and reduced diversity or biomass, nor any indication of a more metal tolerant community being selected for. There is data on the specific changes of soil microbial communities subjected to heavy metal stress, and once more these studies present conflicting data. For example, PFLA and DGGE data has indicated that fungi, Proteobacteria and Gram negative increase under such conditions, whereas, Actinobacteria and Firmicutes decrease (Lorenz et al. 2006; Kelly et al. 1999; Hinojosa et al, 2005). However, Bååth et al (2005) showed in historically contaminated soil that using PFLA that Gram positive bacteria where more prevalent and that fungi decreased in Pb contaminated soils. It is of interest that many reports identify a decrease in Actinobacteria on the addition of heavy metals to soil, we did not measure the change in actinobacterial biomass, however, Lorenz et al. (2006) identified a decrease in the respiratory quinones associated with this group in soil exposed to arsenic and cadmium. However, their DGGE analyses, while also showing a reduction in the number of bands present (and by inference the diversity of the community), indicated that bands were lost predominantly from the upper part of the gel, these would be low G+C taxa. Therefore, it may be worth undertaking more focussed studies to identify whether when Actinobacteria are seen to decline whether this can be attributed to a decline in abundance but not in diversity. As in our study no significant decrease in actinobacterial diversity in contaminated soils was observed.

The second field site at Garrigill, was studied to characterise the microbial community across a metallophytic grassland site with a soil subject to differences in both metal concentration and vegetation cover. The soil contained quantities of several heavy metals, the most abundant being Pb and Zn, as the effects of heavy metals on biological communities co-vary and demonstrate a high co-linearity which would as a result confound any attempts to analyse the effects of individual metals (Feris et al. 2003). As a result the contamination index (CI) has been employed to overcome this and to express a relevant measure of metal contamination. The characterisation of the site into 8 areas had been done on the basis of the predominant vegetation cover, it is known that the microbial community in soil, particularly in the rhizosphere, can alter as a result of changes in the quantity and composition of root exudates that plants produce (Baudoin et al 2003). Therefore, it might be expected that the microbial communities will demonstrate significant differences as a result of this rhizodeposition of organic carbon across the site. A crude measure of organic carbon in the soil showed that, area 5 had the most organic C and total N in the soil. This area also had the highest diversity in the metabolically active eubacterial community (table 4). The presence of high concentrations of heavy metals often inhibits plant growth and as a result will decrease soil organic C, the most contaminated area on the site had the lowest level of organic C in its soil and relatively sparse vegetation (table 1; table 4). However, other areas with high CI also have high levels of organic C present in their soils. In particular, areas 6 and 7 show high organic C and total N levels, the biodiversity among the eubacterial community are also higher than the average across the whole site in these areas. Both have significant coverage by metallophytic flora (table 1) unlike area 1. metal tolerant plants are known to harbour heavy metal tolerant bacteria in their rhizospheres (Abou-Shanab et al. 2006), thererefore, heavy metal tolerant flora may enhance both soil C and biodiversity quite significantly. Analyses of the community profiles from the different areas showed a very complex relationship between the eubacterial and actinobacterial community profiles and the other parameters (CI, organic C and total N, vegetation cover) (figure 6). A cluster analyses of the similarity of the profiles (figure 7) demonstrates that the metabolically active eubacterial community could be divided into one cluster containing the single area 5 that was 72% similar to the other cluster which comprised of all the other area profiles whose similarity to each other was between 92-98 % (figure 7). Area 5 had both the highest organic C and total N content in its soil and the highest diversity. In contrast, with the analyses of the actinobacterial communities diversities across the site were less variable, however, again 2 clusters were produced Area 1 with the highest diversity formed a cluster that was 66% similar to the second which contained all the other area profiles. The highest diversity of metabolically active Actinobacteria was therefore in soil with the highest heavy metal concentration and the lowest organic C and total N.

The results of the study on the metabolically active Eubacteria and Actinobacteria at the Garrigill site highlights the difficulties in studying bacterial soil populations *in situ.* A number of factors can influence the development of the metabolically active community, including environmental disturbance and resource availability. Our data indicate that the diversity of the active eubacterial population is strongly linked to the availability of organic C in the soil the highest organic C content equating to the highest diversity in area 5. The heavy metal concentration did appears to exert some effect, in that areas 1, and 4 with high CI values and had the lowest diversity, and area 6, and 7 also with high CI had much reduced community diversity compared to area 5 despite all three areas having soil with high organic C contents. However, Actinobacteria had the most diversity in area 1 with the lowest organic C content. The role of soil organic C in shaping the microbial community has received some attention, in metallophytic grasslands organic C levels are equivalent to those of other temperate grasslands but there is some evidence that plant material, particularly, from the metallophytic flora is only poorly degraded (Balabane et al. 1999). As a result the quality of the organic carbon the soil may be reduced and this favours those microorganisms that can access this refractile carbon, typically these are K-strategists such as Actinobacteria (Fierer et al. 2003). Across the Garrigill site it is possible that the quality of the organic C entering the soil is variable, so that in area 5 with low levels of heavy metal and few metallophytic hyperaccumulators the plant derived organic C is readily degradable by r-strategists adapted to acquiring labile carbon rapidly (Hery et al 2005). This would account for the high eubacterial diversity in this area. In contrast where organic C is low and heavy metal high as in area 1 (and to a lesser extent in areas 4, 6 and 7) Actinobacteria will be favoured over r-strategists as they are capable of accessing organic material returned to the soil by turnover of the metallophytic flora that may be sequestered with high levels of heavy metals (Balabane et al. 1999). That being said there are a number of studies that demonstrate Actinobacteria can be competitive in soils containing high levels of heavy metals even when labile organic carbon is present (Hery et al 2005), the high background levels of Pb and Zn in the Garrigill soils may enable the Actinobacteria to maintain a relatively high diversity across the site even in those areas where organic C appears plentiful.

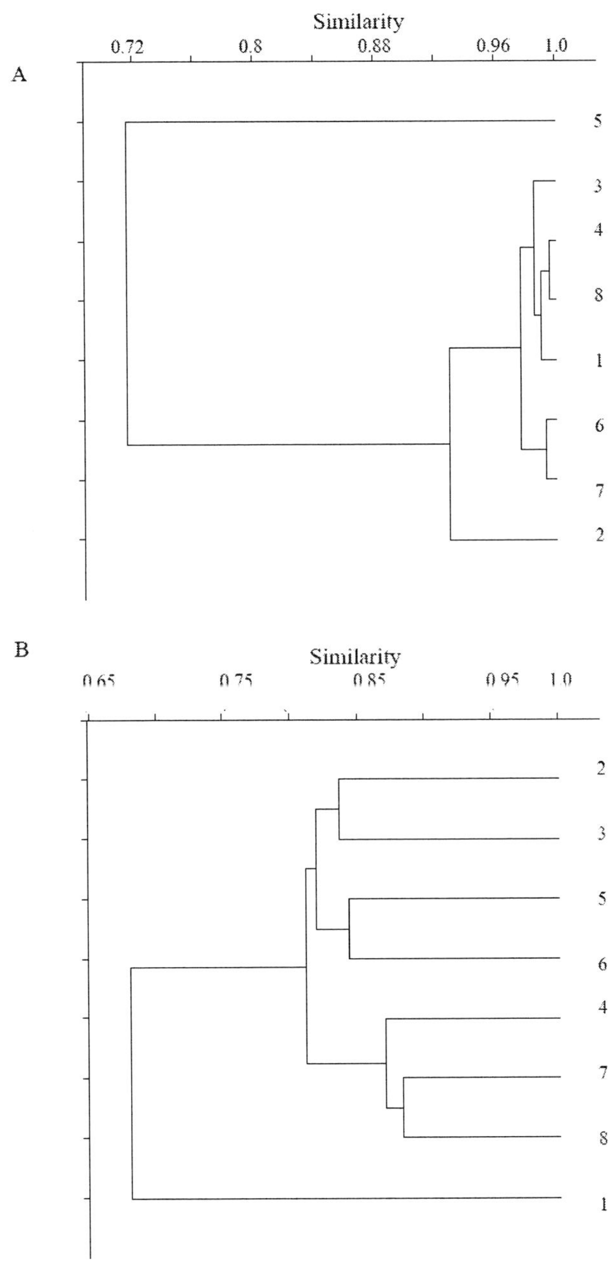

Figure 7. Cluster analyses using UPGMA algorithm of DGGE profiles from the 8 areas sampled on the Garrigill field site. A. Eubacterial community profiles. B. Actinobacterial profiles. The numbers on each branch correspond to the areas of the site from which the profile was derived (figure 1).

Table 4. Characterisation of the soil from the 8 areas of the Garrigill field site in terms of heavy metal, organic C, total N content, and the diversities of the metabolically active eubacterial and actinobacterial communities derived from the DGGE profiles (figure 5)

	Extractable Metals (ppm)			Elemental analyses		Shannon Weiner index (H')	
	Pb	Zn	CI	N (%)	C (%)	Actinobacteria	Eubacteria
Area 1	5285	1617	0.59	0.23	1.76	3.82	0.01
Area 2	1075	621	0.04	0.26	2.23	3.82	1.16
Area 3	1299	638	0.09	0.26	2.19	3.4	0.53
Area 4	3685	1087	0.43	0.18	1.12	3.46	0.34
Area 5	1356	979	0.19	0.73	8.00	3.55	2.76
Area 6	2918	1746	0.48	0.52	6.12	3.55	0.96
Area 7	3924	1984	0.57	0.42	6.19	3.03	0.85
Area 8	909	613	0	0.12	1.05	3.08	0.5

In conclusion, this study has explored the stability and diversity of bacterial communities with an emphasis of the Actinobacteria a significant taxa within the soil flora, in historically contaminated soil from metallophytic grassland sites. We have demonstrated that the eubacterial communities when exposed to additional stress do not exhibit tolerance in terms of maintenance of the diversity of the community. The actinobacterial community, however, appears much more robust in terms of the community structure and diversity in the face of such additional stresses. Subsequently we explored the character of the metabolically active bacterial communities across a field site. Analyses of community structure showed that eubacterial diversity was low and strongly affected by available organic carbon and heavy metal content in the soil. In contrast, the actinobacterial community was much higher, and was lwess susceptible to differences in organic carbon across the site, but did appear to respond to the heavy metal concentration. Such data is important in underpinning our knowledge of scarce metallophytic grassland communities which continue to be threatened by leisure and agricultural activities. more widely, such studies provide important information on how microbial communities respond to heavy metals and can be used to help remediate and restore many of the contaminated sites around the world that have resulted from historical and current industrial activity.

References

Abaye, D. A., Lawlor, K., Hirsch, P. R. and Brookes, P. C. (2005). Changes in the microbial community of an arable soil caused by long-term metal contamination. *European Journal of Soil Science*, *56*, 93-102.

Abou-Shanab, R. A. I., Angle, J. S. and Chaney, R. L. (2006). Bacterial inoculants affecting nickel uptake by Alyssum murale from low, moderate and high Ni soils. *Soil Biology and Biochemistry*, *38*, 2882-2889.

Almås, A. R., Bakken, L. R. and Mulder, J. (2004). Changes in tolerance of soil microbial communities in Zn and Cd contaminated soils. *Soil Biology and Biochemistry*, *36*, 805-813.

Amir, H. and Pineau, R. (1998). Effects of metals on the germination and growth of fungal isolates from New Caledonian ultramafic soils. *Soil Biology and Biochemistry*, *30*, 2043-2054.

Amoroso, M. J., Schubert, D., Mitscherlich, P., Schumann, P. and Kothe, E. (2000). Evidence for high affinity nickel transporter genes in heavy metal resistant *Streptomyces* spec. *Journal of Basic Microbiology*, *40*, 295-301.

Bååth, E. (1989). Effects of heavy metals in soil on microbial processes and populations (a review). *Water Air and Soil Pollution*, *47*, 335-379.

Bååth, E., Diaz-Ravina, M. and Bakken, L. R. (2005). Microbial biomass, community structure and metal tolerance of a naturally Pb-enriched forest soil. *Microbial Ecology*, *50*, 496-505.

Balabane, M., Faivre, D., van Oort, F. and Dahmani-Muller, H. (1999). Mutual effects of soil organic matter dynamics and heavy metals fate in a metallophyte grassland. *Environmental Pollution*, *105*, 45-54.

Bamborough, L. and Cummings, S. P. (2006). The impact of historical heavy metal contamination on the microbial community of upland pastures in the North Pennines. *Aspects of Applied Biology*, *80*, 95-100.

Baudoin E., Benizri E. and Guckert A. (2003) Impact of artificial root exudates on the bacterial community structure in bulk soil and maize rhizosphere. *Soil Biology and Biochemistry*, *35*, 1183-1192.

Baxter, J., Garton, N. J. and Cummings, S. P. (2006) The impact of acrylonitrile and bioaugmentation on the biodegradation activity and bacterial community structure of a topsoil. *Folia Microbiologica*, *51*, 591-598.

Baxter, J. and Cummings, S. P. (2006). The impact of bioaugmentation on metal cyanide degradation and soil bacteria community structure. *Biodegradation*, *17*, 207-217.

Becker, J. M., Parkin, T., Nakatsu, C. H., Wilbur, J. D. and A. Konopka, A. (2006). Bacterial activity, community structure, and centimeter-scale spatial heterogeneity in contaminated soil. *Microbial Ecology*, *51*, 220-231.

Brodie, E., Edwards, S. and Clipson, N. (2003), Soil fungal community structure in a temperate upland grassland soil. *FEMS Microbioogyl Ecology*, 45, 105-114.

Brookes, P. C. and McGrath, S. P. (1984). Effects of metal toxicity on the size of the soil microbial biomass. *Journal of Soil Science*, *35*, 341-346.

Bűrgmann H., Widmer F., Sigler W. V. and Zeyer J. (2003). mRNA extraction and reverse transcription-PCR protocol for detection of *nifH* gene expressed by *Azotobacter vinelandii* in soil. *Applied and Environmental Microbiology*, *69*, 1928-1935.

Cavalier-Smith, T. (2002). The neomuran origin of archaebacteria, the negibacterial root of the universal tree and bacterial megaclassification. *International Journal of Systematic and Evolutionary Microbiology*, *52*, 7-76.

Curtis, T. P., Head, I. M., Lunn, M., Woodcock, S., Schloss, P. D. and Sloan, W. T. (2006). What is the extent of prokaryotic diversity? *Philosophical Transactions of the Royal Society B-Biological Sciences*, *361*, 2023-2037.

Dawson, J. J. C., Godsiffe, E. J., Thompson, I. P., Ralebitso-Senior, T. K., Killham, K. S. and Paton, G. I. (2007). Application of biological indicators to assess recovery of hydrocarbon impacted soils. *Soil Biology and Biochemistry*, *39*,164-177.

DeGrood, S. H., Claassen, V. P. and Scow, K. M. (2005). Microbial community composition on native and drastically disturbed serpentine soils. *Soil Biology and Biochemistry*, *37*,1427-1435.

Duffus, J. H. (2002). "Heavy Metals" – A meaningless term? *Pure and Applied Chemistry*, *74*, 793-807.

Euzéby J. P. (2007) List of Prokaryotic names with standing in the nomenclature. http://www.bacterio.cict.fr. Accessed 18th August 2007.

Fierer, N., Schimel, J. P. and Holden P. A. (2003). Variations in microbial
community composition through two soil depth profiles. *Soil Biology and Biochemistry*, *35*, 167-176.

Fierer, N., Jackson, J. A., Vilgalys, R. and Jackson, R. B. (2005). Assessment of soil microbial community structure by use of taxon-specific quantitative PCR assays. *Applied and Environmental Microbiology*, *71*, 4117-4120.

Feris, K. P., Ramsey, P. W., Rillig, M., Moore, J. N., Gannon, J. E. and Holbert, W. E. (2004). Determining rates of change and evaluating group-level resiliency differences in hyporheic microbial communities in response to fluvial heavy-metal deposition. *Applied and Environmental Microbiology*, *70*, 4756-4765.

Feris, K., Ramsey, P., Frazar, C., Moore, J. N., Gannon, J. E. and Holbert, W. E. (2003). Differences in hyporheic-zone microbial community structure along a heavy-metal contamination gradient. *Applied and Environmental Microbiology*, *69*, 5563-5573.

Frey, B., Stemmer, M., Widmer, F., Luster, J. and C. Sperisen, C. (2006). Microbial activity and community structure of a soil after heavy metal contamination in a model forest ecosystem. *Soil Biology and Biochemistry*, *38*, 1745-1756.

Frostegard, A., Tunlid, A. and Bååth, E. (1993). Phosopholipid fatty-acid composition, biomass and activity of microbial communities from two soil types experimentally exposed to different heavy metals. *Applied and Environmental Microbiology*, *59*, 3605-3617.

Gans, J., Wolinsky, M. and Dunbar, J. (2005). Computational improvements reveal great bacterial diversity and high metal toxicity in soil. *Science*, *309*,1387-1390.

Giller, K. E., Witter, E. and McGrath, S. P. (1998). Toxicity of heavy metals to microorganisms and microbial processes in agricultural soils: A review. *Soil Biology and Biochemistry*, *30*, 1389-1414.

Gremion F., Chatzinotas A. and Harms H. (2003). Comparative 16S rDNA and 16S rRNA sequence analysis indicates that *Actinobacteria* might be a dominant part of the metabolically active bacteria in heavy metal-contaminated bulk and rhizosphere soil. *Environmental Microbiology*, *5*, 896-907.

Griffiths, B. S., Bonkowski, M., Roy, J. and Ritz, K. (2001). Functional stability, substrate utilisation and biological indicators of soils following environmental impacts. *Applied Soil Ecology*, *16*, 49-61.

Griffiths, B. S., Hallett, P. D., Kuan, H. L., Pitkin, Y. and Aitken, M. N. (2005). Biological and physical resilience of soil amended with heavy metal-contaminated sewage sludge. *European Journal of Soil Science*, *56*, 197-205.

Griffiths, R. I., Whitely, A. S., O'Donnell, A. G. and Bailey, M. J. (2003). Influence of depth and sampling time on bacterial community structure in an upland grassland soil. *FEMS Microbiology Ecology*, *43*, 35-43.

Hery, M., Nazaret, S., Jaffre, T., Normand, P. and Navarro, E. (2003). Adaptation to nickel spiking of bacterial communities in neocaledonian soils. *Environmental Microbiology*, *5*, 3-12.

Hery, M., Herrera, A., Vogel, T. M., Normand, P. and Navarro, E. (2005). Effect of carbon and nitrogen input on the bacterial community structure of Neocaledonian nickel mine spoils. *FEMS Microbiology Ecology*, *51*, 333-340.

Heuer, H., Krsek, M., Baker, P., Smalla, K. and Wellington, E. M. H. (1997). Analysis of actinomycete communities by specific amplification of genes encoding 16S rRNA and gel-electrophoretic separation in denaturing gradients. *Applied and Environmental Microbiology*, *63*, 3233-3241.

Hinojosa, M. B., Carreira, J. A., Garcia-Ruiz, R. and Dick, R. P. (2005). Microbial response to heavy metal-polluted soils: Community analysis from phospholipid-linked fatty acids and ester-linked fatty acids extracts. *Journal of Environmental Quality*, *34*, 1789-1800.

Holmes, A. J., Bowyer, J., Holley, M. P., O'Donoghue, M., Montgomery, M. and Gillings, M. R. (2000). Diverse, yet-to-be-cultured members of the Rubrobacter subdivision of the Actinobacteria are widespread in Australian arid soils. Fems *Microbiology Ecology*, *33*,111-120.

Hu, Q., Qi, H. Y., Zeng, J. H. and Zhang, H. X. (2007). Bacterial diversity in soils around a lead and zinc mine. *Journal of Environmental Sciences-China*, *19*, 74-79.

Joynt, J., Bischoff, M., Turco, R., Konopka, A. and Nakatsu, C. H. (2006). Microbial community analysis of soils contaminaed with lead, chromium and petroleum hydrocarbons. *Microbial Ecology*, *51*, 209-219.

Jackson, R. B., Fierer, N. and Schimel, J. P. (2007). New directions in microbial ecology. *Ecology*, *88*, 1343-1344.

Kandeler, E., Tscherko, D., Bruce, K. D., Stemmer, M.,Hobbs, P. J., Bardgett, R. D. and Amelung, W. (2000). Structure and function of the soil microbial community in microhabitats of a heavy metal polluted soil. *Biology and Fertility of Soils*, *32*, 390-400.

Kao, P. H., Huang, C. C. and Hseu, Z. Y. (2006). Response of microbial activities to heavy metals in a neutral loamy soil treated with biosolid. *Chemosphere*, *64*, 63-70.

Kefala, M. I., Zouboulis, A. I. and Matis, K. A. (1999). Biosorption of cadmium ions by *Actinomycetes* and separation by flotation. *Environmental Pollution*, *104*, 283-293.

Kelly, J. J., Häggblom, M. M., Tate, R. L. (1999). Changes in soil microbial communities over time resulting from one time application of ainc: a laboratory microcosm study. *Soil Biology and Biochemistry*, *31*, 1455-1465.

Kelly, J. J., Häggblom, M. M. and Tate, R.L. (2003). Effects of heavy metal contamination and remediation on soil microbial communities in the vicinity of a zinc smelter as indicated by analysis of microbial community phospholipid fatty acid profiles. *Biology and Fertility of Soils*, *38*, 65-71.

Knight, B., and McGrath S. P. (1995). A method to buffer the concentrations of free Zn and Cd ions using a cation-exchange resin in bacterial toxicity studies. *Environmental Toxicology and Chemistry*, *14*, 2033-2039.

Kozdroj, J., and van Elsas, J. D. (2001). Structural diversity of microbial communities in arable soils of a heavily industrialised area determined by PCR-DGGE fingerprinting and FAME profiling. *Applied Soil Ecology*, *17*, 31-42.

Kuan, H. L., Fenwick, C., Glover, L. A., Griffiths, B. S. and Ritz, K. (2006). Functional resilience of microbial communities from perturbed upland grassland soils to further persistent or transient stresses. *Soil Biology and Biochemistry*, *38*, 2300-2306.

Lazzaro, A., Hartmann, M., Blaser, P., Widmer, F., Schulin, R. and Frey, B. (2006). Bacterial community structure and activity in different Cd-treated forest soils. *FEMS Microbiology Ecology*, *58*, 278-292.

Lazzaro, A., Schulin, R., Widmer, F. and Frey, B. (2006). Changes in lead availability affect bacterial community structure but not basal respiration in a microcosm study with forest soils. *Science of the Total Environment*, 371, 110-124.

Leonard, J. (2001). A Study of metallophytic grassland in Weardale. MSc thesis. University of Northumbria.

Liao, M. and Xie, X. M. (2007). Effect of heavy metals on substrate utilization pattern, biomass, and activity of microbial communities in a reclaimed mining wasteland of red soil area. *Ecotoxicology and Environmental Safety*, *66*, 217-223.

Li, Z. J., Xu, J. M. Tang, C. X., Wu, J. J., Muhammad, A. and Wang, H. Z. (2006). Application of 16S rDNA-PCR amplification and DGGE fingerprinting for detection of shift in microbial community diversity in Cu-, Zn-, and Cd-contaminated paddy soils. *Chemosphere*, *62*, 1374-1380.

Lorenz, N., Hintemann, T., Kramarewa, T., Katayama, A., Yasuta, T., Marschner, P. and Kandeler, E. (2006). Response of microbial activity and microbial community composition in soils to long-term arsenic and cadmium exposure. *Soil Biology and Biochemistry*, *38*, 1430-1437.

Lunn, M., Sloan, W. T. and Curtis, T. P. (2004). Estimating bacterial diversity from clone libraries with flat rank abundance distributions. *Environmental Microbiology*, *6*, 1081-1085.

Moffett B. F., Nicholson F. A., Uwakwe N. C., Chambers B. J., Harris J. A. and Hill T. C. J. (2003). Zinc contamination decreases the bacterial diversity of agricultural soil *FEMS Microbiology Ecology*, *43*, 13-19.

Mummey, D. L., and Stahl, P. D. (2004). Analysis of soil whole- and inner-microaggregate bacterial communities. *Microbial Ecology*, *48*, 41-50.

Muyzer G., De Waal E. C. and Uitterlinden A. G. (1993). Profiling of complex microbial populations by denaturing gradient gel electrophoresis analysis of polymerase chain reaction-amplified genes coding for 16S rRNA. *Applied and Environmental Microbiology*, *59*, 695-700.

Mertens, J., Springael, D., De Troyer, I., Cheyns, K., Wattiau, P. and Smolders, E. (2006). Long-term exposure to elevated zinc concentrations induced structural changes and zinc tolerance of the nitrifying community in soil. *Environmental Microbiology*, *8*, 2170-2178.

Nakatsu, C. H., Carmosini, N., Baldwin, B., Beasley, F., Kourtev, P. and Konopka, A. (2005). Soil microbial community responses to additions of organic carbon substrates and heavy metals (Pb and Cr). *Applied and Environmental Microbiology, 71*, 7679-7689.

Narang, R. and Dunbar, J. (2004). Modeling bacterial species abundance from small community surveys. *Microbial Ecology, 47*, 396-406.

Nies D.H. (2003). Efflux-mediated heavy metal resistance in prokaryotes. *FEMS Microbiology Reviews, 27*, 313-339.

Odum, E. P. The effects of stress on the trajectory of ecological succession. In Barrett, G.W. and Rosenburg R. (eds) Stress effects on natural ecosystems. London. J. Wiley and Sons; 1981; pp 43-47.

Paradkar, A., Trefzer, A., Chakraburtty, R. and D. Stassi, D. (2003). Streptomyces genetics: A genomic perspective. *Critical Reviews in Biotechnology, 23*,1-27.

Prokop, Z., Cupr, P., Zlevorova-Zlamalikova, V., Komarek, J., Dusek, L. and Holoubek, I. (2003). Mobility, bioavailability, and toxic effects of cadmium in soil samples. *Environmental Research, 91*,119-126.

Prosser, J. I., Bohannan, B. J. M. Curtis, T. P. Ellis, R. J. Firestone, M. K. Freckleton, R. P. Green, J. L. Green, L. E. Killham, K. Lennon, J. J. Osborn, A. M. Solan, M. van der Gast, C. J. and Young, J. P. W. (2007). Essay - The role of ecological theory in microbial ecology. *Nature Reviews Microbiology, 5*, 384-392.

Ranjard, L., Echairi, A., Nowak, V., Lejon, D. P. H., Nouaim, R. and Chaussod, R. (2006). Field and microcosm experiments to evaluate the effects of agricultural Cu treatment on the density and genetic structure of microbial communities in two different soils. *FEMS Microbiology Ecology, 58*, 303-315.

Rasmussen D. L. and Sørensen S. J. (2001). Effects of mercury contamination on the culturable heterotrophic, functional and genetic diversity of the bacterial community in soil. *FEMS Microbiology Ecology, 36*,1-9.

Ravel J., Amoroso M. J., Colwell R. R. and Hill R. T. (1998). Mercury-resistant *Actinomycetes* from the Chesapeake Bay. *FEMS Microbiology letters, 162*, 177-184.

Renella, G., Chaudri, A. M. and Brookes, P. C. (2002). Fresh additions of heavy metals do not model long-term effects on microbial biomass and activity. *Soil Biology and Biochemistry, 34*,121-124.

Richards, J. W., Krumholz, G. D., Chval, M. S. and Tisa L. S. (2002). Heavy metal resistance patterns of *Frankia* strains. *Applied and Environmental Microbiology, 68*, 923-927.

Rho, J. Y., and Kim, J. H. (2002). Heavy metal biosorption and its significance to metal tolerance of Streptomycetes. *Journal of Microbiology*, 40, 51-54.

Rowan, A.K., Snape, J.R., Fearnside, D., Barer, M.R., Curtis, T.P. and Head, I.M. (2003). Composition and diversity of ammonia-oxidising bacterial communities in wastewater treatment reactors of different design treating identical wastewater. *FEMS Microbiol Ecol, 43*, 195-206.

Sandaa, R. A., Torsvik, V. and Enger, O. (2001). Influence of long-term heavy-metal contamination on microbial communities in soil. *Soil Biology and Biochemistry, 33*, 287-295.

Sandaa, R. A., Torsvik, V., Enger, O., Daae, F. L., Castberg, T. and Hahn, D. (1999). Analysis of bacterial communities in heavy metal-contaminated soils at different levels of resolution. *FEMS Microbiology Ecology*, *30*, 237-251.

Schmidt, A., Haferburg, G., Sineriz, M., Merten, D., Buchel, G. and Kothe, E. (2005). Heavy metal resistance mechanisms in actinobacteria for survival in AMD contaminated soils. *Chemie Der Erde-Geochemistry*, *65*, 131-144.

Sigler, W.V. and Turco, R.F. (2002). The impact of chlorothalonil application on soil bacterial and fungal populations as assessed by denaturing gradient gel electrophoresis. *Applied Soil Ecology*, *21*, 107-118.

Sharma, S., Aneja, M. K., Mayer, J., Munch, J. C. and Schloter, M. (2005). Characterization of bacterial community structure in rhizosphere soil of grain legumes. *Microbial Ecology*, *49*, 407-415.

Sloan, W. T.,Woodcock, S., Lunn, M., I. M. Head, I. M. and Curtis, T. P. (2007). Modeling taxa-abundance distributions in microbial communities using environmental sequence data. *Microbial Ecology*, *53*, 443-455.

Stacktbrandt E., Rainey F.A. and Ward-Rainey N.L. (1997). Proposal for a new hierarchic classification system, *Actinobacteria classis* nov. *International Journal of Systematic Bacteriology*, *47*, 179-491.

Tobor-Kaplon, M., Bloem, A. J., Romkens, P. and de Ruiter, P. C. (2005). Functional stability of microbial communities in contaminated soils. *Oikos*, 111,119-129.

Usman, A., Kuzyakov, Y. and Stahr, K. (2005). Effect of clay minerals on immobilization of heavy metals and microbial activity in a sewage sludge-contaminated soil. *Journal of Soils and Sediments*, *5*, 245-252.

Valls, M., Atrian, S., de Lorenzo, V. and Fernández, L.A. (2000). Engineering a mouse metallothionein on the cell surface of *Ralstonia eutropha* CH34 for immobilization of heavy metals in soil. *Nature Biotechnology*, *18*, 661-665.

Vaasquez-Murrieta, M., Migueles-Gardunno, I., Franco-Hernandez, O., Govaerts, B. and Dendooven, L. (2006). C and N mineralization and microbial biomass in heavy-metal contaminated soil. *European Journal of Soil Biology*, *42*, 89-98.

Wang, Y. P., Shi, J. Y., Wang, H., Lin, Q., Chen, X. C. and Chen, Y. X. (2007). The influence of soil heavy metals pollution on soil microbial biomass, enzyme activity, and community composition near a copper smelter. *Ecotoxicology and Environmental Safety*, *67*, 75-81.

Wardle, D. A. and Zackrisson, O. (2005). Effects of species and functional group loss on island ecosystem properties. *Nature*, *435*, 806-810.

Wertz, S., Degrange, V. Prosser, J. I., Poly, F., Commeaux, C. Freitag, T., Guillaumaud, N. and Le Roux, X. (2006). Maintenance of soil functioning following erosion of microbial diversity. *Environmental Microbiology*, *8*, 2162-2169.

Witter, E., Gong, P., Baath, E. and Marstorp, H. (2000). A study of the structure and metal tolerance of the soil microbial community six years after cessation of sewage sludge applications. *Environmental Toxicology and Chemistry, 19*, 1983-1991.

Wu, S. C., Luo, Y. M., Cheung, K. C. and Wong, M. H. (2006). Influence of bacteria on Pb and Zn speciation, mobility and bioavailability in soil: A laboratory study. *Environmental Pollution, 144*, 765-773.

Zhang, C. B., Huang, L. N. Luan, T. G. Jin, J. and Lan, C. Y. (2006). Structure and function of microbial communities during the early stages of revegetation of barren soils in the vicinity of a Pb/Zn Smelter. *Geoderma, 136*, 555-565.

In: Microbial Ecology Research Trends
Editor: Thijs Van Dijk, pp. 207-220
ISBN: 978-1-60456-179-1

Chapter IX

Physiolgical Diversity in Widely Distributed Microzooplankton: Digestion in the Ciliate *Euplotes vannus*

***John R. Dolan*[1*] *and D. Wayne Coats*[2]**

[1]Marine Microbial Ecology, Laboratoire d'Océanographie, Université Paris6 CNRS UMR 7093, Station Zoologique, B.P. 28, 06230 Villefranche-sur-Mer, France

[2]Smithsonian Environmental Research Center, P. O. Box 28, Edgewater MD, USA

Abstract

Many planktonic microbes exhibit nearly global distributions. Apparent adaptability to wide ranges of environmental conditions in cosmopolitan species could involve physiological diversity among individuals. However, variability between individuals among microorganisms is rarely considered. We examined individual variability in the common marine ciliate *Euplotes vannus*. Digestion was followed in individual cells subjected to different levels of ambient food concentration. Average digestion times were not significantly different because, in the absence of food or in the presence of super-saturating food concentrations, there was a large variability between individuals. Some ciliates completed food vacuole processing under 1 h while others took as long as 17 h. The range of digestion times, which increased under extreme conditions, may correspond to range of growth rates or growth potentials among individuals.

* dolan@obs vlfr.fr

Introduction

Limits of Averages

Variation between individuals is a component of biological diversity which has received very little attention; commonly attempts are made in to minimize, rather than characterize, individual variability (Spicer & Gaston 1999). Establishing and comparing "averages" is the standard means of describing populations or species. This is true despite the fact that employing averages has familiar pitfalls such as masking distinct differences within a group and thus representing a descriptor of little value. Obviously, averages are acknowledged to be mathematical representations, not literal descriptions, but the possibility that averages may be unrealistic representations can be easy to neglect. Consider for example, estimating population reproductive rates with averages and a hypothetical human population with an "average" birth rate of 2.5 children per female. Let the number 2.5 represents the average of two sets of individuals (equal in abundance and mortality rates) but with different fertilities, say one with 1 child and the other with 4 children per female. Not only does t the "average" female does exist in this population but more importantly, if any factors differentially influence the two distinct sets of individuals, it would be impossible to predict the net effect using a single arithmetic average value of fertility. Thus, consideration of average rates may be not only unrealistic in the sense the "average does not exist", but can lead to erroneous conclusions. It is then of some significant interest to consider physiological diversity within a population. This has long been recognized and most recently underlined by Spicer & Gaston (1999). For any particular species, its capability to colonize new habitats, or react to shifts in climatic conditions, or adapt to changes in resources are the types of topics difficult if not impossible to address without some knowledge of the physiological diversity found within the species in terms of diversity among individuals.

With regard to planktonic organisms, physiological diversity between individuals is very rarely assessed. Firstly, as most planktonic organisms are microscopic observations of individual microorganisms are techniquely difficult. Secondly, they live in time and space scales very distinct from those with which we are familiar. Perhaps though the most important reason is that individual variability among microbes are simply not considered to be of much importance. Microbes and most planktonic organisms are characterized by very large population sizes, and as large sample or population size is often equated with 'normality' of distribution, average values are thought to be good descriptors.

Here we report on physiological diversity in the planktonic protist ciliate *Euplotes vannus*. Data on the variability of digestion with feeding conditions, ambient food levels, was gathered. The experiments were not designed originally to establish variability or ranges of rates, but rather to produce averages. We were able to examine physiological diversity only because we followed individual ciliates. However, we show the inadequacy of using averages to describe a physiological rate in the ciliate. Below, we place planktonic ciliates in their ecological rôle and briefly describe the nature of previous studies on *Euplotes vannus*, on which a great many studies have been conducted but very few of which documented differences between individuals.

Ecological Role of Plankton Ciliates and Euplotes Vannus

In the plankton, primary production is carried out by organisms occupying 3 different size-classes: 1) picoplankton-size organisms (0.2 - 2.0 µm in size) autotrophic prokaryotes such as *Synechococcus* and *Prochlorococcus* and small eukaryotic flagellates, 2) nanoplankton-sized organisms (2.0 - 20 µm) mainly eukaryotic flagellates and small diatoms, and 3) microplankton-size organisms (20 -200 µm in size) primarily large dinoflagellates and diatoms. Microzooplankton, composed of planktonic ciliates and heterotrophic dinoflagellates between the sizes of 20 and 200 microns, are the consumers of much of this primary production. Calbet & Landry (2004) estimated that a surprisingly constant 70 % of total planktonic primary production is consumed by ciliates and dinoflagellates across marine systems. This consistency and magnitude of microzooplankton grazing has been challenged (Dolan & McKeon 2005) but there is little doubt that microzooplankton are important in plankton food webs. Microzooplankton, are the link between nanoplankton, and to a lesser extent pico-plankton, and the higher trophic levels which prey on microzooplankton in the form of copepods, larval fish, etc. Among the ciliate component of the microzooplankton, distinct trophic guilds (groups of ciliate species whose diets appeared specialized to some degree) can de distinguished (Dolan 1991). Ciliates whose prey are mainly pico-plankton sized prey are microphagous, those feeding on larger prey are macrophagous and ciliates capable of feeding on other ciliates are predatory. *Euplotes vannus* feeds ciliates as well as other prey types and so is considered a predatory ciliate.

Euplotes Vannus

We isolated *Euplotes vannus* from the Chesapeake Bay. However, the ciliate is very widely distributed, as are many species of the microzooplankton. Euplotes are described as "marine cosmopolitan" and indeed *Euplotes vannus* is found in a wide range of ecosystems from the Chesapeake Bay, the origin of the cultures used in the present work (Dolan & Coats 1991a), to the coastal waters of Denmark (Fenchel 2004), Japan, Italy and California (e.g., Dini & Nyberg 1999). In common with most species of *Euplotes*, while often found in the plankton, it is ordinarily found among the benthic ciliates. In size it is about 100 microns in length. *Euplotes vannus* appears to feed upon algae, other ciliates as well as bacteria. It can be among the ciliates which colonize marine snow particles (Artolozaga et al. 1997).

Euplotes vannus has been the subject of many investigations. The species was used in some of the first studies on nutrient regeneration by protists (Johannes 1965; Gast & Hörstman 1983). The relationships between food type and food concentration and population growth have been investigated (Capriuolo et al. 1988; Dini & Nyberg 1999; Dolan & Coats 1991a). Selective feeding has been examined (Premke & Arndt 2000; Lewitus et al. 2006; Scott et al. 2003). Behavioral experiments have been conducted with *Euplotes vannus* (Tomaru et al. 2001; Fenchel 2004). Taken as a typical coastal ciliate, it has been subjected to several toxicology studies (Stebbing et al. 1990; Coppellotti 1988; Ricci et al. 1997; Xu et al. 2004; Mori et al. 2003). Many membrane physiology investigations have focused on *Euplotes vannus* (Stock et al. 1977; Kruppel & Wissing 1996; Kruppel et al. 1995; Lueken et al. 1996). Molecular and phylogenetic data have become available in recent years (Petroni et al. 2003; Shang et al. 2002; Cheng & Song 2002; Petroni et al. 2002). While diversity, including physiological, between strains or mating types has been considered (Walton et al. 1995; Jones & Gates 1994; Caprette & Gates 1994) physiological diversity between individuals, to our knowledge, has not been addressed.

Use of Digestion Data

The interest in examining digestion in ciliates may appear obscure but it is potentially very useful in the estimation of feeding in natural populations. Estimating feeding rates in field populations is generally problematic with regard to planktonic organisms. One manner, possible in principle, is using 'gut contents' combined with digestion rate data. The ingestion rate, assuming steady-state conditions, equals food contents multiplied by digestion rate. The interest in the approach lies in the possibility of combining experimentally derived digestion rate data with estimates of average food contents from *in situ* or natural populations. Food content data can often be obtained relatively easily by analyzing freshly caught animals with little manipulation. This approach has been applied to benthic ciliates (Goulder 1972, 1973; Fenchel 1975), planktonic ciliates (Kopylov & Tumantseva 1987; Dolan & Coats 1991b) and heterotrophic nanoflagellates feeding on Synechococcus (Dolan & Simek 1999). Thus digestion rates, and their variability with ambient food levels and food type have been studied in ciliates fed algae and other ciliates (Capriuolo & Degan 1991; Dolan & Coats 1991a, Dolan & Simek 1997) and have concluded that digestion rate is insensitive to ambient food concentrations and food type. However, these studies sampled populations with time and did not attempt to asses individual variability.

Significance of Physiological Diversity

We report the results of a set of experiments designed to determine the effects of ambient food levels on digestion in the ciliate *Euplotes vannus*. The experimental design of following individual cells is a design rarely employed with regard to microbes and allowed us to assess physiological diversity in *Euplotes vannus*. Our results show that a population of *Euplotes vannus* contains individuals with distinct reactions to a given set of conditions. We followed digestion of labeled prey by monitoring individual ciliates digesting prey under different conditions of ambient food levels. Comparison of average values suggested no significant effect of ambient food levels on digestion rates. However, examination of the distribution of digestion times among individual ciliates showed that when food was abundant, individual *Euplotes vannus* cells reacted in different fashions- including notably rapid or prolonged digestion. The different durations of digestions among individuals likely correspond with different growth rates. Thus, *Euplotes vannus* populations are composed of individuals which will probably exhibit distinct growth responses when exposed to high prey concentrations. This hypothesis would not have been supported based on the evidence of no differences in digestion with ambient food level, a conclusion based on population averages.

Methods and Materials

Isolation, cultivation and generation of growth curves of *Euplotes vannus* and its food organism, the bactivorous scuticociliate *Metanophrys* sp. are given in detail in Dolan & Coats (1991a). For each of the 4 experiments, log-phase cultures of *Euplotes vannus* and *Metanoprys* were used. Prey ciliates (*Metanophrys* sp.) were labeled with fluorescent microspheres and fed to *Euplotes vannus* as described in Dolan & Coats (1991a). *Euplotes vannus* were washed free of labeled *Metanophrys* and individual cells were micro-pipetted into, and isolated in, 1 ml well plates. Ciliates in the well plates were examined using an inverted microscope equipped with epifluorescence permiting visual verification of ciliate cell contents. For each experiment, a set of 30 individual *Euplotes vannus* containing *Metanophrys* labeled with fluorescent microspheres (Figure 1) assembled. Elapsed time from separating *Euplotes vannus* from labeled *Metanophyrys* to composing a set of 30 Euplotes, each isolated in an individual well of a well plate, was about 20 minutes.

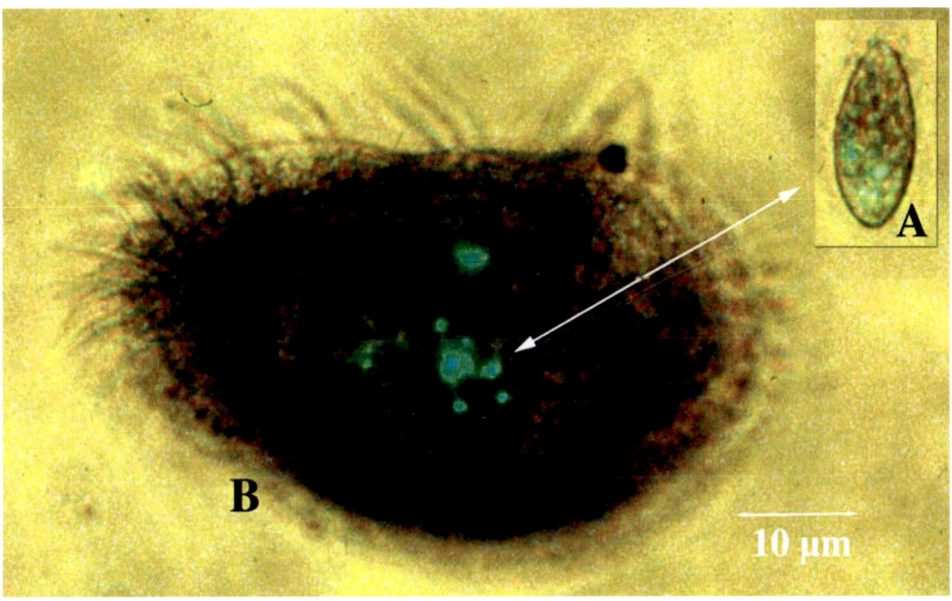

Figure 1. Photomicrographs of a *Metanophrys* (A) labeled with fluorescent microspheres and a *Euplotes vannus* containing several labeled Metanophyrys. The photomicrographs were obtained by using transmitted phase-contrast and epifluorescence illumination.

For each experiment, individual *Euplotes vannus* were examined at 30 minute intervals until the fluorescent microspheres contained within the labeled prey were expelled, marking the end of prey digestion. We thus obtained a digestion time for each *Euplotes* cell. All cultures, labeling and observations were conducted at 20° C in a temperature-controlled chamber.

In the first experiment *Euplotes vannus* were monitored in wells containing culture medium filtered free of food organisms. Thus the ciliates were digesting prey while being starved. In the second, third and fourth experiments, the solutions in the well plates contained, respectively, 100,1000 and 10,000 *Metanophyrys* ml^{-1}. In the latter 3 experiments,

Euplotes vannus cells were transferred to fresh prey solutions approximately every 2 hours. It should be noted that feeding rates reach a maximum in *Euplotes vannus* preying on *Metanophrys* saturates at a food concentration of about 100 *Metanophrys* ml^{-1} (Dolan & Coats 1991). Thus, the 1000 and 10000 prey concentrations represented super-saturated food concentrations for *Euplotes vannus*.

Results

Digestion time, in terms of averages, while appearing to increase with ambient food level, were not significantly different using a t-test comparing results of the 4 experiments due to the relatively large standard deviations (Figure 2). At 20° C, the 'average' Euplotes digesting *Metanophrys* while held in prey-free water and therefore not feeding, digested its food vacuole contents and expelled the inert fluorescent microsphere which labeled its food item, in about 5 hours. The 'average' value for Euplotes digesting in the presence of 100 prey per ml was also about 5 hours. Population averages for ciliates held digesting in the presence of high concentrations of food increased to about 10 h but with a considerable amount of variability. Considering the averages alone and their associated standard deviations, no effect of ambient food level on the digestion time of *Euplotes vannus* can be shown. Thus, digestion time appeared fixed and not responsive to the availability of food.

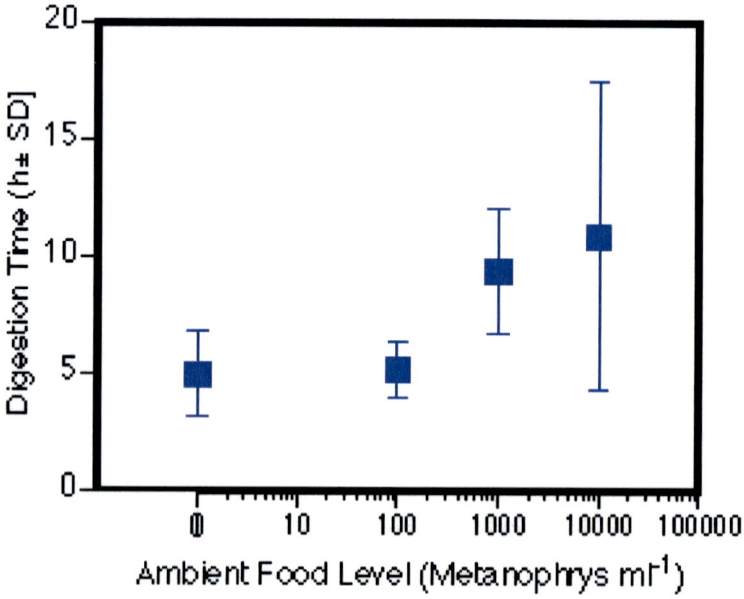

Figure 2. Average digestion times (±standard deviations) in Euplotes vannus digesting Metanophrys as a function of ambient concentrations of Metanophrys. Note that wityh the relatively large standard deviations, no significant differences in digestion times based on a comparison means could be shown.

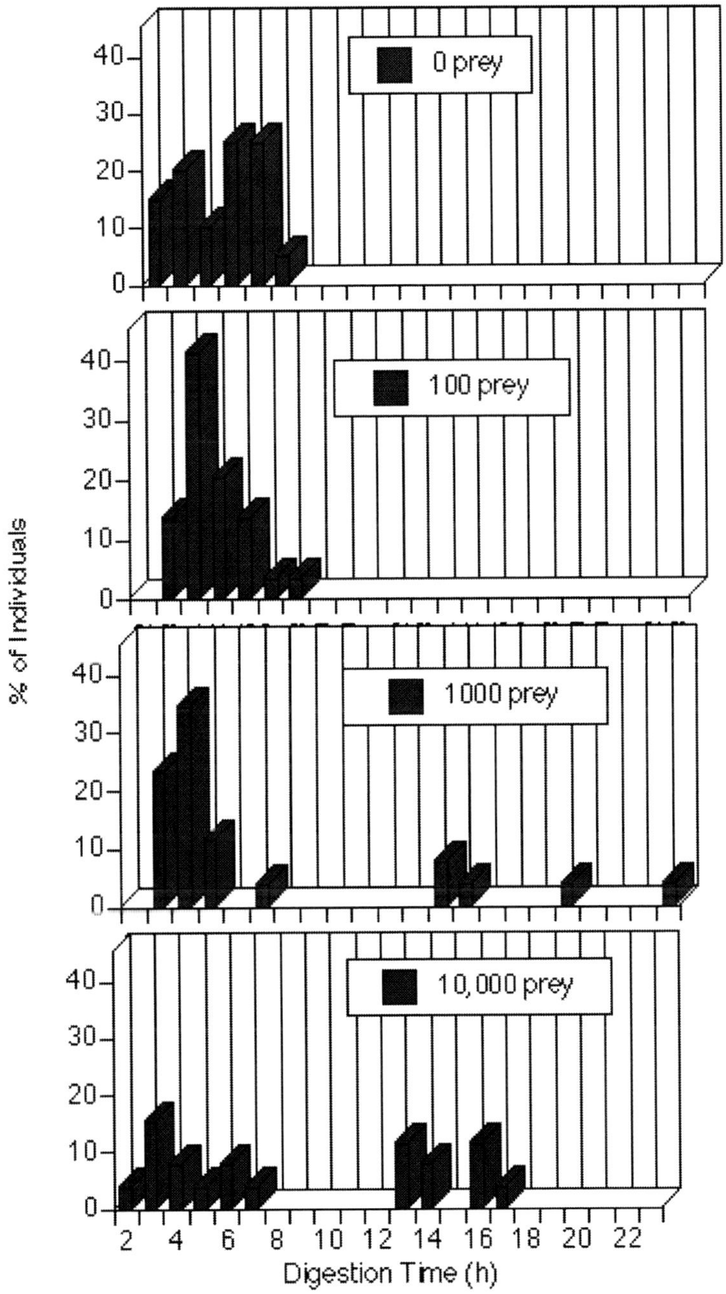

Figure 3. Distribution of digestion times in the 4 experiments. Note that rapid digestion times (1 - 2 h) were detected only in experiments with *Euplotes* digesting while being starved (top panel) or while digesting in the presence of super-abundant prey. Long digestion times (> 9 h) were observed in presence of high food concentrations.

However, distributions of digestion times, in contrast to average values, showed distinct differences with ambient food levels (Figure 3). In non-feeding cells, digestion times appeared to show a bi-modal distribution. Approximately equal and relatively large proportions (about 30 %)of individuals ciliates were rapid digesters ($\leq$ 2 h) and others slow digesters (≥ 6h). The small numbers of cells used precludes a rigorous test of distribution. In *Euplotes vannus* held in prey concentrations of 100 prey per ml, the distribution of digestion times approached normality with a modal digestion time of about 3 h accounting for about 45 % of the individuals. Despite the differences in distributions between digestion in starving and ciliates feeding in 100 prey ml-1, average digestion times for the entire populations were very similar (Figure 3). With high prey concentrations of 1000 or 10,000 prey ml^{-1}, some *Euplotes vannus* showed digestion times similar to ciliates held in lower food concentrations of 2 or 3 h, while others showed very long digestion times, approaching 24 h. At 1000 prey ml-1, there was a relatively high proportion, about 60% of 'fast digesters' , showing digestion times ≤ 2 h). Thus, the existence of two distinct populations, or individuals showing distinct digestion durations, were apparent in *Euplotes vannus* when subjected to super-abundant prey levels and perhaps starved (held in food-free water).

Discussion

Physiological Diversity

The manner in which we gathered data on digestion in *Euplotes vannus*, following individual ciliates, allowed us to see the physiological diversity among individual ciliates. In experiments sampling with time an entire population (e.g., Capriuolo & Degnan 1991; Dolan & Coats 1991a,b) this information can not be extracted. To return to our example in the "Introduction" concerning rates birth rates in a human population, - net rates of change can be estimated by sampling a population over time and simply noting shifts in abundances, but important demographic details will be masked. Our experimental data showed that the ciliate population was not homogenous but rather it appears that ciliates can react in 2 distinct fashions to the presence of super-saturating food concentrations and perhaps a lack of food.

Euplotes vannus digesting prey in a solution free of food showed a dispersed population, relative to those processing food in the presence of 100 prey ml^{-1}, a concentration supporting exponential population growth (Dolan & Coats 1991a). Under very high food concentrations, 1000 or 10,000 *Metanophrys* ml^{-1}, the tendency was towards a dispersal of digestion times within a population was exaggerated. Thus, under perhaps unusual circumstances of no food or very abundant food, the some individual ciliates tended to either process food rapidly or very slowly. Presumably the distinct digestion times recorded would correspond with different digestion efficiencies, biomass production and different rates of cell division. Before considering if there are indeed different patterns employed by different individuals, or rather in some individuals the digestion process simply malfunctions when challenged, it is worth briefly reviewing food vacuole formation in ciliates. The following account is based on studies of food vacuole formation and membrane processing in *Paramecium* (Allen & Fok 1993; Fok & Allen 1990,1993; Fok & Schockley 1985; Fok et al. 1982), Kaneshiro et al. (1992), Plattner & Kissmehl 2003; Ramoino (1996).

One of the characteristics which distinguish most ciliates from members of many other protist groups is the existence of a differentiated mouth region where food vacuole formation begins (e.g., Hausmann & Radek 1993). Thus, in contrast to amboeba, in ciliates the food vacuole membrane is formed from disk-shaped vesicles in the mouth region, rather than simply from the plasma membrane. In addition to discodial vesicles, acidosomes, at about 1 μm in dia considerably larger than discodial vesicles, are also found at the mouth region. As the food vacuole forms from the discoidal vesicles, the acidosomes 'dock' but do not fuse with the forming vacuole. Acidosomes, as the name implies are low pH vesicles. When the food vacuole reaches a certain size, or yet some other unknown cellular trigger is reached, the vacuole is pinched off from the mouth region and is transported to the interior of the cell. The acidosomes then fuse with the vacuole and the original vacuole membrane is retrieved with an end result of the nearly entire replacement of the membrane with that derived from the acidosomes. The food vacuole, once formed of acidosome membrane is then associated with docking of lysosomes. These membrane bound sacs contain acidic hydrolyses. It is possible that a second membrane replacement occurs with the fusion of lysosomes. The lysomes appear to simultaneously fuse with the vacuole and actual digestion begins. After a short time (in *Paramecium* 20 minutes) another change in membrane characteristics occurs. Lysosome membrane appears to be removed from the vacuole. At this point, the food vacuole may be 'defecation competent' or another change in membrane properties may be necessary before binding at the cytoproct region is possible. The vacuole is transported to the cytoproct region and at this point is acid phosphatase negative. The cytoproct opens and the undigested matter is expelled but the vacuole membrane remains inside the cell. Membrane from the food vacuole is tubularized and the tubules released into the cell. From this scenario it is clear that membrane supply and flow is primordial in the formation and processing of food vacuoles. Most if not all membranes appear to be recycled. Thus, the original food vacuole membrane formed from the discodial vesicles are themselves derived, at least in part, from re cycled food vacuole membrane transported back to the mouth region after indigested matter is expelled from the cell. Likewise, acidosomes, based on antigen results, seem to have a heterogeneous origin but recycling of membrane is suspected to occur.

The complex sequence of events based on studies of *Paramecium* may or may not characterize most ciliates. Food vacuole processing has been examined in few species in any detail and there is at least one report of a different pattern. *Pseudomicrothorax dubius* feeds on filamentous blue-green algae and in this ciliate, lysosomes fuse with food vacuole as it is formed in the mouth region (Peck & Hausmann 1980). The processing of food vacuoles in *Euplotes vannus*, if strictly sequential and following that known from *Paramecium*, can be short-circuited as we observed a few instances in which prey-containing food vacuoles merged. Indeed, a breakdown in the sequence is an explanation for long digestion times.

In the scenario based on *Paramecium*, a food vacuole will remain in the cell if it is not first fused with acidosomes and the membrane replaced and/or not subsequently fused with lysosomes. The longest digestion times in *Euplotes vannus* were observed in cells held in solutions of 1,000 prey ml^{-1} (Figure 3). In circumstances of over-abundant prey one may hypothesize an imbalance between the formation of early stage food vacuoles and the supply of acidosomes and lysosomes. For example, some of the membrane generally used for the fabrication of lysosomes or acidosomes was perhaps monopolized in the form of early stage food vacuoles. It is difficult to explain the retention of ingested matter in the presence of abundant food as a useful strategy. Therefore, it seems most likely that in some individuals there was a breakdown in coordination among the distinct components of the digestion process. What is noteworthy is the fact that long digestion times characterized only a portion of the ciliates subjected to high food concentrations while digesting.

Our data show that a treatment effect of ambient food level can be discerned on a portion of the ciliates, a portion hidden in the averages, a comparison of which suggested no significant differences among treatments. We conclude then perhaps re-stating the obvious (but often ignored or under-appreciated): individuals vary. In the ciliate *Euplotes vannus*, we found considerable variability in terms of patterns of digestion found in even the small populations examined. Such an apparent physiological diversity may in part explain its wide distribution.

Using Food Contents and Digestion Rates to Estimate Ingestion Rates

Our results indicate that while average population rates of digestion appeared insensitive to ambient food levels, parts of the ciliate populations displayed long digestion times. The portion of the ciliate population thus effected by food level could be substantial, with nearly 40% of individuals showing long (> 12h digestion rates (Figure 3) but the phenomenon masked by individual to individual variability. If long digestion times are associated with inefficient exploitation of food, a population feeding under conditions of high prey availability will possibly have a lower average growth rate. However, as we noted, a substantial portion of the populations showed very rapid digestion times perhaps with corresponding generation times.

Conclusion

Previous studies have underlined the differences between strains of the same ciliate species or clones derived from a single strain, largely in terms of growth rates or temperature responses (e.g., Wiesse & Montagnes 1998; Weisse & Rammer 2006). Differences between individuals, while suspected to exist, have received very little, if any, attention even with regard to basic activities of ecological interest such as feeding rates (Weisse 2002). However, differences between individual cells can exist, as we have shown. Some *Euplotes vannus* showed digestion times in the presence of over-abundant prey similar to those found in simply optimal prey concentrations, others showed faster digestion or slower digestion rates. In the absence of food, some individuals slowed while other increased digestion times. Such physiological diversity among individuals is likely in part an explanation for the wide geographic distribution of *Euplotes vannus*.

Acknowledgements

Financial support for this work was originally provided by the National Science Foundation (USA) through grant OCE 8800076. Motivation to re-examine old data was provoked by Spicer & Gaston's book "Physiolgical Diversity and Its Ecological Implications" and the French ANR Biodiversity project AQUAPARADOX. Financial support was also provided by the Marplan project, part of the MARBEF Network of Excellence (Contract No. GOCE-CT-2003-505446). This is contribution MPS-07100 of MARBEF.

References

Allen, R.D., Fok, A.K. 1993. Endosomal membrane traffic of ciliates. Advances in Cell and Molecular Biology of Membranes, vol 2B, *Membrane Traffic in Protozoa,* pp 283-309.

Artolozaga, I., Santamaría, E., López, A., Ayo, B., Iriberri, J. 1997. Succession of bacterivorous protists on laboratory-made marine snow. *Journal of Plankton Research,* 19:1429-1440.

Calbet, A., Landry, M.R. 2004. Phytoplankton growth, microzooplankton grazing and carbon cycling in marine systems. *Limnology and Oceanography*, 49:51-57.

Caprette, C.L., Gates, M.A. 1994. Quantitative analysis of interbreeding in populations of vannus-morphotype *Euplotes*, with special attention to the nominal species *Euplotes vannus* and *Euplotes crassus*. *Journal of Eukaryotic Microbiology,* 41: 316-324.

Capriulo, G.M.,, Schreiner, R.A., Dexter, B.L. 1988. Differential growth of *Euplotes vannus* fed fragmented versus unfragmented chains of *Skeletonema costatum*. *Marine Ecology Progress Series,* 47:205-209.

Capriulo, G. M. and C. Degnan. 1991. Effect of food concentration on digestion and vacuole passage time in the heterotrichous marine ciliate *Fabrea salina. Marine Biolology,* 110:199-202.

Chen Z.G., Song W.B. 2002. Phylogenetic positions of *Aspidisca steini* and *Euplotes vannus* within the order euplotida (Hypotrichia : Ciliophora) inferred from complete small subunit ribosomal RNA gene sequences. *Acta Protozoologica*, 41:1-9.

Coppellotti O. 1988. Sensitivity to copper in a ciliate as a possible component of biological monitoring in the Lagoon of Venice. *Archives of Environmental Contamination and Toxicology,* 35:417-425.

Dini, F., Nyberg, D. 1999. Growth rates of marine ciliates on diverse organisms reveal ecological specializations within morphospecies. *Microbial Ecology,* 37:13-22.

Dolan, J.R. 1991. Guilds of ciliate microzooplankton in the Chesapeake Bay. *Estuarine Coastal and Shelf Science,* 33:137-152.

Dolan, J.R., Coats, D.W. 1991a. A study of feeding in predacious ciliates using prey ciliates labeled with fluorescent microspheres. *Journal of Plankton Research,* 13:609-627.

Dolan, J.R., Coats, D.W. 1991b. Preliminary prey digestion in a predacious estuarine ciliate and the use of digestion data to estimate ingestion. *Limnology and Oceanography,* 36:558-565.

Dolan, J.R., Simek, K. 1997. Processing of ingested matter in *Strombidium sulcatum*, a marine ciliate (Oligotrichida). *Limnology and Oceanography,* 43:393-397.

Dolan, J.R., Simek, K. 1998. Ingestion and digestion of an autotrophic picoplankter, *Synechococcus*, by a heterotrophic nanoflagellate, *Bodo saltans*. *Limnology and Oceanography,* 43:1740-1746.

Dolan, J.R., Simek, K. 1999. Diel periodicity in *Synechococcus* populations and grazing by heterotrophic nanoflagellates: analysis of food vacuole contents. *Limnology and Oceanography,* 44:1565-1570.

Dolan, J.R., McKeon, K. 2005. The reliability of grazing rate estimates from dilution experiments: have we over-estimated rates of organic carbon consumption by microzooplankton? *Ocean Science,* 1:1-7.

Fenchel, T. 1975. The quantitative importance of benthic microfauna of an arctic tundra pond. *Hydrobiologia,* 46:445-464.

Fenchel, T. 2004. Orientation in two dimensions: chemosensory motile behaviour of *Euplotes vannus*. *European Journal of Protistology,* 40:49-55.

Fok, A.K., Allen, R.D. 1990. The phagosome-lysosome membrane system and its regulation in *Paramecium*. *International Review of Cytology,* 123:61-94.

Fok, A.K., Allen, R.D. 1993. Membrane flow in the digestive cycle of *Paramecium*. Advances in Cell and Molecular Biology of Membranes, vol 2B, *Membrane Traffic in Protozoa,* pp 311-337.

Fok, A.G., Lee, Y., Allen, R.D. 1982. The correlation of digestive vacuole pH and size with the digestive cycle in *Paramecium caudatum*. *Journal of Protozoology,* 29:409-414.

Fok, A.G., Schockley, B.U. 1985. Processing of digestive vacuoles in *Tetrahymena* and the effects of dichloroisoproternol. *Journal of Protozoology,* 32:6-9.

Gast, V., Horstmann, U. 1983. N-remineralization of phyto- and bacterioplankton by the marine ciliate *Euplotes vannus*. *Marine Ecology Progress Series,* 13:55-60.

Goulder, R. 1972. Grazing by the ciliated protozoon *Loxodes magnus* on the alga *Scenedesmus* in a eutrophic pond. *Oikos,* 23:109-115.

Goulder, R. 1973. Observations over 24 hour on the quantity of algae inside grazing ciliated protozoa. *Oecologia,* 13:117-182.

Hausmann, K., Radek, R. 1993. A comparative survey on phagosome formation in protozoa. Advances in Cell and Molecular Biology of Membranes, vol 2B, *Membrane Traffic in Protozoa,* pp 259-282.

Johannes, R.E. 1965. Influence of marine protozoa on nutrient regeneration. *Limnology and Oceanography,* 10:434-442.

Jones, T.C., Gates, M.A. 1994. A morphometric study of euryhalanity in marine populations of the ciliate genus *Euplotes*. *Journal of Eukaryotic Microbiology,* 41: 303-316.

Kaneshiro, E.S., Reuter, S.F., Quattrone, F.J., Morris, R.E. 1992. Sustained food vacuole formation by axenic *Paramecium tetraurelia* and the inhibition of membrane recycling by alcian blue. *Journal of Protozoology,* 39:713-718.

Kopylov, A.I., Tumantseva, N.I.1987. Analysis of the contents of tintinnid food vacuoles and evaluation of their contribution to the consumption of phytoplankton production off the Peru coast. *Oceanology,* 27:343-347.

Kruppel T., Wissing, F. 1996. Characterisation of the voltage-activated calcium current in the marine ciliate *Euplotes vannus*. *Cell Calcium,* 19: 229-241.

Kruppel, T., Rabe, H., Dummler, B., Lueken, W. 1995. The Depolarizing mechanoreceptor potential and Ca/Mg receptor current of the marine ciliate *Euplotes vannus*. *Journal of Comparative physiology A- Sensory Neural and Behavioral Physiology,* 177:511-517.

Lewitus, A.J., Wetz, M.S., Willis, B.M., Burkholder, J.M., Parrow, M.W., Glasgow, H.B. 2006. Grazing activity of *Pfiesteria piscicida* (Dinophyceae) and susceptibility to ciliate predation vary with toxicity status. Harmful Algae, 5:427-434.

Lueken W., Ricci N., Kruppel T. 1996. Rhythmic spontaneous depolarizations determine a slow-and-fast rhythm in walking of the marine hypotrich *Euplotes vannus*. *European Journal of Protistology,* 32:47-54.

Mori, G., Erra F., Cionini K. 2003. Sublethal doses of heavy metals and Slow-Down pattern of *Euplotes crassus* (Ciliophora, Hypotrichia): a behavioural bioassay. *Italian Journal of Zoology,* 70:23-30.

Peck, R.K., Hausmann, K. 1980. Primary lysosomes of the ciliate *Pseudomicrothorax dubius*: cytochemical identification and role in phgocytosis. *Journal of Protozoology,* 32:501-508.

Petroni, G. Rosati, C. Vannini, L. Modeo, F., Dini, F. Verni, B. 2003. In situ identification by fluorescently labeled oligonucleotide probes of morphologically similar, closely related ciliate species. *Microbial Ecology,* 45:156-162.

Petroni G., Dini F., Verni F. 2002. A molecular approach to the tangled intrageneric relationships underlying phylogeny in *Euplotes* (Ciliophora, Spirotrichea). *Molecular Phylogenetics and Evolution,* 22:118-130.

Petroni, G., Rosati G., Vannini C., 2003. In situ identification by fluorescently labeled oligonucleotide probes of morphologically similar, closely related ciliate species. *Microbial Ecology,* 45:156-162.

Plattner, H., Kissmehl, R. 2003. Molecular aspects of membrane trafficking in *Paramecium*. *International Review of Cytology,* 232:185-216.

Premke, K., Arndt, H. 2000. Predation on heterotrophic flagellates by protists: Food selectivity determined using a live-staining technique. *Archiv für Hydrobiologie,* 150:17-28.

Ramoino, P. 1996. Membrane supply and food vacuole formation in *Paramecium primaurelia*. *Archiv für Protistenkunde,* 147:323-329.

Ricci N., Luvera G., Cacciatori M., Banchetti R., Lueken W. 1997. The effects of 2 μM Hg^{++} on the ethogram of *Euplotes vannus* (Ciliata, Hypotrichida). *European Journal of Protistlogy,* 33:63-71.

Lueken W, Ricci N, Kruppel T. 1996. Rhythmic spontaneous depolarizations determine a slow and fast rhythm in walking of the marine hypotrich ciliate *Euplotes vannus*. *European Journal of Protistology,* 32: 47-57.

Scott, G., Gransden, S.G., Lewitus, A.J. 2003. Grazing of two euplotid ciliates on the heterotrophic dinoflagellates *Pfiesteria piscicida* and *Cryptoperidiniopsis* sp. *Aquatic Microbial Ecology,* 33:303-308.

Shang H.M., Chen Z.G., Song W.B. 2002. Species separation among seven *Euplotes* spp. (Protozoa : Ciliophora : Hypotrichida) using PCR/RFLP analysis of nuclear ribosomal DNA. *Journal of Zoology* 258:375-379.

Spicer, J.I., Gaston, K.J. 1999. *Physiological Diversity and Its Ecological Implications.,* London: Blackwell Science, 241pp. Book.

Stebbing, A.R.D., Soria, S., Burt, G.R., Cleary, J.J. 1990. Water quality bioassays in two Bermudan Harbours using the ciliate *Euplotes vannus*, in relation to tributyltin distribution. *Journal of Experimental Marine Biology and Ecology, 1*38:159-166.

Stock C., Kruppel T., Lueken W., Key, G. 1997. Congruence of electrical properties in two Antarctic and two middle-latitude marine species of *Euplotes* (Ciliata, Hypotrichida). *Journal of Eukaryotic Microbiology* 44:427-433.

Tomaru A, Matsuoka T, Kogure K. 2001. Photoreaction of a colorless marine ciliate, *Euplotes vannus*. *Archiv für Hydrobiologie* 152:647-660.

Walton, B.M., Gates, M.A., Kloss, A., Fisher, J. 1995. Intraspecific variability in the thermal-dependance of locomotion, population-growth, and mating in the ciliated protist *Euplotes vannus*. *Physiological Zoology,* 68:98-113.

Weisse, T. 2002. The significance of inter- and intraspecific variation in bacterivorous and herbivorous protists. *Antonie van Leeuwenhoek,* 81:327-341.

Weisse, T., Montagnes, D.J.S. 1998. Effect of temperature on inter- and intraspecific isolates of *Urotricha* (Prostomatida, Ciliophora). *Aquatic Microbial Ecology,* 15:285-291.

Weisse, T., Rammer, S. 2006. Pronounced ecophysiological clonal differences of two common freshwater ciliates, *Coleps spetai* (Prostomatida) and *Rimostrombidium lacustris* (Oligotrichida), challenge the morphospecies concept. *Journal of Plankton Research,* 28:55-63.

Xu, H.L, Song W.B., Warren A., 2004. An investigation of the tolerance to ammonia of the marine ciliate *Euplotes vannus* (Protozoa, Ciliophora). *Hydrobiologia,* 519:189-195.

In: Microbial Ecology Research Trends
Editor: Thijs Van Dijk, pp. 221-230

ISBN: 978-1-60456-179-1

Chapter X

Polyfunctional SAS of Microbial Origin for Agriculture

M.G. Sagdieva, G.V. Cherkasova, G.I. Djumaniyazova, A.E. Kurbanbaeva, S.D. Gusakova and Zh.S. Ayropetova
Institute of Microbiology, Uzbekistan Academy of Sciences, Uzbekistan, Tashkent, 100128, A. Kadyriy street, b. 7[b]

Abstract

On the basis of screening 19 strains of nitrogen-fixing and 9 strains of phosphorous-mobilizing bacteria, isolated from rhizosphere of cotton, alfalfa and sugar-beet, two active strains - producers of biosurfactants (BS) - have been isolated and identified as *Pseudomonas aeruginosa CH5* and *Bacillus subtilis 26*. The value of surface tension of cultural broth (CB) of the mentioned bacteria varied within the limits 57.2 – 58.0 mN/m. The optimal cultivation conditions for active production of BS by *P. aeruginosa CH5* and *B. subtilis 26* have been determined, at which the surface tension of CB had value 36.6 – 40.5 mN/m. From CB of *P. aeruginosa CH5* biomass, grown in BEB (beef-extract broth) with addition of sunflower or cotton oil, the BS complex (0.2-2.1%) has been selected. In composition of BS no less than three rhamnolipids and lipophilic pigments of phenazine nature have been found, the amount of which essentially depended on type of vegetative oil in nutrient medium. CB, BS and water-soluble pigments of *P. aeruginosa CH5* displayed growth-stimulant activity towards seeds and seedlings of above-mentioned agricultural crops and antagonistic activity to *Verticillium dahliae, Rhizoctonia solani* and *Fusarium oxysporum* phytopathogenic fungi. This activity remained after BS autoclaving (121°C, 1 atm.), enduring at pH 5.5-8.5 and NaCl concentration up to 20 g/L. CB, BS and water-soluble pigments were stable during storage in usual conditions more than 9 months.

Keywords: *Pseudomonas aeruginosa, Bacillus subtilis, Rhizobacteria, Biosurfactants, Rhamnolipids, Phenazine pigments, Bioremediation.*

Introduction

At present time in connection with considerable contamination of land resources and loss of their aptitude for irrigated agriculture it is necessary to conduct the search of microorganisms, which are able to survive in stress conditions, improve soil properties and fertility and simultaneously to influence on growth and development of plants [1]. The bioremediation with application of microorganisms, producing microbial SAS (biosurfactants, BS) and their enzymes, which are active towards organic and inorganic pollutants, is a modern method to restore poorer and contaminated soils. Microbiological method of clean-up environment and soil restoration should organically combine with natural balance in ecosystems. The main demand of acceptability of soil bioremediation technologies is a complete preservation of population health, inhabiting this area, and also restoration of favorable ecological conditions [2-4].

The advantage of microbial BS over synthetic ones includes their non-toxicity, ecological compatibility, easy utilization, stability to extreme temperatures, overpressure, relatively high values of pH and salinity, possibility of their obtaining by biotechnological way. The microbial BS display immunomodulating and antagonistic properties against phytopatogens [5]. However, the investigations devoted to associative interaction of exometabolites from rhizobacteria including BS with high plants are single.

One of factors limiting wide application of microbial BS is relatively low number of microorganisms-destructors in soils, whereas in rhizosphere zone their number is ten times more. Rhizosphere soil differs on quality and quantity from microflora of control. Since rhizosphere microorganisms excrete bioactive substances into the soil including BS, it is of great interest to use them for soil bioremediation, creation of biological preparations of polyfunctional action, which are able to stimulate plant growth and protect them from phytopathogens [2].

It is known that species of *Pseudomonas* genus *(P. aeruginosa, P. chlororaphis, P. putida)* produce active BS of glycolipid nature – rhamnolipids, amount of which depends on strain and source of carbon [6]. So, cultivation of *P. aeruginosa* in media, containing additions of vegetative oils as source of carbon (corn oil, olive oil, soybean or canola oil), promotes of accumulation of rhamnolipids in CB from 7.65 to 112 g/L [7]. Such exametabolites of bacteria as pigments can be extracted from biomass together with BS complex. Different strains of *P. aeruginosa* synthesize two main blue phenazine pigments – pyocyanin (1-hydroxy-5-N-methyl-phenazine) and phenazine-1-carboxylic acid. Pyocyanin belongs to high-active antibiotics, it has toxicity and practical application, because it is active against pathogenic bacteria. The close correlation between secondary metabolites, synthesized with bacteria, has been recorded, in particular, if pigments are synthesized, then production of both antibiotics and other bioactive substances occur [8].

The objective of investigations was a search of local industrial-perspective strains of nitrogen-fixing and phosphorous-mobilizing bacteria – producers of BS, as well as selection of optimal conditions of cultivation and characteristic of BS.

Materials and Methods

Microorganism

Isolation of rhizobacteria – producers of BS was conducted on general accepted methods of soil microbiology from rhizosphere of wheat, cotton, alfalfa and sugar-beet. The species identification of microorganisms was determined on Bergey's Manual of Determi-native Bacteriology [9]. Active strains were selected on the basis of analyzing results of CB surface tension value measurement by methods of Rebinder and Wilhelmy plate technique [10]. The active strains were isolated in pure culture in a way of successive passages to agarized BEB (beef-extract broth) at temperature 32-37°C.

Conditions of Cultivation

The BEB, BEA and Chapek`s media were used. According to the value of surface tension of CB, the optimal nutritive medium (BEB + 1% cotton or sunflower oil) was selected. The conditions of cultivation in this medium were as follows – (orbital shaker, 180 rpm, 37°C) and age of culture (7 days and more) – when the surface tension of CB had the lowest value. Phytopathogenic fungi were cultivated in Chapek`s medium.

Determination of Activity

The influence of both CB and BS, isolated from active cultures, was studied in different concentrations on the seeds of above-mentioned cultures in Petri dishes with filter paper. To obtain this, the seeds were sterilized with concentrated H_2SO_4 and introduced into Petri dishes on 50 pieces into each dish, followed with introduction on 10 ml of CB in concentration 10^6cells/ml or BS in concentration 0,01- 0,00001%. The seeds, impregnated with 10 ml of sterile tap water, were served as control. The germination and rate of sprouting seeds were registered during 4 - 6 days.

Antagonistic activity of bacteria to phytopathogenic fungi *Verticillium dahliae, Rhizoctonia solani* and *Fusarium oxysporum* was determined with method of holes in agar. 7-daily suspension of phytopatogenic fungi was sowed as lawn to the surface of agarized Chapek`s media. 7- or 15-daily CB (10^9 cells/ml) or 1% emulsion of BS were introduced into each hole cut out in the center of agar plate (d = 0.7 cm). The incubation was conducted at 28°C.

Rhamnolipids Extraction and Characterization

250 ml of CB of *P. aeruginosa CH5* (pH 7.0, σ = 36.6-40.4 mN/m) were centrifuged at 7,000 g for 30 min. In the first experiments the supernatant was acidified with 10% HCl up to pH 2.0-3.0, it was undergone +5°C during night and further it was extracted 3 times on 70 ml

of mixture of chloroform and methanol (2:1, v/v) [11]. The united extracts were dried up over dry Na_2SO_4 and steamed on rotary evaporator. Because of absence of well-formed sediment of rhamnolipids within acid solution the next samples of supernatant were extracted without acidification.

The complex of BS in the form of brown pasty substance was analyzed with thin-layer chromatography (TLC) method on glass plates (9 x 12 cm) with silica gel of grade L5/40 and addition of 13% Ca_2SO_4 (Cavalier, Czechoslovakia) in system of solvents $CHCl_3$-MeOH-AcOH (65:15:2, v/v/v). The spots were developed with iodine fumes, and also spraying of chromatogram with 0.5% solution of α-naphthol in H_2O-MeOH (1:1, v|v), then after spraying with 50% H_2SO_4 and heating up to appearance of violet color of glycolipids spots [12]. The spots of rhamnolipids were identified on the basis of comparison of R_f with literature data [13,14]. The quantitative estimate of the main types of rhamnolipids of 15-daily culture of *P. aeruginosa CH5*, grown up on sunflower oil, was conducted on data of double TLC under mentioned conditions, for 7-daily culture grown up on cotton oil – by method column chromatography (CC) [11]. After removal of lipophilic BS from supernatant its surface tension was measured, then 50 ml was sampled, dried until dry and water-soluble BS content was determined gravimetrically.

Results and Discussion

Since in rhizosphere soil there are microorganisms by order higher and more diverse types than in control soil [2], i.e. the search of perspective microorganisms – producers of BS was conducted in rhizosphere zone of cotton, alfalfa, wheat and sugar-beet. 19 strains of nitrogen-fixing and 9 strains of phosphorous-mobilizing bacteria have been revealed, from which two strains from wheat rhizosphere have been found as active producers of BS. They were identified as *Pseudomonas aeruginosa CH5* and *Bacillus subtilis 26*. The CB of the mentioned bacteria had value of surface tension within limits 57.2 – 58.0 mN/m. *P. aeruginosa CH5* strain was cultivated in choosed optimal conditions in the medium with BEB + 1% cotton or sunflower oil during 7-15 days. The obtained CB was analyzed on content and composition of BS.

General Characteristics of BS Complex from *P. Aeruginosa* CH5

Growing *Pseudomonas* genus strains in the nutritive media with addition of vegetative oils promotes to synthesis of rhamnolipids [7]. For cultivation of *P. aeruginosa CH5* strain we used cotton and sunflower oils as carbon source. The latter was chosen as the most economical for industrial cultivation of strain. The accumulation of BS in CB was estimated in dynamics of culture growth on value of surface tension. The lowest index was found in CB of 7-daily and 15-daily culture, grown up on sunflower oil, and in 7-daily culture grown on cotton oil. During storage of CB, obtained on cotton oil, during 20 days no noticeable changes of σ were observed. From supernatant of these samples the lipophilic BS (LBS) and water-soluble BS (WBS) were isolated. The results of experiments are represented in table 1, from data of which it is seen that under equal age of cultures (7 days) and comparable σ values of CB and supernatant, the LBS and WBS content is higher than in supernatant from cotton oil. The removal of LBS from supernatants does not increase essentially their surface tension. Content of exometabolites (LBS+WBS) in CB increases with growing age of culture (7-20 days), mainly at the expense of WBS.

Table 1. Characteristic of BS complex from *P. aeruginosa CH5* under growing of culture in the media with addition of vegetative oils

Age of culture	σ, mN/m		Output of LBS from supernatant		σ of super-natant after extraction,	Output of WBS from supernatant		Total output of LBS and WBS	
	CB	Supernatant	g/L	%	mN/m	g/L	%	g/L	%
Sunflower oil									
7-daily	36,60	36,80	0,67	0,07	40,02	8,1	0,81	8,77	0,88
15-daily	36,90	-	2,14	0,21	-	-	-	-	-
Cotton oil									
7- daily	37,19	38,06	0,44	0,04	37,58	6,2	0,63	6,64	0,66
20-daily	40,50	40,10	0,42	0,04	40,45	15,5	1,55	15,92	1,59

BS of *P. Aeruginosa CH5* Culture, Grown in Sunflower Oil

CB and supernatant of these samples had brown colour. According to results of analytical TLC, LBS of 15-daily culture were a mixture of no less than 5 types of rhamnolipids with pigments. Spots of pigments were coloured in yellow, green and brown colors.

On the basis of specific violet color of glycolipids spots after development of chromatogram with α-naphthol and comparison of R_f values with literature data [13,14], rhamnolipid RL-4 (R_f 0.16; 2 molecules of rhamnose, 1 molecule of fatty acid 3-OH-10:0) was identified as main component, rhamnolipids RL-3 (0.33; 1 molecule of rhamnose, 1 molecule of fatty acid 3-OH-10:0) and RL-2 (0.41; 2 molecules of rhamnose, 2 molecules of fatty acid 3-OH-10:0) – as minor components. Other minor components with R_f 0.07 and 0.27 were not identified. Amount of rhamnolipids in complex with LBS (data of preparative TLC) comprised 49%, from them RL-4 – 38.6; RL-3 – 6.6; RL-2 – 3.8% .

Thus, RL-4 and RL-3 are the main rhamnolipids of 15-daily culture of *P. aeruginosa CH5*. In literature the structures of 4 types of rhamnolipids with residuum of one or two molecules of L-rhamnose and one or two molecules of 3-hydroxydecanoic fatty acid (3-OH-10:0) are described. Two types are mains, and two types are accumulated in dependence on cultivation conditions for microorganisms [7, 13]. It has been noted that both RL-4 and RL-3 rhamnolipids are synthesized biosynthetically only under some conditions of cultivation [13].

BS of *P. Aeruginosa CH5* Culture, Grown in Cotton Oil

CB and supernatant of these samples had dark-brown color. The composition of rhamnolipids of 7-daily and 20-daily culture in comparison to analogous one of he culture, grown in sunflower oil, was more simple (analytical TLC) and included mainly both RL-4 and RL-3 with prevalence of the latter one. The set of lipophilic pigments (R_f 0.7-0.9) differed with complexity and color diversity (pink, blue, green, different tints of brown color). According data CC, content of RL-4 and RL-3 rhamnolipids in 7-daily CB comprised 4.8% and 6.8% correspondently, but pigments content – more 85%. It is known that the main blue phenazine pigment of *P. aeruginosa* – pyocyanin – can exist in oxidated and reduced forms. The reduced form is unstable free radical and it reacts quickly with molecular oxygen with formation of coloured compounds [15, 16]. The accumulation of phenazine pigments in *P. aeruginosa CH5* culture, grown in cotton oil, can be explained both specificity of strain itself and composition of lipophilic components of cotton oil, for example gossypol.

Biological Activity of CB and BS from *P. Aeruginosa CH5*

The action of CB and BS in concentrations from 0.01 to 0.00001% on germination and sprouting seeds as well as to antagonistic activity to phytopathogenic fungi was studied. Comparison of activity before and after autoclavating showed that autoclaving did not influence on properties of CB and BS, because their activity remained. Data of these experiments are shown in tables 2 and 3.

Table 2. Seed germination of cotton and wheat, treated with 15-daily CB of *P. aeruginosa CH5*

Sample	Time of incubation, (days)	Control-1 (H_2O), %	Control-2 (BEB of 6th dilution), %	CB 10^6 cells/ml, %
		Cotton		
Seeds	1	10	52	30
	2	50	80	66
	3	72	85	86
	4	88	85	94
Seedlings (length, cm)		2,14	2,0-3,0	4,81
		Wheat		
Seeds	1	56	40	82
	2	76	72	90
	3	78	75	96
	4	82	75	100
Seedlings (length, cm)		0,5-5	0,5 - 5	9,0 – 9,5

Table 3. Seed germination of cotton and wheat, treated by initial and autoclaved BS complex of 15-daily *P. aeruginosa CH5* culture

Sample	Time of incubation, days	BS concentration before autoclaving, %		BS concentration after autoclaving, %			
	1	0,01	0,001	0,01	0,001	0,0001	0,00001
Cotton seeds		60	80	65	85	85	80
	2	80	100	75	100	96	100
	3	80	100	80	100	96	100
	4	100	100	80	100	100	100
Cotton seedlings (length, mm)		1,5-24	1,5-24	10-12	20-21	24-26	24-26
	1	95	85	65	55	80	80
Wheat seeds	2	95	100	90	90	100	100
	5	95	100	90	95	100	100
	6	100	100	100	100	100	100
Wheat seedlings (length, mm)		83,6	84,2	71,7	88	94,6	90,6

Influence of Extreme Effects on Properties of BS from P. Aeruginosa CH5

Autoclaving

The activity of CB and 1% BS solution of 15-daily culture remained after their autoclaving (121°C, 1 atm.). CB, BS and water-soluble pigments were stable in usual conditions more than 9 months.

Alkaline and Acidic Treatments

Alfalfa seeds were treated with alkaline (pH 8.0-10.0) or subacidic (pH 4.5 - 6.5) solutions of BS. The solutions with pH 10.0 and 4.5 completely suppressed seed germination. The seeds, treated with BS with pH 5.0-8.0, developed normally and, hence BS complex keeps its activity within this pH range.

Treatment with Saline Solutions

BS solutions were prepared with NaCl concentrations 0.85-2.5%. Seeds germinated and actively developed after treatment with BS saline solutions with concentrations 0.85-2.0%, not differening from control.

The data on antagonistic activity of CB and BS towards phytopathogenic fungi are submitted in table 4.

From data of table 4 it is seen, that antagonistic activity of CB and BS is practically equal. The most fungicidal activity the samples display towards to *Verticillium dahliae* phytopathogene.

Table 4. Zones of growth inhibition of phytopathogenic fungi under the influence of CB and BS from *P. aeruginosa CH5*

Fungus culture	Inhibition zone, mm	Inhibition zone, mm
	CB	BS
Verticillium dahliae	75-85	75-87
Rhizoctonia solani	30-40	30-40
Fusarium oxysporum	25-26	26-27

Conclusion

From rhizosphere zone of cotton, alfalfa, wheat and sugar-beet 19 strains of nitrogen-fixing and 9 strains of phosphorous-mobilizing bacteria were isolated, among which two strains originated from wheat rhizosphere were found as active producers of BS. They were identified as *Pseudomonas aeruginosa CH5* and *Bacillus subtilis 26*. On the value of surface tension the optimal nutritive medium was selected, as well as conditions of cultivation in this medium and age of culture (7 days and more), under which the surface tension of CB had the lowest value 36.6 – 40.5 mN/m. From CB of *Ps. aeruginosa CH5* strain, grown up in the medium with addition of sunflower or cotton oil, the complex of lipophilic BS, containing rhamnolipids and phenazine pigments, has been isolated. The content and composition of BS depended on age of culture and nature of vegetative oil, used as source for carbon nutrition. In BS of this strain RL-4 and RL-3 types of rhamnolipids have been identified, in the medium with sunflower oil RL-4 type predominates, in the medium with cotton oil - RL-3 type of rhamnolipid. Autoclaving of CB and BS from *Ps. aeruginosa CH5* can be used as preparations of polyfunctional action for sanitation of sowing material, increase of crop capacity of agricultural cultures, and also as method for bioremediation of sowing lands, contaminated with different pollutions.

Acknowledgements

This research was supported in part in grant No 3200 from the Science and Technology Centre of Ukraine.

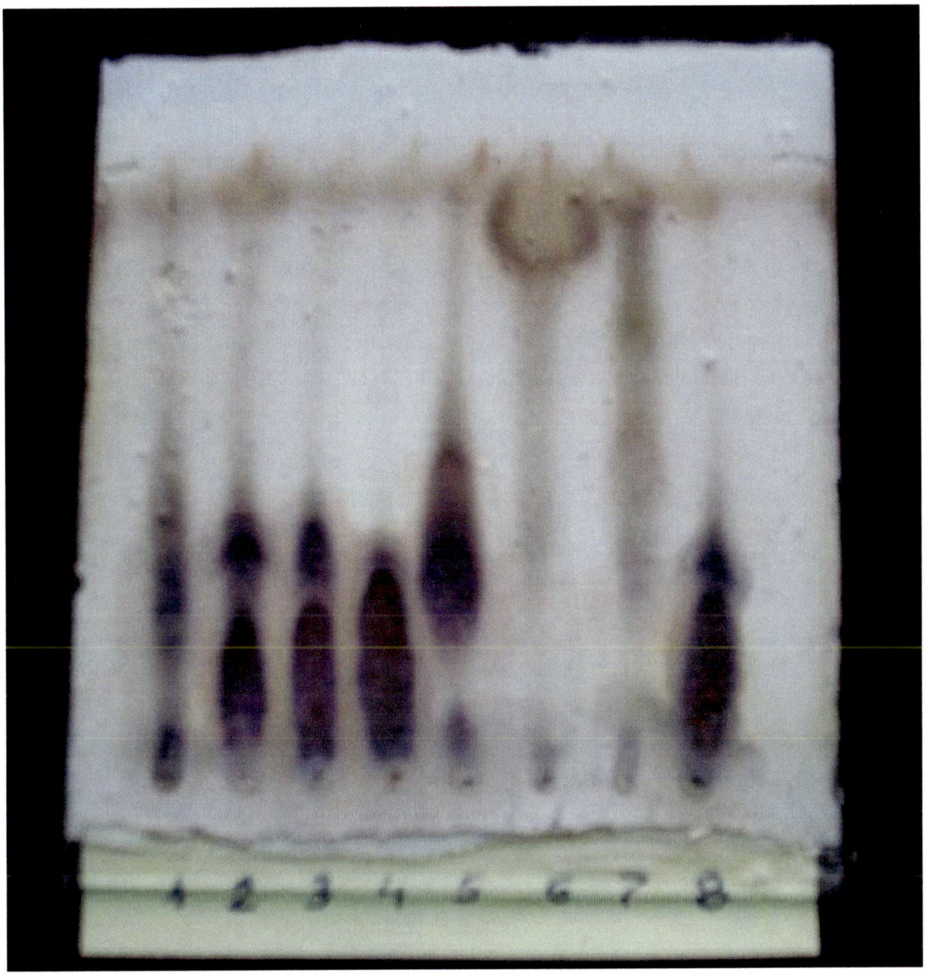

Photo ATLC. Samples: 1 - LBS of 15-daily culture of *P. aeruginosa CH5* (sunflower oil); Fractions of rhamnolipids after preparative TLC LBS: 2, 3, 8 - mixture of rhamnolipids; 4 – RL-4; 5 – RL-3; 6 – main pigment; 7 – mixture of pigments.

References

[1] Vorobiyova L.I. (1989) *Industrial microbiology.* M. MSU, 182-187.

[2] Boronik K.A.M. (2002) Biotechnology of soil remediation on the base of microbial-plant interaction. Proceed. 1st Internation. Congress *"Biology – condition and prospects of development".* M. Russia, 14-18 October, 138.

[3] Gamayurova I.S., Krynitckaya L.Yu., Elizarova Yu.V. (2002) Peculiarity of technology of preparation production for plants protection «Phytospas». Proceed. 1st Internation. Congress *"Biology – condition and prospects of development".* M. Russia, 14-18 October, 137.

[4] Katrukha G.S., Fedorova G.B., Galkina N.N. «Bactofit» – biological fungicide and bactericide of wide spectrum of action. Proceed. 1st Internation. Congress *"Biology – condition and prospects of development"*. M. Russia, 14-18 October, 218.

[5] Desai J.D., Danat I.M. (1997) Microbial production of surfactants and their commercial potential. *Microbiol. Molecul. Biol. Reviews.* 61, #1, 47-64.

[6] Gunter N.W. IV, Nunez A., Fett W., Solaiman D.K.Y. (2005) Production of rhamnolipids by Pseudomonas chlororaphis, a nonpathogenic bacterium. *Appl. Environ. Microbiol.* 71, #5, 2288 -2293.

[7] Lang S., Wullbrandt D.(1999) Rhamnose lipids – biosynthesis, microbial production and application potential. *Appl. Microbiol. Biotechnol.* 51, 22-32.

[8] Prescott L.M., Harley J.P., Klein D.A. (1993) *Microbiology.* Sec.Ed. Wm.C.Brown Publishers. 912.

[9] Bergey's Manual of Determinative Bacteriology (1997). Ed. D.Hout, N.Kreeg, P. Sneet et al. 9th edition. B.1-2.. M., «Mir».

[10] Abramzon A.A. (1981) Surface active substances. L. *«Khimiya»,* 304 p.

[11] Tahzibi A., Fatemeh R., Mahnaz M.A. (2004) Improved production of rhamnolipids by a Pseudomonas aeruginosa mutant. *Iran. Biomed. J.* 8(1), 25-31.

[12] Keyts M. (1975) Technique of lipidology. Moscow: *«Mir»,* 322.

[13] Shenk T., Schuphan I., Schmidt B. (1995) High-performance liquid chromatographic determination of the rhamnolipids produced by Pseudomonas aeruginosa. *J. Chromatogr.* A 693, 7-13.

[14] Arino S., Marchal R., Vandecasteele J.-P. (1996) Identification and production of a rhamnolipidic biosurfactant by a Pseudomonas species. *Appl. Microbiol. Biotechnol.* 45, 162-168.

[15] Hasset D.J., Charniga L., Bean K. et all.(1992) Respons of Pseudomonas aeruginosa to pyocyanin: mechanisms of resistance, antioxidant defenses, and demonstration of a manganese-cofactored superoxide dismutase. *Infect. Immun.* 60,#2, 328-336.

[16] Fernandez R.O., Pizzaro R.A. (1997) High-performance liquid chromatographic analysis of Pseudomonas aeruginosa phenazines. *J. Chromatogr.* A 771, 99-104.

In: Microbial Ecology Research Trends
Editor: Thijs Van Dijk, pp. 231- 255

ISBN: 978-1-60456-179-1

Chapter XI

Changes of Low Molecular Weight Organic Acids, Soil Organic Matter, Nitrogen Transformation and Microbial Populations in Rhizosphere and Bulk Soils in a Temperate Rain Forest

Ming K. Wang[1*], *Po Neng Chiang*[1], *Shun Yao Zhuang*[1], *Hen Biao King*[2] *and Jeen Lian Hwong*[2]

[1]Department of Agricultural Chemistry,
National Taiwan University,Taipei, Taiwan, 106
[2]Taiwan Forestry Research Institute, Taipei, Taiwan, 106

Abstract

The ecosystem research in a temperate rain forest was not fully understood. Thus, the aims of this chapter are illustrated the changes of low molecular weight organic acids (LMWOAs), stable isotope ratios of soil organic matter, forest fire influenced the changes of vegetation communities, nitrogen transformation and microbial populations study in rhizosphere and bulk soils, selecting Ta-Ta-Chia long-term ecological research (LTER) site, Taiwan in a temperate rain forest as an example. Low molecular weight organic acids and the nature of organic matter were determined by gas chromatograph (GC) and CPMAS ^{13}C nuclear magnetic resonance analysis. Changes in the meadow-forest boundary at this LTER site detected by stable isotope ratios of soil organic matter of δ^{13}C analysis of plants and soils. The buried bag incubation method was used to study temporal pattern of net N mineralization and nitrification in soils. Several types of microorganisms were counted for their spatial and temporal variations. The recovery percentage of LMWOAs determination by GC shows a high correlation of r^2 = 0.99. All recovery of LMWOAs was above 83.6%. The proportion of organic functional groups in

*Corresponding author: Ming K. Wang, (0118862) 2363-0231 ext. 2491 or 3066 (O); (0118862) 2366-0751 (Fax); E-mail: mkwang@ntu.edu.tw.

the rhizosphere and bulk soils quantified by ^{13}C NMR analysis followed the general order: alkyl-C > O-alkyl-C > N-alkyl-C > acetal-C > aromatic-C > carboxylic-C > phenolic-C. $\delta^{13}C_{PDB}$ values in pedons 1 and 2 tended to decrease with increasing with soil depth. The values of EB, E and Bw horizons were close to $\delta^{13}C_{PDB}$ values of C_3 plants. The data indicated that SOM was a mixture of *transmorrisonensis* (C_4), *nittakeyamensis* (C_3) and *Tsuga* (C_3) plant residues.$\delta^{13}C_{PDB}$ values obtained from deeper meadow pedons 1 and 2 of soil layers indicated that the vegetation community has changed over the years. *M. transmorrisonensis* (C_4) plants replaced C_3 plants (e.g., *Tsuga*). Extractable NO_3^- concentrations ranged from 0.2 to 2.1 mg N kg^{-1} soil in the meadow zone and from 0.2 to 13.0 mg N kg^{-1} soil in the forest zone. No seasonal changes in soil nitrate concentration were present. Extractable NH_4^+ concentrations ranged from 1.0 to 12.4 mg N kg^{-1} soil in the meadow zone and from 2.8 to 25.0 mg N kg^{-1} soil. The NH_4^+ concentrations showed highly temporal variations. The net of mineralization and nitrification rates showed spatial and temporal variations, depending on microbial activities and populations. Bacteria were most abundant, followed by cellulolytic microbes, phosphate-solubilizing microbes, nitrogen-fixing microbes, fungi and actinomycetes. Microbial populations showed high correlation with microbial biomass carbon and organic acid contents.

Keywords: *$\delta^{13}C$ analysis, forest fire, low molecular weight organic acids (LMWOAs), net mineralization and nitrification rates, ^{13}C nuclear magnetic resonance, microbial populations.*

1. Introduction

In natural ecosystems, soil organic carbon is derived predominantly from the residues of native vegetation, thus the stable carbon isotope ratio of the vegetation has a direct influence on$\delta^{13}C$ of soil organic matter. It has been shown that the natural isotope abundance of SOM corresponds closely to that of the vegetation cover from which it originates [1-3]. Therefore, the characteristics$\delta^{13}C$ of the vegetation serves as a marker to indicate the origin of SOM [4]. Mass-balance calculations based on$\delta^{13}C$ values of the geological sample and assumed end-member$\delta^{13}C$ values for C_3 and C_4 plants have been used to estimate the relative proportion of C_3 and C_4-derived carbon in organic carbon [5,6]. As a consequence of different photosynthesis pathway, C_3 plants have less ^{13}C than C_4 plants.

Plant roots continuously release nutrients, such as sugar, amino acids, organic acids and vitamins, which serve as substrate for microorganisms. As a consequence, a number of soil bacteria and fungi become closely associated with the developing rhizosphere during plant growth. Furthermore, the LMWOAs of root exudates in the tree rhizosphere will vary with tree species, which in turn varies with elevation [7-10]. Nitrogen mineralization and nitrification are the dominant microbial processes affecting inorganic N availability in forest ecosystems. Various experimental approaches have been used to quantify net rates of mineralization and nitrification in soils from a diverse range of forest ecosystems. A comprehensive review by Binkley and Hart [11] discussed a number of methods that can be used to compare N availability among forest soils. Despite this method's labor intensive nature, soil incubations in buried plastic bags is among the more commonly used methods due to its simplicity and relative low cost. The earliest use of the buried bag's method was in agricultural soils by Eno [12]. The changes of $\delta^{13}C$ values and nature of organic matter, mineralization and nitrification rates, LMWOAs exudates from roots and microbial populations at a temperate forest was not fully understood. Thus, the aims of this chapter are demonstrated changes of LMWOAs, stable isotope ratios of soil organic matter, forest fire influenced the changes of vegetation communities, nitrogen transformation and microbial populations study in rhizosphere and bulk soils, selecting Ta-Ta-Chia LTER site, Taiwan in a temperate rain forest as an example, showing the temporal and spatial variations.

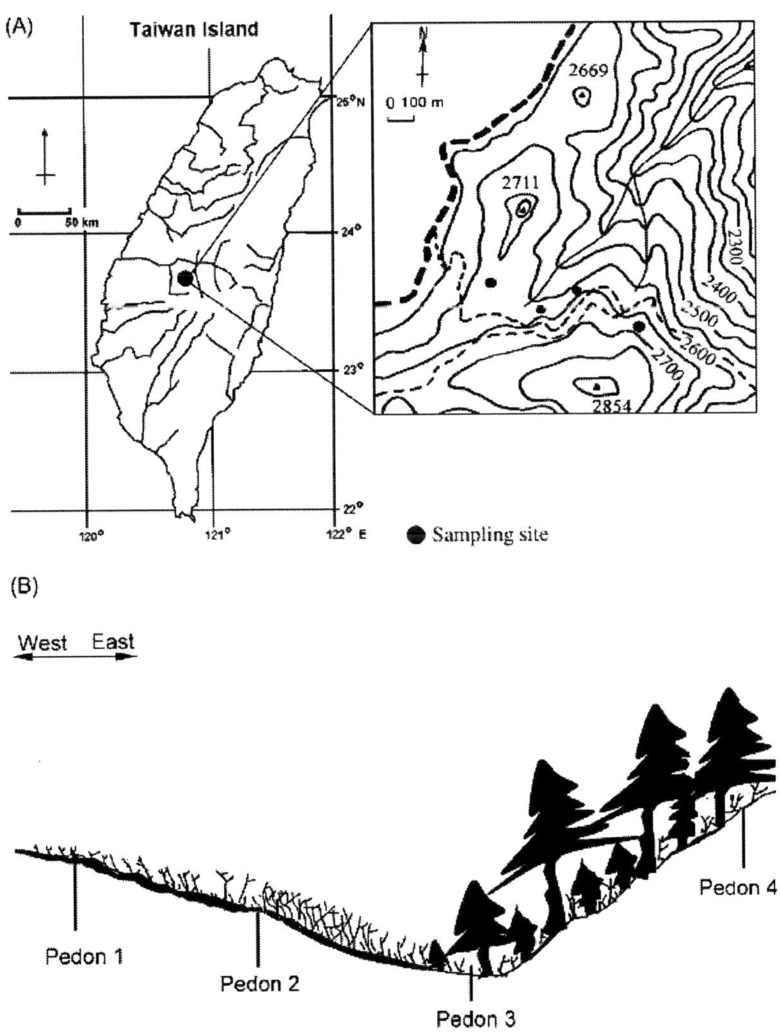

Figure 1. (A) Location maps of Ta-Ta-Chia site in Taiwan, and (B) study pedons across the meadow-forest boundary at Ta-Ta-Chia site (Reproduced with permission from Chiang et al. (2004) Changes in the grassland-forest boundary at Ta-Ta-Chia long term ecological research (LTER) site detected by stable isotope ratios of soil organic matter. Chemosphere 54: 217-224). [19]).

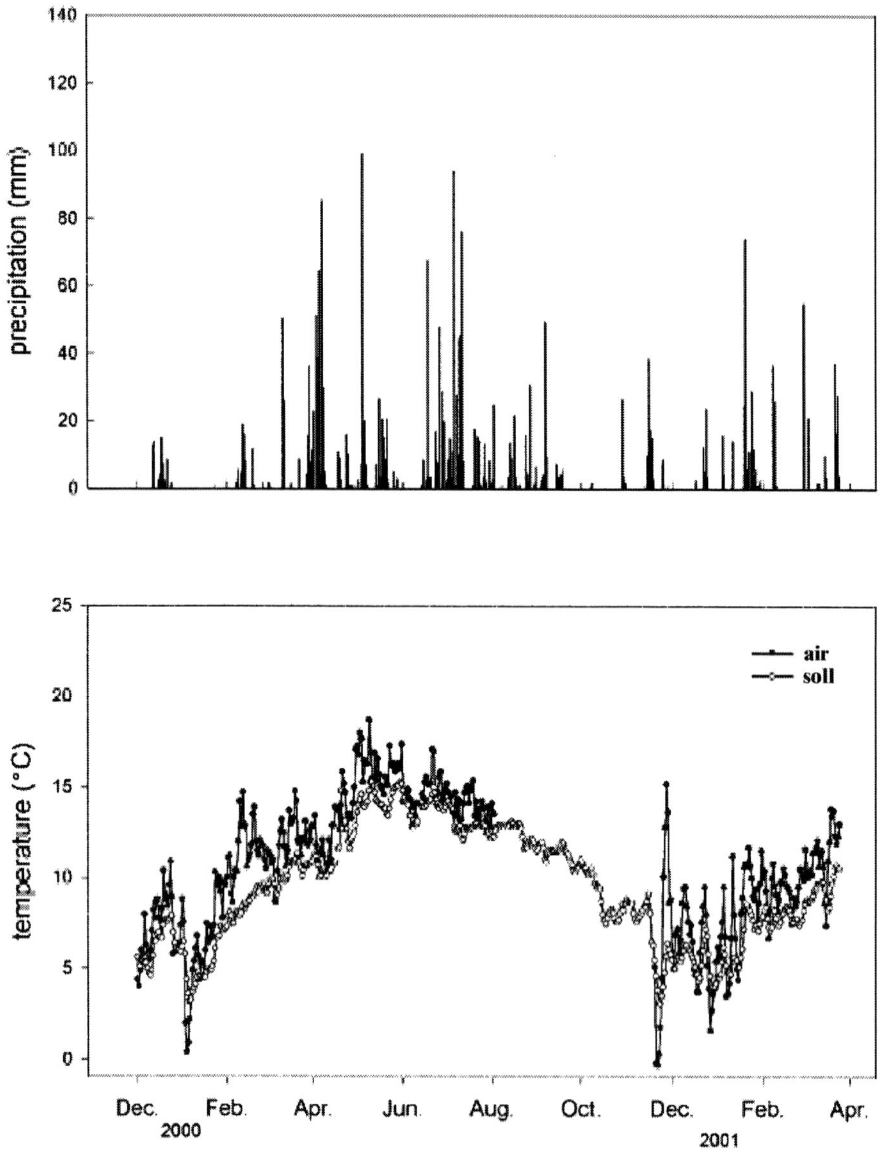

Figure 2. (A) Mean daily precipitation, and (B) air (closed circle) and soil (open circle) temperatures at 0.1 m depth during the study period (December 1999 to February 2001). Data are from a permanent weather station located approximately 50 m from the meadow site (Reproduced with permission from Owen et al. (2003b) Comparison of soil nitrogen mineralization and nitrification in a mixed grassland and forest Ecosystem in central Taiwan. Plant and Soil. 251: 167-174). [44]).

2. Environmental Conditions of Study Site

The Ta-Ta-Chia Long-Term ecological Research (LTER) site is one of the five sites in Taiwan at the highest elevation (i.e., 2000-3952 m) (Figure 1A). The total area of this LTER site is about 600 ha. There are meadow, hemlock, spruce and pinus zones. The dominant vegetations at meadow zones are *Miscanthus transmorrissonsis*, *Yushania niitakayamensis,*

Rhododendron rubropitosuin, and scrub; hemlock zones of *Tsuga chinensis varformosana* stand mixed with scattered dwarf bamboo; spruce zone of *Picea morrisonicola* mixed with *Pinus armandi var. masteriana and Chamaecyparis formosensis*; and pinus zone of *Pinus taiwanensis*. The meadow zone cover 20 ha, hemlock zone of 220 ha, spruce of 160 ha and pinus zone of 200 ha.

The bed rock of the entire site consists of interbeded Tertiary shale and slate [13,14]. Pedons of meadow are located in the range of 23°28'54"N and 120°52'52"E, and elevation of 2650 m. The hemlock zone is located in the range of 23°28'56"N and 120°53'9"E, and elevation of 2680 m. The spruce zone is located at 23°28'57"N and 120°53'19"E and an elevation about 2600 m. The pinus zone ranges from 23°28'52"N and 120°52'05"E, elevation about 2300 m. The whole Ta-Ta-Chia site is in the territory of the Yu-Shan National Park in central Taiwan (Figure 1). Soil classification is sandy loam, mixed, mesic, Dystric Cryochrepts [15]. The study site was on the mountain ridges with annual precipitation of about 2030 mm. The average relative humidity is 85% and the annual average temperature is 10.3°C. The minimum air and soil temperature at December 1999 - April 2001 were 5.0 and 1.0°C, and maximum air and soil temperatures were 10.6 and 10.5°C, respectively (Figure 2) [16].

Plant tissues and soil profile from the meadow, hemlock and spruce zones were selected for changes of stable isotope ratios of soil organic matter, low molecular weight organic acids (LMWOAs), nitrogen transformation and microbial populations study in rhizosphere and bulk soils in a temperate rain forest. There are four experiments were designed and discussed in this chapter, including (1) Low molecular weight organic acids and organic functional groups in rhizosphere and bulk soils of *Tsuga* and *Yushania* in hemlock zone; (2) Changes in the meadow-forest boundary at Ta-Ta-Chia LTER site detected by stable isotope ratios of soil organic matter; (3) Soil nitrogen mineralization and nitrification rates in a mixed meadow and forested ecosystem; and (4) Microbial population of spruce soil at this site.

3. General Physical and Chemical Properties

3.1. Laboratory Analysis of Soil Physical and Chemical Properties

Soil samples were air-dried and crushed to pass through a 2 mm sieve. Sub-sample was prepared for different chemical and physical analyses. Soil pH was determined with 1:1 soil to water suspensions using a pH meter. Organic carbon and nitrogen were determined by automated dry combustion gas chromatograph using a Carlo-Erba NA-1500 elemental analyzer (Carlo-Erba, Milan, Italy). Cation-exchange capacity and particle size analyses were determined according to the method described by Jackson [17]. The exchangeable K, Na, Ca and Mg cations were extracted by 0.1 M ammonium acetate solution. The extracts of cations were determined by flame photometer and atomic absorption spectrophotometry (Hitachi, AAS 180-30). The silt was separated from the sand by wet-sieve (53 μm) techniques. The cay fractions were dialyzed against distilled water and freeze-dried. The deferrated clays were saturated with Mg and K, and mounted as slurries on glass slides for x-ray diffraction (XRD) [17]. The Mg-satuarted clays were examined at 25°C before and after glycerol solvation. The

K-saturated clays were examined at 25°C and after heating at 110, 350 and 550°C for two hours. The oriented clay mineral aggregates were examined with an x-ray diffractometer (Rigaku Miniflex) with CuKα radiation. The XRD patterns were recorded in the range of 3-50°2θwith a scan rate of 2°2θmin^{-1}.

3.2. Soil Physical and Chemical Properties

Spruce soil showed higher C and N contents, C/N ratio, CEC and base saturation than those in meadow and hemlock soils. The texture of three soils is silty loam (Table 1). Exchangeable potassium (K) and magnesium (Mg) were high in rhizosphere soils than that in bulk soils. The availability of aluminum (Al), iron (Fe), manganese (Mn), copper (Cu) and zinc (Zn) in rhizosphere was influenced by the ionic species and concentrations, which depend on the pH change and chemical compositions of root exudates [18]. The clay mineralogical compositions were vermiculite, illite, kaolinite and trace amount of gibbsite [19].

4. Low Molecular Weight Organic Acids and Organic Functional Groups in Rhizosphere and Bulk Soils of *Tsuga* and *Yushania* in Hemlock Zone

Two study blocks were located within a hemlock stand mixed with scattered dwarf bamboo bushes and few short broad-leaved trees. Three samples were collected in each block [20]. There are three sites in each block of different slopes of tree. Rhizosphere and bulk soils from the A horizon (0-15 cm) were sampled below *Tsuga* and *Yushania.* The moist color was 10YR2/2 for all blocks [20,21]. The soil that remained adhered to the roots after gentle shaking was sampled as rhizosphere soils, by operational definition [22]. Root exudation primarily take place from fine roots near-root tips, which are usually closely spaced in the A horizon soil.

Table 1. Selected Surface Soil (0-20 cm) Physical and Chemical Properties

Zone	pH	Total C	Total N	C/N Ratio	Na	Exchangeable K	Ca	Mg	CEC	Base Saturation	Texture
	(1:1)	(%)	(%)		-----------cmol kg^{-1} soil-----------					(%)	
Meadow	4.84± 0.12	1.62± 0.15	0.15± 0.01	11.0± 0.86	3.52	18.5	5.46	4.03	14.09	2.24	SL[a]
Hemlock	4.49± 0.24	1.88± 0.13	0.14± 0.01	13.5± 1.19	4.75	19.2	6.33	4.80	14.74	2.38	SL
Spruce	4.81± 0.14	7.54± 1.17	0.54± 0.10	14.2± 1.25	10.2	30.9	85.6	14.5	20.14	7.01	SL

[a]SL: Sandy loam

The bulk soils were sampled from a location about 15 cm in distance from the main root surface. Fresh roots and leaves of *Tsuga* and *Yushania* were also collected as comparison. The samples were immediately preserved with dry ice. Soil and plant materials were ground and passed through a sieve (< 2 mm), and then stored in a deep freezer at –24°C.

4.1. Determination of Low Molecular Weight Organic Acids (LMWOAs) in Root Exudates by Gas Chromatograph (GC)

GC analysis of methylated organic acids was performed on a China chromatograph 8700 F series GC equipped with a flame ionization detector aid as AJandW scientific GC column (30 m x 0.5 mm i.d.) coated with DB-FFAP was used. The injector, column, and detector temperatures were 125, 200 and 250°C, respectively. Helium was used as the carrier gas at a flow rate of 0.63 kg cm^{-2}. 2 μL of sample were injected at a sensitivity setting of 8 x 10^{-10} afs, and the chromatograms were integrated by a Hitachi Model D-2500 chromato-integrator. Standards for the low molecular weight organic acids (oxalic, malonic, succinic, fumaric, and maleic) were obtained from Sigma Chemical Co., while methanol and chloroform used for sample preparation were analytical-grade reagents. Methyl-malonic acid, not found in root exudates of *Tsuga* and *Yushania* was used as an internal standard and added to both standard mixtures and to samples before methylation.

4.2. Acid Extraction and Concentration from Roots, Leaves and Liter Exudates

Fresh material corresponding to 1 gm of dry weight of ground roots, leaves and litters and 15 gm of soils were extracted in a 250 mL centrifuge tube for extractions with 20 mL 0.5 M HCl in methanol (MeOH) 1:1 ratio by shaking for 1 h, and then centrifuges at 10,000 xg. The detailed acid extraction and concentration from roots, leaves and liter exudates, recovery of the extraction/concentration and derivation procedure and sample stability tests were reported in Chen et al. [10].

4.3. CPMAS ^{13}C Nuclear Magnetic Resonance Analysis

Ground fresh roots, leaves and soils were examined their organic functional groups by CPMAS ^{13}C NMR spectroscopy [23], using a Bruker MSL-200 NMR instrument. Samples were spun at the magnetic angle in the boron nitride rotor with a Kel-F cap. Data acquisition conditions were: spectrometer frequency, 50.33 MHz; spectra width, 20,000 Hz; number of scans, 28,000; spinning speed, 3500 Hz; contact time, 1 ms; pulse delay time, 1 s. the NMR spectra were divided into the following chemical shift regions: alkyl-C (0-45 ppm), N-alkyl-C (46-65 ppm), O-alkyl-C (65-90 ppm), acetal-C (90-110 ppm), aromatic-C (110-140 ppm), phenolic-C (140-160 ppm) and carboxyl (160-200 ppm) [23,24]. Areas of the chemical shift regions were measured by cutting and weighing, and were expressed as percentages of total area (relatively intensity) [24].

4.4. Low Molecular Organic Acids Analyzed by GC

All recovery of low molecular weight organic acids (LMWOAs) was above 83.6%. Recovery of acids from the extraction/concentration and methylation procedure was examined by spiking the *Tsuga* rhizosphere soil (TRS), *Yushania* rhizosphere soil (YRS), and bulk soils (BS) with 100 μg of selected acids. Very good recovery and reproducibility were obtained for the saturated organic acids, oxalic, malonic and succinic acids. It has been reported that the methylation of unsaturated dicarboxylic acids may be accompanied by undesirable side reactions, as well as cis-trans isomerization [25]. The results of methodology evaluation demonstrate that the method developed in this study is accurate and precise. The LMWOAs content in various rhizosphere soils of TRS, YRS and BS are shown in Table 2. The proportion of oxalic acid contents follows the order YRS > TRS > BS with significant differences ($P < 0.05$).

Oxalic, malonic, furmaric and succinic acids in the *Tsuga* fresh roots, *Tsuga* fresh twigs and leaves; *Yushania* fresh roots and *Yushania* twigs and leaves were identified on the basis of retention time. It should be noted that the reproducibility of retention time was very good as indicated by the recovery rate of over 90%.

Table 2. Low Molecular Weight Organic Acids in the TRS, YRS, and BS[a] (Reproduced with permission from Chen et al. (2001) Determination of low molecular weight dicarboxylic acids and organic functional groups in rhizosphere and bulk soils of Tsuga and Yushania in a temperate rain forest. Plant and Soil 231: 37-44). [10])

Sample[b]	Organic acid concentration (mg kg^{-1} soil)			
	Oxalic acid	Malonic acid	Fumaric acid	Succinic acid
TRS	67b	9b	2b	22b
YRS	95a	18a	4a	35a
BS	46c	7c	2b	14c

[a] Average followed by the same letter are not significantly different (P<0.05) as determined by Duncan's multiple range (n = 6)
[b] TRS: *Tsuga* rhizosphere soil, YRS: *Yushania* rhizosphere soil, BS: Bulk soil.

Table 3. Low Molecular Weight Organic Acids in Plants. (Reproduced with permission from Chen et al. (2001) Determination of low molecular weight organic acids and organic functional groups in rhizosphere and bulk soils of *Tsuga* and *Yushania* in a temperate rain forest. Plant and Soil 231: 37-44. [10]

Sample[b]	Organic acid concentration (mg kg^{-1} dry weight of plant)[a]			
	Oxalic acid	Malonic acid	Fumaric acid	Succinic acid
Tsuga fresh roots	282	85	□[b]	62
Tsuga fresh twigs and leaves	2411	1590	□	1320
Yushania fresh roots	3789	314	117	113
Yushania fresh twigs and leaves	1470	140	61	350

[a] Recovery (> 90%) of low molecular weight organic acids acids from HCL/MeOH extraction of plant as determinated by GC after sample methylaton.
[b] Not detectable (MDL = 1 mg kg^{-1}).

High oxalic acid was present in fresh twigs and leaves of *Tsuga* (2411 mg kg^{-1}), *Yushania* fresh roots (3789 mg kg^{-1}) and *Yushania* fresh twigs and leaves (1470 mg kg^{-1}). High malonic acid (1590 mg kg^{-1}) and succinic acid (1320 mg kg^{-1}) contents were also present in *Tsuga* fresh twigs and leaves. The *Yushania* rhizosphere soils (YRS), *Yushania* fresh roots, *Yushania* fresh twigs and leaves showed high concentrations of oxalic acids (Table 3). Oxalic acid is a metabolic product of microorganisms and also a component of root exudates [26]. The amount of malonic, furmaric and succinic acids in YRS are greater than in TRS and BS. Fumaric acid was not detected in the *Tsuga* fresh roots and *Tsuga* fresh twigs and leaves. However, 61 and 117 mg kg^{-1} of fumaric acid were found in decidous environment in twigs, leaves and *Yushania* fresh roots, respectively. The amounts of succinic acids are 62, 1320, 113 and 350 mg kg^{-1} with respect to *Tsuga* fresh roots, *Tsuga* fresh twigs, and leaves; *Yushania* fresh roots and *Yushania* fresh twigs and leaves. No maleic acid was detected by GC analysis and its minimum detected limit was 1 mg kg^{-1}. In general, the amount of LMWOAs in YRS was greater than TRS and BS [10].

4.5. Nature of Organic Matter

The ^{13}C NMR analysis of SOM and organic carbon of *Tsuga* rhizosphere soil (TRS), *Yushania* rhizosphere soil (YRS) and bulk soil (BS) can be classified into seven groups (Figure 3A). The concentrations of these organic functional groups in the rhizosphere and bulk soils followed the general trend: alkyl-C > O-alkyl-C > N-alkyl-C > acetal-C > aromatic-C > carboxylic-C > phenolic-C (Table 4). The distribution of organic functional groups in *Tsuga* fresh twigs and leaves, *Tsuga* litters, *Tsuga* fresh roots (Figure 3B), *Yushania* fresh twigs and leaves, *Yushania* litters and *Yushania* fresh roots is shown in Figure 3C and Table 5. The amount of O-alkyl-C in the *Tsuga* litters was lower than *Tsuga* fresh twigs and leaves. However, the amounts of alkyl-C, N-alkyl-C, acetal-C, aromatic-C and phenolic-C in the *Tsuga* litters were higher than in the *Tsuga* fresh twigs and leaves. The *Tsuga* fresh plant materials contain more alkyl-C (carbohydrate), lignin (aromatic-C and phenolic-C) and carboxyl-C than the *Yushania* fresh plant materials. However, hemi-cellulose (N-alkyl-C and acetal-C), and cellulose (O-alkyl-C) contents in the *Yushania* fresh plant materials are greater than in the *Tsuga* fresh plant materials [27]. After decomposition, the components of hemi-cellulose, microbial resynthesis carbohydrates and lignin tend to increase. On the other hand, the O-alkyl-C contents in the *Yushania* litters are greater than the *Yushania* fresh twigs and leaves, while alkyl-C, N-alkyl-C and carboxylic-C contents in the *Yushania* litters are slightly lower than in the *Yushania* fresh twigs and leaves. This may be because the *Yushania* plant materials decompose easily. The microbial resynthesis production of mono-saccharide is a major component in the alpine forest soil environments with high annual precipitation [28,29]

In this environment, the degradation of *Tsuga* materials is slower than that of *Yushania* materials [30,31]. Hence, the cellulose contents (O-alkyl-C) in the *Tsuga* rhizosphere soils (TRS) are greater than in the *Yushania* rhizosphere soil (YRS) (Table 2). However, the amount of hemi-cellulose (N-alkyl-C) and carboxylic-C in the YRS is higher than in the TRS.

Table 4. Distribution (%) of Organic Carbon Functional Groups in the TRS, YRS, and BS[a] (Reproduced with permission from Chen et al. (2001) Determination of low molecular weight dicarboxylic acids and organic functional groups in rhizosphere and bulk soils of *Tsuga* and *Yushania* in a temperate rain forest. Plant and Soil 231: 37-44). [10]

Sampleb	Functional group (%)						
	Alkyl-C	N-Alkyl-C	O-Alkyl-C	Acetal-C	Aromatic-C	Phenolic-C	Carboxyl-C
TRS	32.9±1.7 b	16.0±0.9 a	32.7±1.7 a	9.0±0.9 a	4.9±1.4 a	1.6±0.4 b	3.0±0.8 b
YRS	34.9±2.6 a	16.6±0.2 a	28.9±2.7 b	8.3±0.6 a	4.8±1.1 a	2.0±0.5 b	4.6±1.1 a
BS	30.0±0.9 c	15.1±0.4 b	32.5±1.7 a	8.7±0.4 a	5.6±0.7 a	3.0±0.6 a	5.3±0.5 a

[a]Average followed by the same letter are not significantly different ($p < 0.05$) as determined by Duncan's multiple range test. *b*TRS; *Tsuga* rhizosphere soil; YRS; *Yushania* rhizosphere soil, BS: Bulk soil

Table 5. Distribution of Organic Carbon Functional Groups in Plants (Reproduced with permission from Chen et al. (2001) Determination of low molecular weight dicarboxylic acids and organic functional groups in rhizosphere and bulk soils of *Tsuga* and *Yushania* in a temperate rain forest. Plant and Soil 231: 37-44). [10]

Sample	Functional group (%)						
	Alkyl-C	N-Alkyl-C	O-Alkyl-C	Acetal-C	Aromatic-C	Phenolic-C	Carboxyl-C
Tsuga fresh twigs and leaves	17.1	10.3	40.1	10.1	6.4	7.9	8.1
Tsuga litters	21.1	11.5	35.6	11.0	7.0	8.0	5.8
Tsuga fresh roots	16.7	13.6	34.8	10.0	11.1	9.7	4.1
Yushania fresh twigs and leaves	18.3	15.2	44.9	10.7	2.0	2.4	6.5
Yushania litters	14.6	14.6	50.0	12.1	2.1	2.8	3.8
Yushania fresh roots	3.2	16.5	56.5	13.3	4.0	4.2	2.3

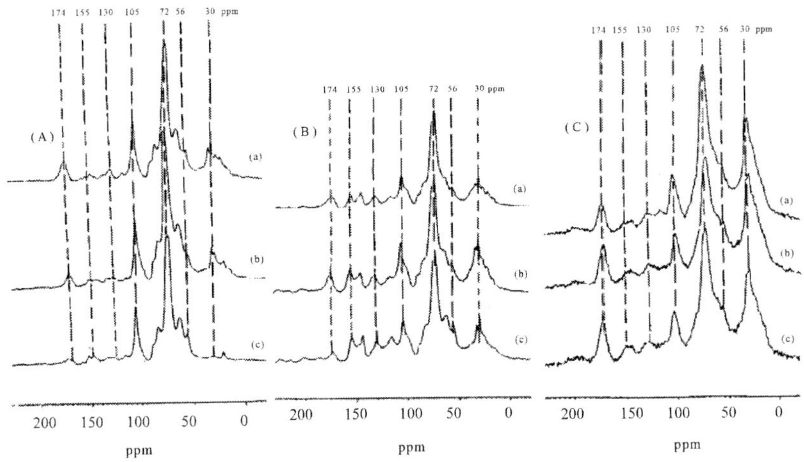

Figure 3. ^{13}C magic angle cross polarization NMR spectra of (A) (a) *Tsuga* rhizosphere soils (TRS), (b) Yushania rhizosphere soils (YRS), and (c) bulk soils (BS); (B) (a) fresh twigs and leaves, (b) litters, and (c) fresh roots of the Taiwan *Tsuga*; and (C) (a) fresh twigs and leaves, (b) litters, and (c) fresh roots of the *Yushania* (Reproduced with permission from Chen et al. (2001) Determination of low molecular weight dicarboxylic acids and organic functional groups in rhizosphere and bulk soils of *Tsuga* and *Yushania* in a temperate rain forest. Plant and Soil 231: 37-44). [10].

The debris of the *Tsuga* is not as easily decomposed under the alpine forest soil environments, so it is a major organic carbon source in the TRS and BS [10].

5. Changes in the Meadow-Forest Boundary at Ta-Ta-Chia LTER Site Detected by Stable Isotope Ratios of Soil Organic Matter

Four pedons were located along a transect from meadow (*M. transmorrisonensis* mixed with *Yushania nittakeyamensis*, pedons 1 and 2), through a transition zone (mixed with C_3 and C_4 plants) (pedon 3), to a hemlock forest (*Tsuga chinesis varformosana* mixed with *Yushania nittakeyamensis*) (pedon 4). The distance between each pedon is around 200 m. The study site was on the mountain ridges. Pedons 1 and 2 (in meadow) are on a south-facing slope, and pedon 4 (forest) is on the north-facing slope (Figure 1B).

5.1. $\delta^{13}C$ Analysis of Plants and Soils

The soil samples that passed through 2 mm sieve and fresh plant samples. The soil samples were first physically treated by examining under a microscope to look for contaminating materials (i.e., fibers, rootlets or other plant materials), which were removed with forceps. Bulk soils were air-dried and sieved through a 0.5 mm sieve. The dried soil samples, fresh leaves and roots of *M. transmorrisonensis, nittakeyamensis* and *Tsuga* were pulverized by agate mortar with a pestle and a machine, respectively. An aliquot of the powder sample was placed in a platinum (Pt) crucible admixed with purified CuO. The crucible containing the mixture was combusted at 1100°C under vacuum in a quartz reaction vessel [3]. The CO_2 was separated from the other combustion products by condensation and evaporation under the control of liquid nitrogen and dry ice. The carbon-isotope ratios of the CO_2 samples were determined using an isotope ratio-mass spectrometer, VG SIRA-10 (Oxford, UK)[32]. The result s are reported in $\delta^{13}C_{PDB}$ values [33]. The overall reproducibility is better than 0.1 per mil at the 90% level of confidence. Each sample was tested in triplicate. The carbon isotope of the organic matter was expressed as

$$\delta^{13}C\ (‰) = [(R_{sample}/R_{PDB}) - 1] \times 1000 \qquad (1)$$

where $R = {}^{13}C/{}^{12}C$. PDB refers to the belemnite carbonate standard of the Peedee Formation, South Carolina, USA. Measurements of SOM were reproducible to ±0.05‰ [3,34].

The following equation was used to calculate the proportion of the soil carbon from C_4 plants:

$$\delta^{13}C_{SOM} = \delta^{13}C_{C4} \times f + \delta^{13}C_{C3} \times (1 - f) \qquad (2)$$

Where $\delta^{13}C_{SOM}$ = carbon isotope ratio of the SOM,
$\delta^{13}C_{C4}$ = carbon isotope ratio of C_4 plants,
$\delta^{13}C_{C3}$ = carbon isotope ratio of C_3 plants,
f = proportion of carbon from C_4 plants, and
f = proportion of carbon from C_3 plants.

5.2. $\delta^{13}C$ of Plant Leaves and Roots

$\delta^{13}C$ values of *M. Transmorrisonensis* leaves and roots are –12.8±0.1‰ and –13.8±0.3‰, respectively. These values are typical of C_4 plants similar to those from other reports (-12.37±0.14‰) [3]. On the other hand, the *nittakeyamensis* and *Tsuga* leaf and root fragments showed $\delta^{13}C$ values in the range of –23.0±0.1‰ to –28.7±0.2‰ (Table 6). The roots of the *nittakeyamensis* had a higher $\delta^{13}C$ value. These values are typical of C_3 plants and close to these reported by Kao (1997) (i.e., *Baeothryon subcapitatum* (*Thwaites T. Koyama*) (-23.33±0.26‰), *Pinus taiwanensis Hayata* (-26.03±0.22‰), *Y. nittakayamensis* (-26.03±0.31‰), and *Lycopodium cernuum Linn* (-29.21±0.13‰). The $\delta^{13}C$ values of these soils and plants were compared and showed reproducible results.

5.3. Distribution of SOM$\delta^{13}C$

$\delta^{13}C$ values of the O horizon in meadow pedons 1 and 2 seem to indicate that the vegetation is to a certain extent mixed with *nittakeyamensis*. However, $\delta^{13}C_{PDB}$ values of O horizon of the *Tsuga* forest leaves pedon 4 were close to those of fresh *Tsuga* leaves and roots.

Table 6. $\delta^{13}C_{PDB}$ Values of Fresh Plants (Reproduced with permission from from Chiang et al. (2004) Changes in the grassland-forest boundary at Ta-Ta-Chia long term ecological research (LTER) site detected by stable isotope ratios of soil organic matter. Chemosphere 54: 217-224). [19])

Species		δ^{13} C_{PDB} (‰)	Vegetation type
M. transmorrisonensis	Leave	-12.8 ± 0.1	C4
	Root	-13.8 ± 0.3	
Y. nittakayamensis	Leave	-27.8 ± 0.3	C3
	Root	-23.0 ± 0.1	
T. chinesis var formosana	Leave	-28.7 ± 0.2	C3
	Root	-26.0 ± 0.3	

Boutton [6] measured five dominant plants in a woodland and indicated that$\delta^{13}C_{PDB}$ values of plant tissue remain relatively constant during the early stages of decomposition. Wedin et al. [35] found small (i.e., 0.6-1.5‰) shifts in the$\delta^{13}C_{PDB}$ values of four grass liters. In addition, some indirect evidence also suggests that fractionation during decomposition is small (from <1‰ to 2‰) especially in terrestrial ecosystems [2,6]. The average$\delta^{13}C_{PDB}$ values of *transmorrisonensis* is about 14 higher than that of C_3 plants [3]. Thus, even if there is discrimination during decomposition, the isotopic discrimination would be very small compared to the large$\delta^{13}C_{PDB}$ difference between C_3 and C_4 plants.

5.4. Distribution of SOM $\delta^{13}C$ in Meadow Pedons 1 and 2, Transition Zone of Pedon 3 and in *Tsuga* Zone of Pedon 4

In summary, $\delta^{13}C_{PDB}$ values in pedons 1 and 2 tended to decrease with increasing with soil depth. The values of EB, E and Bw horizons were close to$\delta^{13}C_{PDB}$ values of C_3 plants. However, $\delta^{13}C_{PDB}$ values in Oe and A horizons ranged between surface O, B and B4 horizons with respect to pedons 1 and 2 (Figure 4A and 4B). The data indicated that SOM was a mixture of *transmorrisonensis* (C_4), *nittakeyamensis* (C_3) and *Tsuga* (C_3) plant residues.$\delta^{13}C_{PDB}$ values obtained from deeper meadow pedons 1 and 2 of soil layers indicated that the vegetation community has changed over the years. *M. transmorrisonensis* (C_4) plants replaced C_3 plants (e.g., *Tsuga*). In the meadow pedons, we observed the presence of the fine charcoals mixed with soil particles. However, the particle size of charcoals is too fine to determine whether they belong to C_3 or C_4 plants, or for the age to be identified by ^{14}C dating study.

$\delta^{13}C_{PDB}$ values of transition zone (pedon 3) ranged from –24.5 to –21.5‰ (Figure 4C). The transition zone located in the valley of pedons 1, 2 and 4 (Figure 4) showed large variation in$\delta^{13}C_{PDB}$ values with soil depth.$\delta^{13}C_{PDB}$ values should be influenced by both C_3 and C_4 plants.$\delta^{13}C_{PDB}$ values of the *Tsuga* forest (pedon 4) ranged from –27.0‰ to –23.5‰ (Figure 4D). The O and A horizons showed an increase in$\delta^{13}C_{PDB}$ values by 4.8% at a depth of between –13 and 11 cm. The minimum$\delta^{13}C_{PDB}$ values ranged from -25.2 to -23.5‰ and remained relatively constant until a depth of about 33-68 cm. All forest soils have similar distribution profiles for SOMδ$^{13}C_{PDB}$ values, regardless of proximity to the meadow-forest boundary. The C_3 plants are the major species in this pedon.

5.5. Factors Influencingδ$^{13}C_{PDB}$ Values

There are several factors influencing δ$^{13}C_{PDB}$ values, such as the relative amount of C_3 and C_4 plant residual compositions in SOM, genesis of soils and plants, decomposition rate of plants, climate change of CO_2, transportation of CO_2 from ambient air to the sites where carboxylation occurs, effects of photo-respiration and respiration, and effect of metabolism. Mariotti and Peterschmitt [36] and Schwarz et al. [37] illustrated that the past dry climate favored the C_4 plant growth, where C_3 plants could grow better in moist climate. In pedon 4 (C_3 plants), variation inδ$^{13}C_{PDB}$ values versus soil depths was 12.5% (Figure 4D). Soil genesis in pedons, mineralization, humification and rate of decomposition of SOM caused slight change inδ$^{13}C_{PDB}$ values versus soil depths of around 1-2%, if the SOM sources came from the same plant [38-40]. Mariotti and Peterschmitt [36], Schwartz et al. [37], and Martin et al. [38] also reported that the SOM decomposition in the tropical and subtropical regimes of forest soil had increased inδ$^{13}C_{PDB}$ values of one to several per mill (‰) with increasing soil depth. The variation of forest pedon 4 with soil depth is due to humification and decomposition of SOM in a temperate rain forest.

5.6. Rate of Humification and Decomposition of SOM

The degree of humification and decomposition of SOM are also affected the precipitation of isotope measurement. Chen et al. [10] identified the nature of organic matter at Ta-Ta-Chia LTER site using ^{13}C NMR techniques techniques.

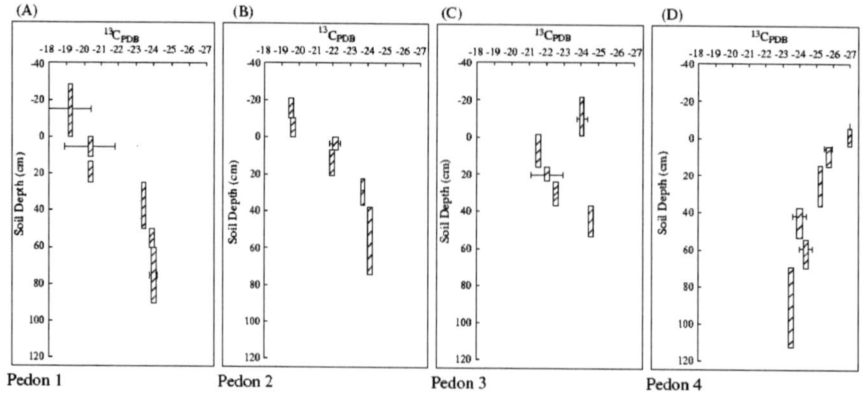

Figure 4. Profiles of δ$^{13}C_{PDB}$ distributions in pedons under plants, each point represents the mean value of triplicated measurements, (A) meadow (pedon 1), (B) meadow (pedon 2), (C) transition zone (pedon 3), and (D) *Tsuga* forest (pedon 4) (Reproduced with permission from Chiang et al. (2004) Changes in the grassland-forest boundary at Ta-Ta-Chia long term ecological research (LTER) site detected by stable isotope ratios of soil organic matter. Chemosphere 54: 217-224). [19].

The bulk and rhizosphere soil of *nittakeyamensis* and *Tsuga*, fresh *nittakeyamensis* and *Tsuga* twigs and leaves, litters and roots were subjected to ^{13}C NMR analysis. In this temperate rain forest environment, the degradation of *Tsuga* materials is slower than that of *nittakeyamensis* [30,31,41]. The debris of the *Tsuga* is not as easily decomposed under the alpine forest soil environment, so it is a major organic carbon in the *Tsuga* rhizosphere (pedon 4) and bulk soils. It seems that $\delta^{13}C_{PDB}$ values in pedon 4 (C_3 plants) showed slightly change in plant communities.

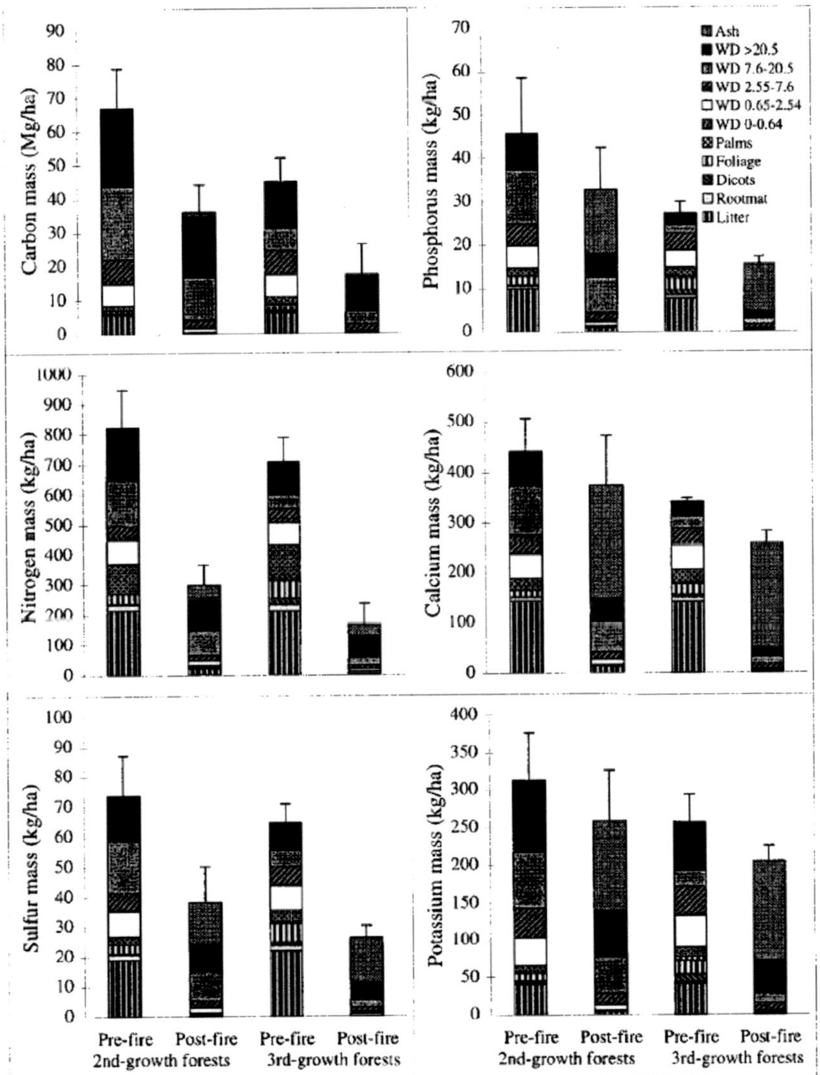

Figure 5. Dynamics of aboveground pools of C, N, S, P, Ca and K in second- and third-growth forests of the Brazilian Amazon before and after fire. Error bars are the SEs of the total above ground pool of each element. WD denotes woody debris, and associated values in each case denote the diameter range class (cm) of the woody debris component. (Reproduced with permission from Hughes et al. (2000) Fire in the Brazilian Amazon. 3. Dynamics of biomass, C, and nutrient pools in regenerating forest. Oecologia 124: 574-588). [43].

5.7. Dynamics of Aboveground Elemental Pools in Response to Forest Fire

Forest fire at this LTER site, particular the historic record at meadow has been reported several times by the Yu-Shan National Park Management Office [42]. Fire frequency at earlier time was probably greater, and more extensive forest may have resulted [34]. Elemental pools can be affected by fire events. Elemental pools can be decreased owing to volatilization or aerosol transport, transform from organic to inorganic form and deposit as ash, or remain on site in the form of residual, un-combusted debris. The fates of elemental pools during the fire events measured to be influenced at least factors: (1) levels of biomass consumption; (2) the relative distribution of elemental mass within biomass categories with varying degree of combustion; and (3) the temperatures of volatilization of the respective elements [43]. In the case of nitrogen mass, the coupled factors of volatilization and disproportionately high concentration of N in readily combusted fine fuels resulted in an average combustion of 76% of pre-fire N mass and an average site loss of 70% of pre-fire N mass (Figure 5). On the other hand, even though the quantity of biomass consumed represented 74, 66, and 77% of pre-fire aboveground pools of P, K, and Ca, much smaller proportions of these elements were actually lost from the site due to the relatively high temperature of volatilization of those elements. Combustion of S mass averaged 75% of pre-fire pools, and the site losses of S across second- and third-growth forests (54%) were intermediate between the high losses of N and the lower losses of P, K and Ca [43].

6. Soil Nitrogen Mineralization and Nitrification Rates in a Mixed Meadow and Forested Ecosystem

Two different vegetation zones were selected for this study. There are meadow and hemlock zones. Study plots were separated by approximately 100 m. At each zone, we located and marked five plots in each vegetation community. Study zones are separately by 50 m and plot size was 2 x 2 m. Plots were random located within each zone. Soil samples (0.5 m diameter) were collected from the O/A horizon (0.1-0.2 cm depth). Net N mineralization and nitrification rates were estimated *in situ* using the buried bag technique [11,12,16,44]. Four or five paired samples were collected from random locations inside each plot; one sample was used to determine initial concentrations of extractable NH_4^+ and NO_3^- and the other was placed in a polyethylene bag and buried *in situ*. The average incubation period was 42 days and ranged from 32 to 65 days.

6.1. Soil Extraction and Chemical Analyses

Fresh soil samples were kept chilled during transport back to the laboratory. Soils were sieved (4 mm) to homogenize the sample and a sub-sample was used to determine moisture content. These samples were dried at 105°C until constant weight and moisture data are presented as percent dry weight. Fresh soil samples (10 gm) were extracted with 100 mL of 2 M KCl for 60 min. Soil extracts were filtered with Whatman No. 1 filter paper and blank corrections were used. Extracts were frozen until analyzed. Ammonium concentration was measured using a manual indophenol colorimetric method [45]. Nitrate concentration was measured using the manual Cd reduction method [46]. Net N mineralization was calculated as the difference between NH_4^+ and NO_3^- concentrations in the incubated sample and the NH_4^+ and NO_3^- in the initial samples. Net nitrification was calculated as the difference between NO_3^- concentrations in the incubated sample minus NO_3^- in the initial samples. Annual rates of net N mineralization and nitrification were estimated by summing all monthly values. We estimated soil bulk density using the core method [47]. Nitrogen mineralization and nitrification rates were converted to mass per area per unit time using these bulk density estimates. Linear regression analysis w as carried out using the GLM procedure with SAS statistical package.

6.2. Extractable NH_4^+ and NO_3^- Concentrations Versus Net Mineralization and Nitrification Rates

Extractable NH_4^+ concentrations ranged from 1.0 to 12.4 mg N kg^{-1} soil in the meadow zone and from 2.8 to 25.0 mg N kg^{-1} soil. The lowest NH_4^+ concentrations were found during April and June in the forest zone and during Junc in the meadow. The highest NH_4^+ concentrations were found during October in both of the meadow and forest zones. Subsequently, soil NH_4^+ concentrations decreased and remained low during the following winter [16]. Extractable NO_3^- concentrations ranged from 0.2 to 2.1 mg N kg^{-1} soil in the meadow zone and from 0.2 to 13.0 mg N kg^{-1} soil in the forest zone (Figs. 6, 7). No seasonal changes in soil nitrate concentration were present.

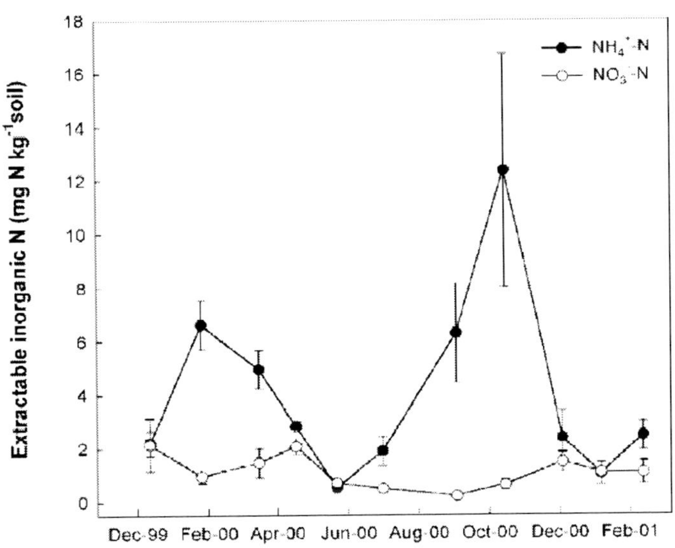

Figure 6. Soil extractable ammonium and nitrate concentrations in the meadow soil (± SE, n = 5) at 0.1–0.2 m depth at Ta-Ta-Chia during the study period (Reproduced with permission from Owen et al. (2003b) Comparison of soil nitrogen mineralization and nitrification in a mixed grassland and forest Ecosystem in central Taiwan. Plant and Soil. 251: 167-174). [44]).

Net mineralization rates in the meadow were generally highest in the summer and yet were nearly as high during January and February 2001. Net mineralization rate s in the meadow soil varied between –0.1 and 0.24 mg N kg^{-1} $soil^{-1}$ day^{-1} (Figure 8). The temporal pattern for net mineralization in the forest zone was similar, with mineralization rates highest during in the summer and moderate rates during the following winter. Net nitrification rates were generally low in the meadow and forest soils and often near zero (Figs. 8, 9). Mineralization rates in the forest zone were higher than that in the meadow zone. Net N mineralization, as indicated by negative mineralization and nitrification rates, occurred during the sampling interval with the highest NH_4^+ concentration (October).

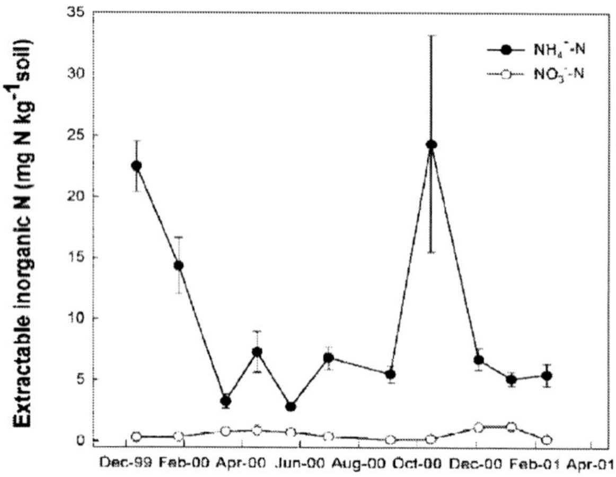

Figure 7. Soil extractable ammonium and nitrate concentrations in the forest soil (± SE, $n = 5$) at 0.1–0.2 m depth at Ta-Ta-Chia during the study period (Reproduced with permission from Owen et al. (2003b) Comparison of soil nitrogen mineralization and nitrification in a mixed grassland and forest Ecosystem in central Taiwan. Plant and Soil. 251: 167-174). [44]).

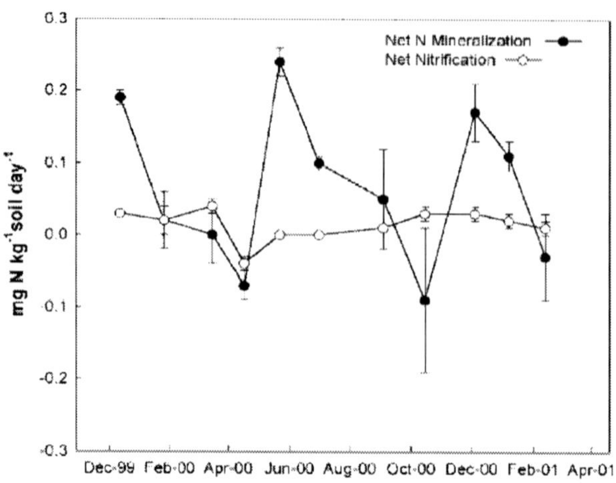

Figure 8. Net N mineralization and nitrification rates for each incubation period in the meadow soil (±SE, n = 5) at 0.1–0.2 m depth at Ta-Ta-Chia (Reproduced with permission from Owen et al. (2003b) Comparison of soil nitrogen mineralization and nitrification in a mixed grassland and forest Ecosystem in central Taiwan. Plant and Soil. 251: 167-174). [44]).

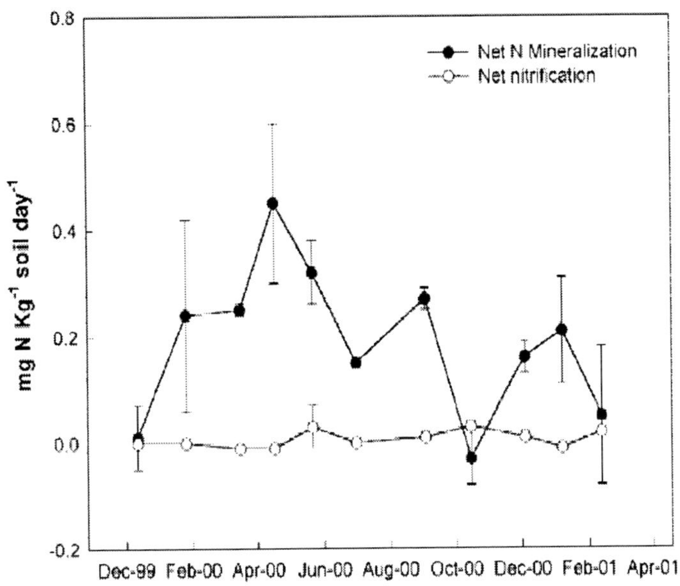

Figure 9. Net N mineralization rates and nitrification rates, for each incubation period in the forest soil (± SE, $n = 5$) at 0.1–0.2 m depth at Ta-Ta-Chia (Reproduced with permission from Owen et al. (2003b) Comparison of soil nitrogen mineralization and nitrification in a mixed grassland and forest Ecosystem in central Taiwan. Plant and Soil. 251: 167-174). [44]).

6.3. Nitrogen Mineralization and Nitrification in Forest and Meadow Zones: in Most Forest Soils

NH_4^+ concentrations are higher than NO_3^- concentrations because of greater leaching losses of NO_3^- [48]. In both the *Tsuga* forest zone and meadow zone at Ta-Ta-Chia, extractable NH_4^+ accounted for nearly all the extractable inorganic soil nitrogen. We found no evidence for temporal changes in the pool of extractable NO_3^- at either zone. However, extractable NH_4^+ concentrations did exhibit temporal changes and were lowest during the growing season in both meadow and forest zones.

More complex relationships have been found between mineralization rate and litter quantity and quality [49]. Elevated N mineralization rates can occur in plant communities with litter that contains higher N concentrations since litter with higher N concentration can decompose faster than lower quality litter and contribute to higher mineralization rates. Epstein et al. [50] showed that plant cover type affected rates of net N mineralization in soils. They found higher net N mineralization rates in C_3 plant communities. Our results form Ta-Ta-Chia are in agreement with this finding since we also found lower rates of N mineralization in the *Miscanthus-Yushania* meadow zone (*Miscanthus* is a C_4 plant) compared to the *Tsuga-Yushania* (C_3 vegetation) forest zone.

6.4. Estimates of Annual N Mineralization Rates

We compared our estimates of annual rates of N mineralization with other published estimates of annual N mineralization from forested ecosystems in Asia (Table 7). At Ta-Ta-Chia, annual net N mineralization was 10.5 and 33.5 kg ha^{-1} in the meadow and forest zones, respectively [44]. Theses estimates are among the lower annual rates of N mineralization in the region. This is probably due to climatic differences between mountainous area of central Taiwan and the location of other studies in Japan and India with higher precipitation and temperatures. Well-planned multi-site comparison, both within Taiwan and the East-Asia region, would further enhance our understanding regional patterns in nitrogen cycling.

7. Microbial Population of Spruce Soil

The organic layer, surface (0-20 cm) and subsoil (21-40 cm) samples in spruce zone were collected for this study.

7.1. Culture Media and Conditions

Bacteria were counted at 25°C after 5 days incubation on nutrient agar containing (g L^{-1}) beef extract 3.0, peptone 5.0, and agar 15.0 at pH 6.8±0.1. Actinomycetes were cultivated at 25°C for 7 days on glycerol-yeast extract medium comprised of (g L^{-1}) glycerol 5.0, yeast extract 2.0, K_2HPO_4 1.0, and agar 15.0 at pH 7.0±0.1. Streptomycin and cycloheeximide were added to inhibit the growth of bacteria and fungi at a final concentration of 10 μg cm^{-3} [51]. Fungi were grown at 25°C for 5 days on Rose Bengal medium containing (g L^{-1}) glucose 10.0, peptone 5.0, K_2HPO_4 1.0, $MgSO_4.7H_2O$ 0.5, Rose Bengal 0.033 and agar 15.0 at pH 6.8±0.1. Cellulolytic microbes were counted at 25°C after 7 days incubation on Mandels-Reese medium with carboxymethylcellulose (Sigma Co.) as the sole carbon source and sprayed with Congo red to show clear zone around the colonies. Phosphate-solublizing microbes were measured at 25°C after 5 days on rock phosphate medium and had clear zone around the colonies. Nitrogen-fixing microbes were characterized at 25°C after 7 days incubation on nitrogen-free mannitol medium [51]. All experiments were carried out in triplicate.

7.2. Microbial Populations

Bacteria were the most abundant, followed by cellulolytic microbes, phosphate-solubilizing microbes, nitrogen-fixing microbes, fungi, and actinomycetes. Total organic carbon and nitrogen contents were high in Spruce soil [52]. The microbial populations, microbial biomass carbon and nitrogen contents were also high on surface soils and low in subsoils, and showed spatial variations. Microbial populations were significantly different between surface soils and subsoils, and showed spatial variations [53]. The ratios of cellulolytic microbes to the total viable count were high due to the heavy coverage of organic matter lying on the surface soils. Malic and succinic acids were the major organic acids in soils, while acetic and formic acids were also found in organic layer. Microbial populations showed high correlation with microbial biomass carbon and organic acid contents [54].

Table 7. Comparison of Annual Rates of N Mineralization (NMIN) from Related Studies in Asia (Reproduced with permission from Owen et al. (2003b) Comparison of soil nitrogen mineralization and nitrification in a mixed grassland and forest Ecosystem in central Taiwan. Plant and Soil. 251: 167-174). [44])

Dominant species	Location	NMIN (kg N ha^{-1} $year^{-1}$)	References
Cryptomeria japonica	Japan	67	Wu et al. [55]
Chamecyparis obtuse		77	
Cryptomeria japonica 15 year	Japan	58	Murakami et al. [56]
Chamaecyparis obtuse 15 year		34	
Chamaecyparis obtuse 35 year		22	
Tsuga chinensis var. *formonsana*	Taiwan	34	Owen et al. [16]
Miscanthus transmorrisonensis Hayata		11	Owen et al. [16]
Pinus kesiya 7 year	India	85	Maithani et al. [57]
Quercus dealbata 13 year		106	
Q. dealbata/Rhododendron arboreum 16 year		115	
Acacia catechu/Emblica officinalis hillbase	India	210	Raghubanshi [58]
A. catechu/lannea coromandelica midslope		170	
Boswellia serrata/A. catechu hilltop		200	

Conclusion

$\delta^{13}C$ values observed from deeper soil horizons of meadows 1 and 2 indicated the vegetation type had been changed over the years. Pedons 1 and 2 are on south-facing slope, which is drier than that of pedon 4 on a north-facing slope in forest. It may be caused by forest fire of natural or human disturbance affected the change of plant community in meadow zone. In this temperate rain forest environment, the degradation of *Tsuga* materials is slower than that of *nittakeyamensis*. The degree of humification and decomposition of SOM are also affected the precipitation of isotope measurement. Decomposition of organic residues is really depending on microbial growing environments and populations.

The highest soil NH_4^+ concentrations generally occurred at the upper location in the O horizon (range 14.5-58.5 mg N kg^{-1}). The spatial pattern in extractable NO_3^- concentration showed a different trend with the highest concentrations at the upper locations (2.6-20.1 mg N kg^{-1}). Total extractable inorganic N concentrations in the O horizon ranged from 16.8 to 63.4 mg N kg^{-1}. In the mineral soil, the range was 1.9-11.2 mg N kg^{-1}. The net of mineralization and nitrification rates showed spatial and temporal variations, depending on microbial activities and populations. The microbial populations and biomass carbon and nitrogen contents were high in surface soils and low in subsoils. Bacteria were most abundant, followed by cellulolytic microbes, phosphate-solubilizing microbes, nitrogen-fixing microbes, fungi and actinomycetes. The ratios of cellulolytic microbes to the total viable count were high due to the heavy coverage of organic matter lying on the surface soils.

Microbial populations showed high correlation with microbial biomass carbon and organic acid contents. The nutrient cycling and forest fire influence the dynamics of biomass, C, P, Ca, Mg, K, Na and other anions on this LTER site merit further study.

Acknowledgements

We thank the National Science Council, Taiwan, Republic of China, for financial support under the projects of forest sFoils (NSC #85-2621-B002-017, 86-2621-B002-022-A07, and 87-2621-B002-053-A10).

References

[1] Keeling CD, Carter AF, Mook WG: Seasonal latitudinal, and secular variations in the abundance and isotopic ratios of atmospheric CO_2. 2. Results from oceanographic cruises in the tropical Pacific Ocean. *J. Geophys. Res.* 89: 4615-4628, 1984

[2] Nadelhoffer KJ, Fry B: Controls on natural nitrogen-15 and carbon-13 abundances in forest soil organic matter. *Soil Sci. Soc. Am. J.* 52: 1633-1640, 1988

[3] Kao WY: Contribution of *Miscanthus transmorrisonensis* t soil organic carbon in a mountain grassland: Estimated from stable carbon isotope ratio. *Bot. Bull. Acad. Sinica* 38: 45-48, 1997

[4] Kessel C, Farrel RE, Pennock DJ: Carbon-13 and nitrogen-15 natural abundance in crop residues and soil organic matter. *Soil Sci. Soc. Am. J.* 58: 382-389, 1994

[5] Bird, MI, Haberle SG, Chivas AR: Effect of altitude on the carbon isotope composition of forest composition of forest and grassland soils from Papua New Guinea. Global Biogeochem. *Cycles*. 8: 13-22, 1994

[6] Boutton TW: Stable carbon isotope ratios of soil organic matter and their use as indicators of vegetation and climate change. Mass Spectrometry of Soils. *Boutton TW and Yamasaki TW eds.* Marcel Dekker, Inc., New York, pp. 47-82, 1996

[7] Huang PM, Schnitzer M: Interactions of soil minerals with natural organics and microbes. *Soil Sci. Soc. Am.*, Madison, WI, 1986

[8] Zhang H, Thompson ML, Sandor JA: Compositional differences in organic matter among cultivated and uncultivated Argiudolls and Hapludalfs derived from loess. *Soil Sci. Soc. Am. J.* 25: 216-222, 1988

[9] Zech W, Haumaier L, Guggenberger G, Gill-Sotres F, Arai S: Changes in carbon species distribution of humic substances with depth in mineral soils on various origin. *Humic Substances in the Global Environmental and Implications in Human Health.* Senesi N, Miano MT. eds. Elsevier, Amsterdam. pp. 445-450, 1994

[10] Chen MC, Wang MK, Chiu CY, Huang PM, King HB: Determination of low molecular weight dicarboxylic acids and organic functional groups in rhizosphere and bulk soils of *Tsuga* and *Yushania* in a temperate rain forest. *Plant Soil.* 231: 37-44, 2001

[11] Binkley D, Hart SC: The components of nitrogen availability assessments in forest soils. *Adv. Soil Sci.* 10: 57-112, 1989

[12] Eno CF: Nitrate production in the field by incubating the soil in polyethylene bags. *Am. Soil Sci. Soc. Proc.* 24: 277-279, 1960

[13] Ho CS: An Introduction to the Geology of Taiwan: Explanatory Text of the Geologic Map of Taiwan. *2nd ed. Ministry of Economic Affairs*, Taipei, Taiwan. pp. 192, 1988

[14] King HB, Dystrochrept JK, Bowen LH, Wang C: Iron concretions in a Typic Dystrochrept in Taiwan. *Soil Sci. Soc. Am. J.* 54: 462-468, 1990

[15] Soil Survey Staff: Keys to Soil Taxonomy. Eight ed. *United State Department of Agriculture and Natural Resources Conservation Service*, Washington, DC, 1998

[16] Owen JS, Wang MK, Wang CH, King HB, Sun HL: Net N mineralization and nitrification rates in a forested ecosystem in northeastern Taiwan. *For. Ecol. Manage.* 176: 519-530, 2003a

[17] Jackson ML: Soil Chemical Analysis. *Advanced Course, 2nd ed.* University of Wisconsin, Madison, WI. pp. 497, 1979

[18] Chiu CH, Wang MK, Hwong JL, King HB: Physical and chemical properties in rhizosphere and bulk soils of *Tsuga* and *Yushania* in a temperate rain forest. *Comm. Soil Sci. Plant Analy.* 33(11and12): 1723-1735, 2002

[19] Chiang PN, Wang MK, Chiu CY, King HB, Hwong JL: Changes in the grassland-forest boundary at Ta-Ta-Chia long-term ecological research (LTER) site detected by stable isotope ratios of soil organic matter. *Chemos.* 54: 217-224, 2004

[20] Chen MC: Comparison of physicochemical properties between the rhizosphere and bulk soils in the Hemlock forest of the Ta-Ta-Chia saddle, Yushan. *MSc Thesis.* National Taiwan University, Taiwan, pp. 83, 1998

[21] Munsell Soil Color Charts: Revised edition. Macbeth, Division of Kollmorgen Instruments Corp., New York, 1992

[22] Lynch JM: Soil Rhizosphere. John Wiley and Sons, New York, 1990

[23] Oades JM: Recent advances in organo-mineral interactions: Implication for carbon cycling and soil structure. Vol. 1. Natural and Anthropogenic Organics. Huang PM, Berthelin J, Bollag J-M, McGill WB, Page AL eds. CRC Press/Lewis Publishers, *Boca Raton, FL.* pp. 119-134, 1995

[24] Preston CM, Sollins P, Sayer BG: Changes in organic compounds for fallen logs in old-growth Doulas-fir forests monitored by ^{13}C nuclear magnetic resonance spectroscopy. *Can. J. For. Res.* 20: 1382-1391, 1990

[25] Moghimi A, Tate ME, Oades JM: Characterization of rhizosphere products especially 2-keto-gluconic acid. *Soil Biol. Biochem.* 10: 283-287, 1978

[26] Fox TR, Comerford NB: Low-molecular weight organic acids in selected forest soils of Southeastern USA. *Soil Sci. Soc. Am. J.* 54: 1763-1767, 1990

[27] Hatcher PG.: Chemical structure studies of natural lignin dipolar dephasing soild-state ^{13}C nuclear magnetic resonance. *Org. Geochem.* 11: 31-39, 1987

[28] Chen JS, Chiu CY: Characterization of soil organic matter in different particle-size fractions in humid subalpine soils by CP/MAS ^{13}C NMR. *Geoderma* 117: 129-141, 2001

[29] Chen JS, Chiu CY: Characteristics of humic substances in humid subalpine forest soils in central Taiwan. *Acta Agrophysica* 50: 41-47, 2001

[30] Kögel-Knabner I: Biodegradation and humification processes in forest soils. *Soil Biochemistry, Vol. 8*. Bollag J-M, Stotzky C eds. Marcel Dekker, New York. pp. 101-137, 1993

[31] Stevenson FJ: Humus Chemistry. John Wiley and Sons, New York, 1994

[32] McKinney CR, McCrea JM, Epstein S, Allen HA, Urey HC: Improvements in mass spectrometers for the measurement of small differences in isotope abundance ratios. *Rev. Sci. Instrum*. pp. 724, 1950

[33] Craig H: Isotopic standards for carbon and oxygen and correction factors for mass-spectrometric analysis of carbon dioxide. *Geochim. Cosmochim. Acta*. 12: 133-149, 1957

[34] Desjardins T, Carneirofilho A, Mariotti A, Chauvel A, Girardin C: Changes of the forest-savanna boundary in Brazilian Amazonia during the Holocene revealed by stable isotope ratios of soil organic carbon. *Oecologia* 108: 749-756, 1996

[35] Wedin DA, Tieszen LL, Vewey B, Pastor J: Carbon dynamics during grass decomposition and soil organic matter formation. *Ecology* 75: 1383-1392, 1995

[36] Marriotti A, Pterschmitt E: Forest savanna ecotone dynamics in India as revealed by carbon isotope ratios of soil organic matter. *Oecologia* 97: 475-480, 1994

[37] Schwarz D, Mariotti A, Trouvé C, Ven Den Borg K, Guillet B: Etude des profiles isotopiques ^{13}C et ^{14}C dun sol ferrallitique sableux du littoral congolais. Implications sur la dynamique de la matiére organique et l'histore de la végétation, *CR Acad. Sci.* Paris II. 315: 1411-1417, 1992

[38] Martin A, Mariotti A, Balesdent J, Kacelle P, Vuattoux R: Estimate of organic carbon matter turnover rate in a savanna soil by ^{13}C natural abundance measurements. *Soil Biol. Biochem*. 22: 517-523, 1990

[39] Marriotti A: Le carbone 13 en abundance naturelle traceur de la dynamique de la matiére organique des sols et da l'évolution des paléoenvironments continentauz. *Cah. Orstom. Ser. Pedol.* 26: 299-313, 1991

[40] McPhearson GR, Boutton TW, Midwood AJ: Stable carbon isotope analysis of soil organic matter illustrates vegetation change at the grassland/woodland boundary in south-eastern Arizonia, USA. *Oecologia* 93: 95-101, 1993

[41] Wang Y, Cerlin TE, Effland WR: Stable isotope ratios of soil carbonate and soil organic matter as indicators of forest invasion of prairie near Ames, Iowa. *Oecologia* 95: 365-369, 1993

[42] Yu-Shan National Park Report: Environmental and Ecological Changes at Ta-Ta-Chia, Yu-Shan Forest Fire Residues and their Remediation. Published by Yu-Shan National Park Management Office, *Department of Interior, ROC*, pp. 107, 1994

[43] Hughes RF, Kauffman J.B, Cummings DL: Fire in the Brazilian Amazon. 3. Dynamics of biomass, C, and nutrient pools in regenerating forest. *Oecologia* 124: 574-588, 2000

[44] Owen JS, Wang MK, Sun HL, King HB, Wang CH, Chuang CF: Comparison of soil nitrogen mineralization and nitrification in a mixed grassland and forested ecosystem in central Taiwan. *Plant Soil.* 251: 167-174, 2003b

[45] Dorich RA, Nelson DW: Direct colorimetric measurement of ammonium in potassium chloride extracts of soils. *Soil Sci. Soc. Am. J.* 47: 833-836, 1983

[46] APHA: Cadmium reduction method. Standard Methods for the Examination of Water and Wastewater. *Franson MAH ed.* Washington, DC. pp. 4-118, 1998

[47] Blake GR, Hartge KH: Bulk density. Methods of Soil Analysis. Part 1. Physical and Chemical Methods. 2nd Ed. Agronomy Monograph #9., Klute A et al. eds. *Soil Sci. Soc. Am.*, Madison, WI. pp. 363-367, 1986

[48] Laverman AM, Zoomer HR, van Verseveld HW, Verhoef HA: Temporal and spatial variation of nitrogen transformations in coniferous forest soil. *Soil Biol. Biochem.* 32: 1661-1670, 2000

[49] Vinton MA, Burke IC: Interactions between individual plants species and soil nutrient status in shortgrass steppe. *Ecology* 76: 1116-1133, 1995

[50] Epstein HE, Burker IC, Mosier AR: Plant effects on spatial and temporal patterns of nitrogen cycling in shortgrass steppe. *Ecosystems.* 1: 374-385, 1998

[51] Yang SS, Lin YC, Yang CK, Chang EH, Wei CB: Microbial ecology of Hsieh-Ho thermal power plant and its surrounding area. *J. Microbio. Immuno. Infect.* 32: 269-277, 1999

[52] Vance ED, Brookes PC, Jenkinson DS: An extraction method for measuring soil microbial C. *Soil Biology Biochem.* 19: 70-707, 1987

[53] Imberger KT, Chiu CY: Spatial changes of soil fungal and bacterial biomass from a sub-alpine coniferous forest to grassland in a humid, sub-tropical region. *Biol. Fertil. Soils.* 33: 105-110, 2001

[54] Yang SS, Fan HS, Yang CK, Lin IC: Microbial population of spruce soil in Tatachla Mountain of Taiwan. *Chemosph.* 52: 1489-1498, 2003

[55] Wu G, Toda H, Haibara K, Aiba Y: The effects of nitrogen mineralization on water-soluble ions of forests. *J. Jpn. For. Soc.* 80: 21-26, 1998

[56] Murakami M, Taketa H, Iwatsubo G: Seasonal changes in nitrogen mineralization rates in the soils of *Chamaecyparis* obtuse and *Crytomeria* Japonica plantation. Exp. For. Tokyo Univ. 62: 4-54, 1990

[57] Maithani K, Arunachalam A, Tripathi RS, Pandy NH: Nitrogen mineralization as influenced by climate, soil and vegetation in a subtropical humid forest in northeast India. *For. Ecol. Manage.* 109: 91-101, 1998

[58] Raghubanshi AS: Dynamics of soil biomass C, N, and P in a dry tropical in India. *Biology Fertil. Soils.* 12: 55-59, 1991

Index

A

B

C

D

E

F

G

H

I

P

Q

R

S

T

U

V

W

X

Y

Z